Graduate Texts in Mathematics **188**

Editorial Board
S. Axler   F.W. Gehring   K.A. Ribet

Springer
New York
Berlin
Heidelberg
Barcelona
Budapest
Hong Kong
London
Milan
Paris
Singapore
Tokyo

Graduate Texts in Mathematics

1 TAKEUTI/ZARING. Introduction to Axiomatic Set Theory. 2nd ed.
2 OXTOBY. Measure and Category. 2nd ed.
3 SCHAEFER. Topological Vector Spaces.
4 HILTON/STAMMBACH. A Course in Homological Algebra. 2nd ed.
5 MAC LANE. Categories for the Working Mathematician. 2nd ed.
6 HUGHES/PIPER. Projective Planes.
7 SERRE. A Course in Arithmetic.
8 TAKEUTI/ZARING. Axiomatic Set Theory.
9 HUMPHREYS. Introduction to Lie Algebras and Representation Theory.
10 COHEN. A Course in Simple Homotopy Theory.
11 CONWAY. Functions of One Complex Variable I. 2nd ed.
12 BEALS. Advanced Mathematical Analysis.
13 ANDERSON/FULLER. Rings and Categories of Modules. 2nd ed.
14 GOLUBITSKY/GUILLEMIN. Stable Mappings and Their Singularities.
15 BERBERIAN. Lectures in Functional Analysis and Operator Theory.
16 WINTER. The Structure of Fields.
17 ROSENBLATT. Random Processes. 2nd ed.
18 HALMOS. Measure Theory.
19 HALMOS. A Hilbert Space Problem Book. 2nd ed.
20 HUSEMOLLER. Fibre Bundles. 3rd ed.
21 HUMPHREYS. Linear Algebraic Groups.
22 BARNES/MACK. An Algebraic Introduction to Mathematical Logic.
23 GREUB. Linear Algebra. 4th ed.
24 HOLMES. Geometric Functional Analysis and Its Applications.
25 HEWITT/STROMBERG. Real and Abstract Analysis.
26 MANES. Algebraic Theories.
27 KELLEY. General Topology.
28 ZARISKI/SAMUEL. Commutative Algebra. Vol.I.
29 ZARISKI/SAMUEL. Commutative Algebra. Vol.II.
30 JACOBSON. Lectures in Abstract Algebra I. Basic Concepts.
31 JACOBSON. Lectures in Abstract Algebra II. Linear Algebra.
32 JACOBSON. Lectures in Abstract Algebra III. Theory of Fields and Galois Theory.
33 HIRSCH. Differential Topology.
34 SPITZER. Principles of Random Walk. 2nd ed.
35 ALEXANDER/WERMER. Several Complex Variables and Banach Algebras. 3rd ed.
36 KELLEY/NAMIOKA et al. Linear Topological Spaces.
37 MONK. Mathematical Logic.
38 GRAUERT/FRITZSCHE. Several Complex Variables.
39 ARVESON. An Invitation to C^*-Algebras.
40 KEMENY/SNELL/KNAPP. Denumerable Markov Chains. 2nd ed.
41 APOSTOL. Modular Functions and Dirichlet Series in Number Theory. 2nd ed.
42 SERRE. Linear Representations of Finite Groups.
43 GILLMAN/JERISON. Rings of Continuous Functions.
44 KENDIG. Elementary Algebraic Geometry.
45 LOÈVE. Probability Theory I. 4th ed.
46 LOÈVE. Probability Theory II. 4th ed.
47 MOISE. Geometric Topology in Dimensions 2 and 3.
48 SACHS/WU. General Relativity for Mathematicians.
49 GRUENBERG/WEIR. Linear Geometry. 2nd ed.
50 EDWARDS. Fermat's Last Theorem.
51 KLINGENBERG. A Course in Differential Geometry.
52 HARTSHORNE. Algebraic Geometry.
53 MANIN. A Course in Mathematical Logic.
54 GRAVER/WATKINS. Combinatorics with Emphasis on the Theory of Graphs.
55 BROWN/PEARCY. Introduction to Operator Theory I: Elements of Functional Analysis.
56 MASSEY. Algebraic Topology: An Introduction.
57 CROWELL/FOX. Introduction to Knot Theory.
58 KOBLITZ. p-adic Numbers, p-adic Analysis, and Zeta-Functions. 2nd ed.
59 LANG. Cyclotomic Fields.
60 ARNOLD. Mathematical Methods in Classical Mechanics. 2nd ed.
61 WHITEHEAD. Elements of Homotopy Theory.

(continued after index)

Robert Goldblatt

Lectures on the Hyperreals

An Introduction to Nonstandard Analysis

 Springer

Library
Quest University Canada
3200 University Boulevard
Squamish, BC V8B 0N8

Robert Goldblatt
School of Mathematical and Computing Sciences
Victoria University
Wellington
New Zealand

Editorial Board

S. Axler	F.W. Gehring	K.A. Ribet
Mathematics Department	Mathematics Department	Mathematics Department
San Francisco State University	East Hall University of Michigan	University of California at Berkeley
San Francisco, CA 94132	Ann Arbor, MI 48109	Berkeley, CA 94720-3840
USA	USA	USA

Mathematics Subject Classification (1991): 26E35, 03H05, 28E05

Library of Congress Cataloging-in-Publication Data
Goldblatt, Robert.
    Lectures on the hyperreals : an introduction to nonstandard analysis / Robert Goldblatt.
        p.   cm. — (Graduate texts in mathematics ; 188)
    Includes bibliographical references and index.
    ISBN 0-387-98464-X (hardcover : alk. paper)
    1. Nonstandard mathematical analysis. I. Title. II. Series.
QA299.82.G65 1998
515—DC21   98-18388

Printed on acid-free paper.

© 1998 Springer-Verlag New York, Inc.
All rights reserved. This work may not be translated or copied in whole or in part without the written permission of the publisher (Springer-Verlag New York, Inc., 175 Fifth Avenue, New York, NY 10010, USA), except for brief excerpts in connection with reviews or scholarly analysis. Use in connection with any form of information storage and retrieval, electronic adaptation, computer software, or by similar or dissimilar methodology now known or hereafter developed is forbidden.
The use of general descriptive names, trade names, trademarks, etc., in this publication, even if the former are not especially identified, is not to be taken as a sign that such names, as understood by the Trade Marks and Merchandise Marks Act, may accordingly be used freely by anyone.

Production managed by Victoria Evarretta; manufacturing supervised by Jacqui Ashri.
Photocomposed pages prepared from the author's $\text{L}^{\text{A}}\text{T}_{\text{E}}\text{X}$ files.
Printed and bound by Maple-Vail Book Manufacturing Group, York, PA.
Printed in the United States of America.

9 8 7 6 5 4 3 2 1

ISBN 0-387-98464-X Springer-Verlag New York Berlin Heidelberg   SPIN 10659819

Preface

> *There are good reasons to believe that nonstandard analysis, in some version or other, will be the analysis of the future.*
>
> KURT GÖDEL

This book is a compilation and development of lecture notes written for a course on nonstandard analysis that I have now taught several times. Students taking the course have typically received previous introductions to standard real analysis and abstract algebra, but few have studied formal logic. Most of the notes have been used several times in class and revised in the light of that experience. The earlier chapters could be used as the basis of a course at the upper undergraduate level, but the work as a whole, including the later applications, may be more suited to a beginning graduate course.

This preface describes my motivations and objectives in writing the book. For the most part, these remarks are addressed to the potential instructor.

Mathematical understanding develops by a mysterious interplay between intuitive insight and symbolic manipulation. Nonstandard analysis requires an enhanced sensitivity to the particular symbolic form that is used to express our intuitions, and so the subject poses some unique and challenging pedagogical issues. The most fundamental of these is how to turn the *transfer principle* into a working tool of mathematical practice. I have found it unproductive to try to give a proof of this principle by introducing the formal Tarskian semantics for first-order languages and working through the proof of Łoś's theorem. That has the effect of making the subject seem more difficult and can create an artifical barrier to understanding. But the practical use of transfer is more readily explained informally, and typically involves statements that are no more complicated than the "epsilon-delta" statements used in standard analysis. My approach then has been to *illustrate* transfer by many examples, with demonstrations of why those examples work, leading eventually to a situation in which its formulation as a general principle appears quite credible.

There is an obvious analogy with standard laws of thought, such as induction. It would be an unwise teacher who attempted to introduce this to the novice by deriving the principle of induction as a theorem from the axioms of set theory. Of course one attempts to *describe* induction, and *explain* how it is applied. Eventually after practice with examples the student gets used to using it. So too with transfer.

It is sensible to use this approach in many areas of mathematics, for instance beginning a course on standard analysis with a description of the real number system $\mathbb{R}$ as a complete ordered field. The student already has well-developed intuitions about real numbers, and the axioms serve to summarise the essential information needed to proceed. It is rare these days to find a text that begins by explicitly constructing $\mathbb{R}$ out of the rationals via Dedekind cuts or Cauchy sequences, before embarking on the theory of limits, convergence, continuity, etc.

On the other hand, it is not so clear that such a methodology is adequate for the introduction of the hyperreal field $^*\mathbb{R}$ itself. In view of the controversial history of infinitesimals, and the student's lack of familiarity with them, there is a plausibility problem about simply introducing $^*\mathbb{R}$ axiomatically as an ordered field that extends $\mathbb{R}$, contains infinitesimals, and has various other properties. I hope that such a descriptive approach will eventually become the norm, but here I have opted to use the foundational, or constructive, method of presenting an ultrapower construction of the ordered field structure of $^*\mathbb{R}$, and of enlargements of elementary sets, relations, and functions on $\mathbb{R}$, leading to a development of the calculus, analysis, and topology of functions of a single variable. At that point (Part III) the exposition departs from some others by making an early introduction of the notions of internal, external, and hyperfinite subsets of $^*\mathbb{R}$, and internal functions from $^*\mathbb{R}$ to $^*\mathbb{R}$, along with the notions of overflow, underflow, and saturation. It is natural and helpful to develop these important and radically new ideas in this simpler context, rather than waiting to apply them to the more complex objects produced by constructions based on superstructures.

As to the use of superstructures themselves, again I have taken a slightly different tack and followed (in Part IV) a more axiomatic path by positing the existence of a *universe* $\mathbb{U}$ containing all the entities (sets, tuples, relations, functions, sets of sets of functions, etc., etc.) that might be needed in pursuing a particular piece of mathematical analysis. $\mathbb{U}$ is described by set-theoretic closure properties (pairs, unions, powersets, transitive closures). The role of the superstructure construction then becomes the foundational one of showing that universes exist. From the point of view of mathematical practice, enlargements of superstructures seem somewhat artificial (a "gruesome formalism", according to one author), and the approach taken here is intended to make it clearer as to what exactly is the ontology that we need in order to apply nonstandard methods. Looking to the future, if (one would like to say *when*) nonstandard analysis becomes as widely recognised as its standard "shadow", so that a descriptive approach without any need for ultrapowers is more amenable, then the kind of axiomatic account developed here on the basis of universes would, I believe, provide an effective and accessible style of exposition of the subject.

What does nonstandard analysis offer to our understanding of mathematics? In writing these notes I have tried to convey that the answer includes the following five features.

(1) *New definitions of familiar concepts, often simpler and more intuitively natural*

Examples to be found here include the definitions of convergence, boundedness, and Cauchy-ness of sequences; continuity, uniform continuity, and differentiability of functions; topological notions of interior, closure, and limit points; and compactness.

(2) *New and insightful (often simpler) proofs of familiar theorems*

In addition to many theorems of basic analysis about convergence and limits of sequences and functions, intermediate and extreme values and fixed points of continuous functions, critical points and inverses of differentiable functions, the Bolzano–Weierstrass and Heine–Borel theorems, the topology of sets of reals, etc., we will see nonstandard proofs of Ramsey's theorem, the Stone representation theorem for Boolean algebras, and the Hahn–Banach extension theorem on linear functionals.

(3) *New and insightful constructions of familiar objects*

For instance, we will obtain integrals as hyperfinite sums; the reals $\mathbb{R}$ themselves as a quotient of the hyperrationals $^*\mathbb{Q}$; other completions, including the p-adic numbers and standard power series rings as quotients of nonstandard objects; and Lebesgue measure on $\mathbb{R}$ by a nonstandard counting process with infinitesimal weights.

(4) *New objects of mathematical interest*

Here we will exhibit new kinds of number (limited, unlimited, infinitesimal, appreciable); internal and external sets and functions; shadows; halos; hyperfinite sets; nonstandard hulls; and Loeb measures.

(5) *Powerful new properties and principles of reasoning*

These include transfer; internal versions of induction, the least number principle and Dedekind completeness; overflow, underflow, and other principles of permanence; Robinson's sequential lemma; saturation; internal set definition; concurrence; enlargement; hyperfinite approximation; and comprehensiveness.

In short, nonstandard analysis provides us with an enlarged view of the mathematical landscape. It represents yet another stage in the emergence of new number systems, which is a significant theme in mathematical history. Its rich conceptual framework will be built on to reveal new systems and new understandings, so its development will itself influence the course of that history.

Contents

I Foundations   1

1 What Are the Hyperreals?   3
    1.1   Infinitely Small and Large   3
    1.2   Historical Background   4
    1.3   What Is a Real Number?   11
    1.4   Historical References   14

2 Large Sets   15
    2.1   Infinitesimals as Variable Quantities   15
    2.2   Largeness   16
    2.3   Filters   18
    2.4   Examples of Filters   18
    2.5   Facts About Filters   19
    2.6   Zorn's Lemma   19
    2.7   Exercises on Filters   21

3 Ultrapower Construction of the Hyperreals   23
    3.1   The Ring of Real-Valued Sequences   23
    3.2   Equivalence Modulo an Ultrafilter   24
    3.3   Exercises on Almost-Everywhere Agreement   24
    3.4   A Suggestive Logical Notation   24
    3.5   Exercises on Statement Values   25

	3.6	The Ultrapower	25
	3.7	Including the Reals in the Hyperreals	27
	3.8	Infinitesimals and Unlimited Numbers	27
	3.9	Enlarging Sets	28
	3.10	Exercises on Enlargement	29
	3.11	Extending Functions	30
	3.12	Exercises on Extensions	30
	3.13	Partial Functions and Hypersequences	31
	3.14	Enlarging Relations	31
	3.15	Exercises on Enlarged Relations	32
	3.16	Is the Hyperreal System Unique?	33

4 The Transfer Principle   35

	4.1	Transforming Statements	35
	4.2	Relational Structures	38
	4.3	The Language of a Relational Structure	38
	4.4	∗-Transforms	42
	4.5	The Transfer Principle	44
	4.6	Justifying Transfer	46
	4.7	Extending Transfer	47

5 Hyperreals Great and Small   49

	5.1	(Un)limited, Infinitesimal, and Appreciable Numbers	49
	5.2	Arithmetic of Hyperreals	50
	5.3	On the Use of "Finite" and "Infinite"	51
	5.4	Halos, Galaxies, and Real Comparisons	52
	5.5	Exercises on Halos and Galaxies	52
	5.6	Shadows	53
	5.7	Exercises on Infinite Closeness	54
	5.8	Shadows and Completeness	54
	5.9	Exercise on Dedekind Completeness	55
	5.10	The Hypernaturals	56
	5.11	Exercises on Hyperintegers and Primes	57
	5.12	On the Existence of Infinitely Many Primes	57

II   Basic Analysis   59

6 Convergence of Sequences and Series   61

	6.1	Convergence	61
	6.2	Monotone Convergence	62
	6.3	Limits	63
	6.4	Boundedness and Divergence	64
	6.5	Cauchy Sequences	65
	6.6	Cluster Points	66

6.7	Exercises on Limits and Cluster Points	66
6.8	Limits Superior and Inferior	67
6.9	Exercises on lim sup and lim inf	70
6.10	Series	71
6.11	Exercises on Convergence of Series	71

7 Continuous Functions — 75

7.1	Cauchy's Account of Continuity	75
7.2	Continuity of the Sine Function	77
7.3	Limits of Functions	78
7.4	Exercises on Limits	78
7.5	The Intermediate Value Theorem	79
7.6	The Extreme Value Theorem	80
7.7	Uniform Continuity	81
7.8	Exercises on Uniform Continuity	82
7.9	Contraction Mappings and Fixed Points	82
7.10	A First Look at Permanence	84
7.11	Exercises on Permanence of Functions	85
7.12	Sequences of Functions	86
7.13	Continuity of a Uniform Limit	87
7.14	Continuity in the Extended Hypersequence	88
7.15	Was Cauchy Right?	90

8 Differentiation — 91

8.1	The Derivative	91
8.2	Increments and Differentials	92
8.3	Rules for Derivatives	94
8.4	Chain Rule	94
8.5	Critical Point Theorem	95
8.6	Inverse Function Theorem	96
8.7	Partial Derivatives	97
8.8	Exercises on Partial Derivatives	100
8.9	Taylor Series	100
8.10	Incremental Approximation by Taylor's Formula	102
8.11	Extending the Incremental Equation	103
8.12	Exercises on Increments and Derivatives	104

9 The Riemann Integral — 105

9.1	Riemann Sums	105
9.2	The Integral as the Shadow of Riemann Sums	108
9.3	Standard Properties of the Integral	110
9.4	Differentiating the Area Function	111
9.5	Exercise on Average Function Values	112

10 Topology of the Reals — 113
- 10.1 Interior, Closure, and Limit Points — 113
- 10.2 Open and Closed Sets — 115
- 10.3 Compactness — 116
- 10.4 Compactness and (Uniform) Continuity — 119
- 10.5 Topologies on the Hyperreals — 120

III Internal and External Entities — 123

11 Internal and External Sets — 125
- 11.1 Internal Sets — 125
- 11.2 Algebra of Internal Sets — 127
- 11.3 Internal Least Number Principle and Induction — 128
- 11.4 The Overflow Principle — 129
- 11.5 Internal Order-Completeness — 130
- 11.6 External Sets — 131
- 11.7 Defining Internal Sets — 133
- 11.8 The Underflow Principle — 136
- 11.9 Internal Sets and Permanence — 137
- 11.10 Saturation of Internal Sets — 138
- 11.11 Saturation Creates Nonstandard Entities — 140
- 11.12 The Size of an Internal Set — 141
- 11.13 Closure of the Shadow of an Internal Set — 142
- 11.14 Interval Topology and Hyper-Open Sets — 143

12 Internal Functions and Hyperfinite Sets — 147
- 12.1 Internal Functions — 147
- 12.2 Exercises on Properties of Internal Functions — 148
- 12.3 Hyperfinite Sets — 149
- 12.4 Exercises on Hyperfiniteness — 150
- 12.5 Counting a Hyperfinite Set — 151
- 12.6 Hyperfinite Pigeonhole Principle — 151
- 12.7 Integrals as Hyperfinite Sums — 152

IV Nonstandard Frameworks — 155

13 Universes and Frameworks — 157
- 13.1 What Do We Need in the Mathematical World? — 158
- 13.2 Pairs Are Enough — 159
- 13.3 Actually, Sets Are Enough — 160
- 13.4 Strong Transitivity — 161
- 13.5 Universes — 162
- 13.6 Superstructures — 164

13.7	The Language of a Universe	166
13.8	Nonstandard Frameworks	168
13.9	Standard Entities	170
13.10	Internal Entities	172
13.11	Closure Properties of Internal Sets	173
13.12	Transformed Power Sets	174
13.13	Exercises on Internal Sets and Functions	176
13.14	External Images Are External	176
13.15	Internal Set Definition Principle	177
13.16	Internal Function Definition Principle	178
13.17	Hyperfiniteness	178
13.18	Exercises on Hyperfinite Sets and Sizes	180
13.19	Hyperfinite Summation	180
13.20	Exercises on Hyperfinite Sums	181

14 The Existence of Nonstandard Entities — **183**

14.1	Enlargements	183
14.2	Concurrence and Hyperfinite Approximation	185
14.3	Enlargements as Ultrapowers	187
14.4	Exercises on the Ultrapower Construction	189

15 Permanence, Comprehensiveness, Saturation — **191**

15.1	Permanence Principles	191
15.2	Robinson's Sequential Lemma	193
15.3	Uniformly Converging Sequences of Functions	193
15.4	Comprehensiveness	195
15.5	Saturation	198

V Applications — 201

16 Loeb Measure — **203**

16.1	Rings and Algebras	204
16.2	Measures	206
16.3	Outer Measures	208
16.4	Lebesgue Measure	210
16.5	Loeb Measures	210
16.6	μ-Approximability	212
16.7	Loeb Measure as Approximability	214
16.8	Lebesgue Measure via Loeb Measure	215

17 Ramsey Theory — **221**

17.1	Colourings and Monochromatic Sets	221
17.2	A Nonstandard Approach	223
17.3	Proving Ramsey's Theorem	224

	17.4	The Finite Ramsey Theorem	227
	17.5	The Paris–Harrington Version	228
	17.6	Reference	229

18 Completion by Enlargement — **231**

	18.1	Completing the Rationals	231
	18.2	Metric Space Completion	233
	18.3	Nonstandard Hulls	234
	18.4	p-adic Integers	237
	18.5	p-adic Numbers	245
	18.6	Power Series	249
	18.7	Hyperfinite Expansions in Base p	255
	18.8	Exercises	257

19 Hyperfinite Approximation — **259**

	19.1	Colourings and Graphs	260
	19.2	Boolean Algebras	262
	19.3	Atomic Algebras	265
	19.4	Hyperfinite Approximating Algebras	267
	19.5	Exercises on Generation of Algebras	269
	19.6	Connecting with the Stone Representation	269
	19.7	Exercises on Filters and Lattices	272
	19.8	Hyperfinite-Dimensional Vector Spaces	273
	19.9	Exercises on (Hyper) Real Subspaces	275
	19.10	The Hahn–Banach Theorem	275
	19.11	Exercises on (Hyper) Linear Functionals	278

20 Books on Nonstandard Analysis — **279**

Index — **283**

Part I
Foundations

1
What Are the Hyperreals?

1.1 Infinitely Small and Large

A nonzero number ε is defined to be *infinitely small*, or *infinitesimal*, if

$$|\varepsilon| < \tfrac{1}{n} \text{ for all } n = 1, 2, 3, \ldots.$$

In this case the reciprocal $\omega = \tfrac{1}{\varepsilon}$ will be *infinitely large*, or simply *infinite*, meaning that

$$|\omega| > n \text{ for all } n = 1, 2, 3, \ldots.$$

Conversely, if a number ω has this last property, then $\tfrac{1}{\omega}$ will be a nonzero infinitesimal.

However, in the real number system $\mathbb{R}$ there are no such things as nonzero infinitesimals and infinitely large numbers. Our aim here is to study a larger system, the *hyperreals*, which form an ordered field *$\mathbb{R}$ that contains $\mathbb{R}$ as a subfield, but also contains infinitely large and small numbers according to these definitions. The new entities in *$\mathbb{R}$, and the relationship between *$\mathbb{R}$ and $\mathbb{R}$, provide an intuitively appealing alternative approach to real analysis and topology, and indeed to many other branches of pure and applied mathematics.

1.2 Historical Background

Our mathematical heritage owes much to the creative endeavours of people who found it natural to think in terms of the infinite and the infinitesimal. By examining the words with which they expressed their ideas we can learn much about the origins of our twentieth-century perspective, even if that perspective itself makes it difficult, perhaps impossible, to recapture faithfully the "mind-set" of the past.

Archimedes

An old idea that has never lost its potency is to think of a geometric object as made up of an "unlimited" number of "indivisible" elements. Thus a curve might be regarded as a polygon with infinitely many sides of infinitesimal length, a plane figure as made up of parallel straight line segments viewed as strips of infinitesimal width, and a solid as composed of infinitely thin plane laminas.

The formula $A = \frac{1}{2}rC$ for the area of a circle in terms of its radius and circumference was very likely discovered by regarding the circle as made up of infinitely many segments consisting of isosceles triangles of height r with infinitesimal bases, these bases collectively forming the circle itself. In the third century BC., Archimedes gave a proof of this formula using the *method of exhaustion* that had been developed by Eudoxus more than a century earlier. This involved approximating the area arbitrarily closely by regular polygons. From the modern point of view we would say that as the number of sides increases, the sequence of areas of the polygons converges to the area of the circle, but the Greek mathematicians did not develop the idea of taking the limit of an infinite sequence. Instead, they used an indirect *reductio ad absurdum* argument, showing that if the area was not equal to $A = \frac{1}{2}rC$, then by taking polygons with sufficiently many sides a contradiction would follow.

Archimedes applied this approach to give proofs of many formulae for areas and volumes involving circles, parabolas, ellipses, spirals, spheres, cylinders, and solids of revolution. He wrote a treatise called *The Method of Mechanical Theorems* in which he explained how he discovered these formulae. His method was to imagine geometrical figures as being connected by a lever that is held in balance as the elements of one figure whose magnitude (area or volume) and centre of gravity is known are weighed against the elements of another whose magnitude is to be determined. These elements are as above: line segments in the case of plane figures, with length as the comparative "weight"; and plane laminas in the case of solids, weighted according to area.[1] Archimedes did not regard this procedure as

---

[1] A lucid illustration of the "Method" is given on pages 69–70 of the book

providing a proof, but said of a result obtained in this way that

> *this has not therefore been proved, but a certain impression has been created that the conclusion is true.*

The demonstration of its truth was then to be supplied by the method of exhaustion. The lesson of history is that the way in which a mathematical fact is discovered may be very different from the way that it is proven. Indeed Archimedes' treatise, along with all knowledge of his "method", was lost for many centuries and found again only in 1906.

Newton and Leibniz

In the latter part of the seventeenth century the differential and integral calculus was discovered by Isaac Newton and Gottfried Leibniz, independently. Leibniz created the notation dx for the difference in successive values of a variable x, thinking of this difference as infinitely small or "less than any assignable quantity". He also introduced the integral sign $\int$, an elongated "S" for "sum", and wrote the expression $\int y\,dx$ to mean the sum of all the infinitely thin rectangles of size $y \times dx$. He expressed what we now know as *Leibniz's rule* for the differential of a product xy in the form

$$dxy = x\,dy + y\,dx.$$

To demonstrate this he first observed that

> *dxy is the same thing as the difference between two successive xy's; let one of these be xy, and the other $x + dx$ into $y + dy$.*

Then calculating

$$\begin{aligned} dxy &= (x+dx)(y+dy) - xy \\ &= x\,dy + y\,dx + dx\,dy, \end{aligned}$$

he stated that the desired result follows by

> *the omission of the quantity $dx\,dy$, which is infinitely small in comparison with the rest, for it is supposed that dx and dy are infinitely small.*

Leibniz's views on the actual existence of infinitesimals make interesting reading. In response to certain criticisms, he drew attention to the fact that Archimedes and others

> *found out their wonderfully elegant theorems by the help of such ideas; these theorems they completed with reductio ad absurdum*

---

by C.H. Edwards cited in Section 1.4, showing how it yields the area under the graph of $y = x^2$ between 0 and 1.

6   1. What Are the Hyperreals?

> *proofs, by which they at the same time provided rigorous demonstrations and also concealed their methods,*

and went on to write:

> *It will be sufficient if, when we speak of infinitely great (or more strictly unlimited), or of infinitely small quantities (i.e., the very least of those within our knowledge), it is understood that we mean quantities that are indefinitely great or indefinitely small, i.e., as great as you please, or as small as you please, so that the error that one may assign may be less than a certain assigned quantity ... by infinitely great and infinitely small we understand something indefinitely great, or something indefinitely small, so that each conducts itself as a sort of class, and not merely as the last thing of a class ... it will be sufficient simply to make use of them as a tool that has advantages for the purpose of calculation, just as the algebraists retain imaginary roots with great profit.*

Further indication of this attitude is found in the following passage from an argument in one of his manuscripts:

> *If dx, ddx ... are by a certain fiction imagined to remain, even when they become evanescent, as if they were infinitely small quantities (and in this there is no danger, since **the whole matter can be always referred back to assignable quantities**), then ...*

Newton's formulation of the calculus used a different language and had a more dynamic conception of the phenomena under discussion. He considered *fluents* $x, y, \ldots$ as quantities varying in a spatial or temporal sense, and their *fluxions* $\dot{x}, \dot{y}, \ldots$ as

> *the speeds with which they flow and are increased by their generating motion.*

In modern parlance, the fluxion $\dot{x}$ is the derivative $\frac{dx}{dt}$ of x with respect to time t (or the velocity of x). Newton wrote (1671):

> *The moments of the fluent quantities (that is, their indefinitely small parts, by addition of which they increase during each infinitely small period of time) are as their speeds of flow ... if the moment of any particular one, say x, be expressed by the product of its speed $\dot{x}$ and an infinitely small quantity o (that is by $\dot{x}o$) ... it follows that quantities x and y after an infinitely small interval of time will become $x + \dot{x}o$ and $y + \dot{y}o$. Consequently, an equation which expresses a relationship of fluent*

> *quantities without variance at all times will express that relationship equally between $x + \dot{x}o$ and $y + \dot{y}o$ as between x and y; and so $x + \dot{x}o$ and $y + \dot{y}o$ may be substituted in place of the latter quantities, x and y, in the said equation.*

In other words, if (x, y) is a point on the curve defined by an equation in x and y, then $(x + \dot{x}o, y + \dot{y}o)$ is also on the curve. But this does not seem right: surely $(x + \dot{x}o, y + \dot{y}o)$ should lie on the tangent to the curve, the line through (x, y) of slope $\dot{y}/\dot{x}$, rather than on the curve itself? Moreover, in making the proposed substitution and carrying out algebraic calculations, Newton permitted himself to divide by the infinitely small quantity o while at the same time stating that

> *since o is supposed to be infinitely small so that it be able to express the moments of quantities, terms which have it as a factor will be equivalent to nothing in respect of others. I therefore cast them out ...*

which seems to amount to equating o to zero.

Such perplexities are typical of the confusions caused by the concepts of infinitesimal calculus. In later writing Newton himself tried to explain his theory of fluxions in terms of limits of ratios of quantities. He wrote that he did not (unlike Leibniz)

> *consider Mathematical Quantities as composed of Parts **extreamly small**, but as **generated by a continual motion**,*

and that

> *fluxions are very nearly as the **Augments of the Fluents**.*

His conception of limits is conveyed by the following passages:

> *Quantities, and the ratios of quantities, which in any finite time converge continually to equality, and before the end of time approach nearer to each other than by any given difference, become ultimately equal ... Those ultimate ratios with which quantities vanish are not truly the ratios of ultimate quantities, but limits towards which the ratios of quantities decreasing without limit do always converge; and to which they approach nearer than by any given difference, but never go beyond, nor in effect attain to, till the quantities are diminished ad infinitum.*

Newton considered that the use of limits of ratios provided an adequate basis for his calculus, without ultimately depending on indivisibles:

> *In Finite Quantities so to frame a Calculus, and thus to investigate the Prime and Ultimate Ratios of Nascent or Evanescent Finite Quantities, is agreeable to the Ancients; and I was willing*

8   1. What Are the Hyperreals?

> to shew, that in the Method of Fluxions there's no need of introducing Figures infinitely small into Geometry. For this Analysis may be performed in any Figures whatsoever, whether finite or infinitely small, so they are but imagined to be similar to the Evanescent Figures ...

Euler

The greatest champion of infinitely small and large numbers was Leonhard Euler, said to be the most prolific of all mathematicians. He simply assumed that such things exist and behave like finite numbers. A good illustration of his approach is to be found in the book *Introduction to the Analysis of the Infinite* (1748), where he developed infinite series for logarithmic, exponential, and trigonometric functions from the following basis:

> Let ω be an infinitely small number, or a fraction so small that, although not equal to zero, still $a^\omega = 1 + \psi$, where ψ is also an infinitely small number ... we let $\psi = k\omega$. Then we have $a^\omega = 1 + k\omega$, and with a as the base for the logarithms, we have $\omega = \log(1 + k\omega)$ ... If now we let $j = \frac{z}{\omega}$, where z denotes any finite number, since ω is infinitely small, then j is infinitely large. Then we have $\omega = \frac{z}{j}$, where ω is represented by a fraction with an infinite denominator, so that ω is infinitely small, as it should be.

Euler took it for granted that Newton's formula for the binomial series works for his numbers, and applied it to the expansion of $a^z = a^{\omega j} = (1 + k\omega)^j$ to deduce that

$$a^z = 1 + \frac{kz}{1!} + \frac{k^2 z^2}{2!} + \frac{k^3 z^3}{3!} + \cdots,$$

and hence when $z = 1$ that

$$a = 1 + \frac{k}{1!} + \frac{k^2}{2!} + \frac{k^3}{3!} + \cdots.$$

In fact, since $k\omega = \frac{kz}{j}$, the general term $\binom{j}{n}(k\omega)^n$ of the binomial series for a^z should be

$$\frac{j(j-1)(j-2)\cdots(j-n+1)}{n!} \cdot \frac{k^n z^n}{j^n},$$

but Euler reduced this to $\frac{k^n z^n}{n!}$ by the following extraordinary reasoning:

> Since j is infinitely large, $\frac{j-1}{j} = 1$, and the larger the number we substitute for j, the closer the value of the fraction $\frac{j-1}{j}$ comes to 1. Therefore, if j is a number larger than any assignable number, then $\frac{j-1}{j}$ is equal to 1. For the same reason $\frac{j-2}{j} = 1$, $\frac{j-3}{j} = 1$, and so forth.

1.2 Historical Background

His next step was a natural one:

> Since we are free to choose the base a for the system of logarithms, we now choose a in such a way that $k = 1$ ... we obtain the value for
>
> $$a = 2.71828182845904523536028.$$
>
> When this base is chosen, the logarithms are called natural or hyperbolic. The latter name is used since the quadrature of a hyperbola can be expressed through these logarithms. For the sake of brevity for this number $2.718281828459\cdots$ we will use the symbol e ...

Whereas the modern view is that

$$e = \lim_{n \to \infty} \left(1 + \frac{1}{n}\right)^n,$$

Euler had obtained it by stipulating that $e = \left(1 + \frac{1}{j}\right)^j$, and indeed $e^z = \left(1 + \frac{z}{j}\right)^j$, for infinitely large j. In this way he "proved" that

$$e^z = 1 + \frac{z}{1!} + \frac{z^2}{2!} + \frac{z^3}{3!} + \cdots,$$

$$\log(1+x) = x - \frac{x^2}{2} + \frac{x^3}{3} - \frac{x^4}{4} + \cdots,$$

and also showed that

$$\cos x = 1 - \frac{x^2}{2!} + \frac{x^4}{4!} - \cdots = \frac{e^{ix} + e^{-ix}}{2},$$

$$\sin x = x - \frac{x^3}{3!} + \frac{x^5}{5!} - \cdots = \frac{e^{ix} - e^{-ix}}{2i}$$

by using the equations $\cos \omega = 1$, $\sin \omega = \omega$, and $j = j - 1 = j - 2 = \cdots$ with ω infinitely small and j infinitely large.

Euler's demonstration that the function e^x is equal to its own derivative employed the practice, which, as we saw, was adopted by Leibniz and Newton, of "casting out" higher-order infinitesimals like $dx\, dy$, $(dx)^2$, $(dx)^3$, etc. Applying his series expansion for the exponential function to e^{dx} he argued that

$$\begin{aligned} d(e^x) &= e^{x+dx} - e^x \\ &= e^x(e^{dx} - 1) \\ &= e^x\left(dx + \frac{(dx)^2}{2!} + \frac{(dx)^3}{3!} + \cdots\right) \\ &= e^x\, dx. \end{aligned}$$

Demise of Infinitesimals

The conceptual foundations of the calculus continued to be controversial and to attract criticism, the most famous being that of Berkeley, who wrote (1734) in opposition to the ideas of Newton and his followers:

> And what are these fluxions? The velocities of evanescent increments? And what are these same evanescent increments? They are neither finite quantities, nor quantities infinitely small, nor yet nothing. May we not call them the ghosts of departed quantities?

Eventually infinitesimals were expunged from analysis, along with the dependence on intuitive geometric concepts and diagrams. The subject was "arithmetised" by the explicit construction of the real numbers out of the rational number system by the work of Dedekind, Cantor, and others around 1872. Weierstrass provided the purely arithmetical formulation of limits that we use today, defining $\lim_{x \to a} f(x) = L$ to mean that

$$(\forall \varepsilon > 0)\,(\exists \delta > 0) \text{ such that } 0 < |x - a| < \delta \text{ implies } |f(x) - L| < \varepsilon.$$

Robinson

Three centuries after the seminal discoveries of Newton and Leibniz, infinitesimals were restored with a vengeance by Abraham Robinson, who wrote in the preface to his 1966 book *Non-standard Analysis*:

> In the fall of 1960 it occurred to me that the concepts and methods of contemporary Mathematical Logic are capable of providing a suitable framework for the development of the Differential and Integral Calculus by means of infinitely small and infinitely large numbers.

The progress of symbolic logic in the twentieth century had produced an exact formulation of the syntax of mathematical statements; an account of what it is for a statement to be true of a mathematical system or structure—i.e. for the structure to be a *model* of the statement; and methods for obtaining models of prescribed statements. One such method comes from the *compactness theorem*:

- If a set Σ of statements (of an appropriate kind) has the property that each finite subset Σ' of Σ has a model (a structure of which all members of Σ' are true), then there must be a single structure that is a model of Σ itself.

Now suppose that we take $\Sigma_\mathbb{R}$ to consist of all appropriate statements true of $\mathbb{R}$ (including the axioms for ordered fields amongst other things) together with the infinitely many statements

$$0 < \varepsilon, \quad \varepsilon < 1, \quad \varepsilon < \tfrac{1}{2}, \quad \varepsilon < \tfrac{1}{3}, \quad \ldots, \quad \varepsilon < \tfrac{1}{n}, \quad \ldots.$$

Using the compactness theorem it can be deduced that $\Sigma_\mathbb{R}$ has a model *$\mathbb{R}$, which will be an ordered field in which the element ε is a positive infinitesimal. Moreover, this model will satisfy the *transfer principle*:

- Any appropriately formulated statement is true of *$\mathbb{R}$ if and only if it is true of $\mathbb{R}$.

This is reminiscent of Leibniz's above-quoted remark that

> *the whole matter can be always referred back to assignable quantities,*

and might even suggest that there is no point in considering *$\mathbb{R}$, since it satisfies the same theorems as $\mathbb{R}$. But on the contrary, what it offers is a new methodology for real analysis, because the availability of infinitesimals allows for easier and more intuitively natural proofs in *$\mathbb{R}$ of some theorems that can then immediately be inferred to hold of $\mathbb{R}$ by transfer.

Of course for this to work, the theorems in question must be "appropriately formulated", and explaining what this means is one of our major goals. As we shall see, *$\mathbb{R}$ fails to satisfy Dedekind's *completeness* axiom stipulating that any nonempty set with an upper bound must have a least upper bound, so this is not the sort of assertion to which transfer applies. In order to determine which statements are subject to it we will need the "concepts and methods of contemporary Mathematical Logic" that were available to Robinson, but not to Leibniz, nor indeed to those in the intervening period who tried to work with infinitesimals or construct non-Archimedean extensions of the real number system. Robinson's great achievement was to turn the transfer principle into a working tool of mathematical reasoning. In the last few decades it has been applied to many areas, including analysis, topology, algebra, number theory, mathematical physics, probability and stochastic processes, and mathematical economics.

To those unfamiliar with formal logic, the use of compactness may seem like a kind of sleight of hand. A model of $\Sigma_\mathbb{R}$ is produced, but we do not see where it came from. However, the compactness theorem itself has a proof, and one way to prove it is to use the notion of an *ultraproduct*, an algebraic construction that takes all the assumed models of the finite subsets of Σ and builds a model of Σ out of them. We can apply this construction directly to the structure $\mathbb{R}$ to build *$\mathbb{R}$ as a special kind of ultraproduct called an ultra*power*. This will be our first main task.

1.3 What Is a Real Number?

Consideration of this question provides motivation for the definition of the hyperreal number system. Here are some standard answers.

(1) A real number is an infinite decimal expression, such as

$$\sqrt{2} = 1.4142135623731\cdots,$$

that identifies $\sqrt{2}$ as the sum of the infinite power series

$$1 + \frac{4}{10} + \frac{1}{10^2} + \frac{4}{10^3} + \frac{2}{10^4} + \frac{1}{10^5} + \cdots.$$

(2) A real number is an element of a complete ordered field. Here "complete", often called *Dedekind complete*, means that any nonempty set with an upper bound must have a least upper bound. Any two complete ordered fields are isomorphic, so this notion uniquely characterises $\mathbb{R}$.

(3) A real number is a *Dedekind cut* in the set $\mathbb{Q}$ of rational numbers: a partition of $\mathbb{Q}$ into a pair $\langle L, U \rangle$ of nonempty disjoint subsets with every element of L less than every element of U and L having no largest member. Thus $\sqrt{2}$ can be identified with the cut

$$L = \{q \in \mathbb{Q} : q^2 < 2\}, \quad U = \{q \in \mathbb{Q} : q^2 > 2\}.$$

The set of all Dedekind cuts of $\mathbb{Q}$ can be made into a complete ordered field.

(4) A real number is an equivalence class of Cauchy sequences of rational numbers. A sequence $\langle r_1, r_2, r_3, \ldots \rangle$ is *Cauchy* if its terms get arbitrarily close to each other as we move along the sequence, i.e.,

$$\lim_{n,m \to \infty} |r_n - r_m| = 0.$$

Thus $\sqrt{2}$ is the limit of the rational Cauchy sequence

1, 1.4, 1.41, 1.414, 1.4142, 1.41421, 1.414213, ...

as well as being the limit of any of the subsequences of this sequence, and of other rational sequences besides.

Two Cauchy sequences $\langle r_1, r_2, r_3, \ldots \rangle$ and $\langle s_1, s_2, s_3, \ldots \rangle$ are *equivalent* if their corresponding terms approach each other arbitrarily closely:

$$\lim_{n \to \infty} |r_n - s_n| = 0.$$

This defines an equivalence relation on the set of rational-valued Cauchy sequences, and the resulting set of equivalence classes forms a complete ordered field. Any two equivalent Cauchy sequences will have the same limit, and so represent the same real number. For example, $\sqrt{2}$ corresponds to the equivalence class of the above sequence.

Answer (2) provides the basis for the *axiomatic* or *descriptive* approach to the analysis of ℝ. The object of study is simply described as being a complete ordered field, since all its properties derive from that fact. The axioms for a complete ordered field are listed, and everything follows from that. This is by far the favoured approach in introductory texts on real analysis.

The *constructive* approach takes as given only the rational number system and proceeds to construct ℝ explicitly. There are at least two ways to do this, due respectively to Dedekind (answer (3)) and Cantor (answer (4)).

It would be possible to develop an axiomatic approach to the hyperreals *ℝ by assuming that we are dealing with an ordered field containing ℝ as well as infinitesimals and satisfying the transfer principle "appropriately formulated". However, in view of the controversial history of the notion of infinitesimal, one could be forgiven for wondering whether this is an exercise in fantasy, or whether there does exist a number system satisfying the proposed axioms. The constructive approach is needed to resolve this issue. We will be discussing a construction of *ℝ out of ℝ that is analogous to Cantor's construction of ℝ out of ℚ. Hyperreal numbers will arise as equivalence classes of *real*-valued sequences, and the challenge will be to find an equivalence relation on such sequences that produces the desired outcome.

To conclude this introduction to our subject, let us examine another putative answer to the question "what is a real number?"—namely, that a real number is a point on the number line:

———————————————•———————————————

Now, the intuitive geometric idea of a line is an ancient one, much older than the notion of a set of points, let alone an infinite set. The identification of a line with the set of points lying on that line is a perspective that belongs to modern times. For Euclid a line was simply " length without breadth", and his diagrams and arguments involved lines with a finite number of points marked on them. By applying the field operations and taking limits of converging sequences we can assign a point to each real number, but the claim that this exhausts all the points on the line is just that: a claim. One could seek to justify it by invoking a principle such as the one attributed to Eudoxus and Archimedes that any two magnitudes are such that

the less can be multiplied so as to exceed the other.

This entails that for each real number r there is an integer $n > r$, and that precludes there being any infinitely large or small numbers in ℝ. But then one could say that the Eudoxus–Archimedes principle is just a property of those points on the line that correspond to "assignable" numbers. The hyperreal point of view is that the *geometric* line is capable of sustaining a much richer and more intricate number set than the *real* line.

1.4 Historical References

Amongst the numerous books available, the following are worth consulting for more details on the historical background we have been discussing.

M. E. BARON AND H. J. M. BOS. *Newton and Leibniz.* Open University Press, 1974.

J. M. CHILD. *The Early Mathematical Manuscripts of Leibniz.* Open Court Publishing Co., 1920.

E. J. DIJKSTERHUIS. *Archimedes.* Princeton University Press, 1987.

C. H. EDWARDS. *The Historical Development of the Calculus.* Springer, 1979.

LEONHARD EULER. *Introduction to the Analysis of the Infinite*, Book I, translated by John D. Blanton. Springer, 1988.

2
Large Sets

2.1 Infinitesimals as Variable Quantities

Cauchy (1789–1857) is regarded as one of the pioneers of the precision that is characteristic of contemporary mathematics. He wrote:

> My principal aim has been to reconcile rigor, which I have made a law to myself in my Cours d'analyse, with the simplicity which the direct consideration of infinitely small quantities produces.

His method was to consider infinitesimals as being variable quantities that vanish:

> When the successive numerical values of a variable decrease indefinitely so as to be smaller than any given number, this variable becomes what is called **infinitesimal**, or infinitely small quantity .... One says that a variable quantity becomes infinitely small when its value decreases numerically so as to converge to the limit zero.

Even today there are textbooks containing statements to the effect that a sequence satisfying
$$\lim_{n \to \infty} r_n = 0$$
is an *infinitesimal*, while one satisfying
$$\lim_{n \to \infty} r_n = \infty$$

is an *infinitely large magnitude*. Can we then construct a number system in which such sequences represent infinitely small and large numbers respectively?

According to Cauchy, the sequence

$$1, \tfrac{1}{2}, \tfrac{1}{3}, \tfrac{1}{4}, \ldots$$

is an infinitesimal, as is

$$\tfrac{1}{2}, \tfrac{1}{4}, \tfrac{1}{6}, \tfrac{1}{8}, \ldots$$

If these represent infinitely small numbers, perhaps we should regard the second as being half the size of the first because it converges twice as quickly? Similarly, the sequences

$$1, 2, 3, 4, \ldots,$$
$$2, 4, 6, 8, \ldots$$

both represent infinitely large magnitudes, and arguably the second is twice as big as the first because it diverges to ∞ twice as quickly. On the other hand, the distinct sequences

$$1, 2, 3, 4, \ldots,$$
$$2, 2, 3, 4, \ldots$$

will presumably represent the same infinite number.

These ideas are attractive because they suggest the possibility of using infinitely small and large numbers as measures of rates of convergence. But in the construction of real numbers out of Cauchy sequences (Section 1.3), all sequences converging to zero are identified with the number zero itself, while diverging sequences have no role to play at all. Clearly then we need a very different kind of equivalence relation among sequences than the one used in Cantor's construction of $\mathbb{R}$ from $\mathbb{Q}$.

2.2 Largeness

Let $r = \langle r_1, r_2, r_3, \ldots \rangle$ and $s = \langle s_1, s_2, s_3, \ldots \rangle$ be real-valued sequences. We are going to say that r and s are equivalent if they agree at a "large" number of places, i.e., if their *agreement set*

$$E_{rs} = \{n : r_n = s_n\}$$

is large in some sense that is to be determined. Whatever "large" means, there are some properties we will want it to have:

- $\mathbb{N} = \{1, 2, 3, \ldots\}$ must be large, in order to ensure that any sequence will be equivalent to itself.

- Equivalence is to be a transitive relation, so if E_{rs} and E_{st} are large, then E_{rt} must be large. Since $E_{rs} \cap E_{st} \subseteq E_{rt}$, this suggests the following requirement:

    If A and B are large sets, and $A \cap B \subseteq C$, then C is large.

    In particular, this entails that if A and B are large, then so is their intersection $A \cap B$, while if A is large, then so is any of its supersets $C \supseteq A$.

- The empty set $\emptyset$ is not large, or otherwise by the previous requirement all subsets of $\mathbb{N}$ would be large, and so all sequences would be equivalent.

Requiring $A \cap B$ to be large when A and B are large may seem restrictive, but there are natural situations in which all three requirements are fulfilled. One such is when a set $A \subseteq \mathbb{N}$ is declared to be large if it is *cofinite*, i.e. its complement $\mathbb{N} - A$ is finite. This means that A contains "almost all" or "ultimately all" members of $\mathbb{N}$. Although this is a plausible notion of largeness, it is not adequate to our needs. The number system we are constructing is to be linearly ordered, and a natural way to do this, in terms of our general approach, is to take the equivalence class of sequence r to be *less than* that of s if the set

$$L_{rs} = \{n : r_n < s_n\}$$

is large. But consider the sequences

$$r = \langle 1, 0, 1, 0, 1, 0, \ldots \rangle,$$
$$s = \langle 0, 1, 0, 1, 0, 1, \ldots \rangle.$$

Their agreement set is empty, so they determine distinct equivalence classes, one of which should be less than the other. But L_{rs} (the even numbers) is the complement of L_{sr} (the odds), so both are infinite and neither is cofinite. Apparently our definition of largeness is going to require the following condition:

- For any subset A of $\mathbb{N}$, one of A and $\mathbb{N} - A$ is large.

The other requirements imply that A and $\mathbb{N} - A$ cannot both be large, or else $A \cap (\mathbb{N} - A) = \emptyset$ would be. Thus the large sets are precisely the complements of the ones that are not large. Either the even numbers form a large set or the odd ones do, but they cannot both do so, so which is it to be?

Can there in fact be such a notion of largeness, and if so, how do we show it?

2.3 Filters

Let I be a nonempty set. The *power set* of I is the set
$$\mathcal{P}(I) = \{A : A \subseteq I\}$$
of all subsets of I. A *filter* on I is a nonempty collection $\mathcal{F} \subseteq \mathcal{P}(I)$ of subsets of I satisfying the following axioms:

- Intersections: if $A, B \in \mathcal{F}$, then $A \cap B \in \mathcal{F}$.
- Supersets: if $A \in \mathcal{F}$ and $A \subseteq B \subseteq I$, then $B \in \mathcal{F}$.

Thus to show $B \in \mathcal{F}$, it suffices to show
$$A_1 \cap \cdots \cap A_n \subseteq B,$$
for some n and some $A_1, \ldots, A_n \in \mathcal{F}$.

A filter $\mathcal{F}$ contains the empty set $\emptyset$ iff $\mathcal{F} = \mathcal{P}(I)$. We say that $\mathcal{F}$ is *proper* if $\emptyset \notin \mathcal{F}$. Every filter contains I, and in fact $\{I\}$ is the smallest filter on I.

An *ultrafilter* is a proper filter that satisfies

- for any $A \subseteq I$, either $A \in \mathcal{F}$ or $A^c \in \mathcal{F}$, where $A^c = I - A$.

2.4 Examples of Filters

(1) $\mathcal{F}^i = \{A \subseteq I : i \in A\}$ is an ultrafilter, called the *principal ultrafilter generated by* i. If I is finite, then every ultrafilter on I is of the form $\mathcal{F}^i$ for some $i \in I$, and so is principal.

(2) $\mathcal{F}^{co} = \{A \subseteq I : I - A \text{ is finite}\}$ is the *cofinite*, or *Fréchet*, filter on I, and is proper iff I is infinite. $\mathcal{F}^{co}$ is not an ultrafilter.

(3) If $\emptyset \neq \mathcal{H} \subseteq \mathcal{P}(I)$, then the *filter generated by* $\mathcal{H}$, i.e., the smallest filter on I including $\mathcal{H}$, is the collection
$$\mathcal{F}^\mathcal{H} = \{A \subseteq I : A \supseteq B_1 \cap \cdots \cap B_n \text{ for some } n \text{ and some } B_i \in \mathcal{H}\}$$
(cf. Exercise 2.7(4)). For $\mathcal{H} = \emptyset$ we put $\mathcal{F}^\mathcal{H} = \{I\}$.

If $\mathcal{H}$ has a single member B, then $\mathcal{F}^\mathcal{H} = \{A \subseteq I : A \supseteq B\}$, which is called the *principal filter generated by* B. The ultrafilter $\mathcal{F}^i$ of Example (1) is the special case of this when $B = \{i\}$.

(4) If $\{\mathcal{F}_x : x \in X\}$ is a collection of filters on I that is linearly ordered by set inclusion, in the sense that $\mathcal{F}_x \subseteq \mathcal{F}_y$ or $\mathcal{F}_y \subseteq \mathcal{F}_x$ for any $x, y \in X$, then
$$\bigcup_{x \in X} \mathcal{F}_x = \{A : \exists x \in X \, (A \in \mathcal{F}_x)\}$$
is a filter on I.

2.5 Facts About Filters

(1) The filter axioms are equivalent to the requirement that

$$A \cap B \in \mathcal{F} \quad \text{iff} \quad A, B \in \mathcal{F}.$$

(2) If $\mathcal{F} \subseteq \mathcal{P}(I)$ satisfies the superset axiom, then $\mathcal{F} \neq \emptyset$ iff $I \in \mathcal{F}$. Hence $\{I\} \subseteq \mathcal{F}$ for any filter $\mathcal{F}$.

(3) An ultrafilter $\mathcal{F}$ satisfies

$$\begin{aligned} A \cap B \in \mathcal{F} &\quad \text{iff} \quad A \in \mathcal{F} \text{ and } B \in \mathcal{F}, \\ A \cup B \in \mathcal{F} &\quad \text{iff} \quad A \in \mathcal{F} \text{ or } B \in \mathcal{F}, \\ A^c \in \mathcal{F} &\quad \text{iff} \quad A \notin \mathcal{F}. \end{aligned}$$

(4) Let $\mathcal{F}$ be an ultrafilter and $\{A_1, \ldots, A_n\}$ a finite collection of pairwise disjoint ($A_i \cap A_j = \emptyset$) sets such that

$$A_1 \cup \cdots \cup A_n \in \mathcal{F}.$$

Then $A_i \in \mathcal{F}$ for *exactly one* i such that $1 \leq i \leq n$.

(5) If an ultrafilter contains a finite set, then it contains a one-element set and is principal. Hence a *nonprincipal* ultrafilter must contain all cofinite sets. *This is a critical property used in the construction of infinitesimals and infinitely large numbers (cf. Section 3.8).*

(6) $\mathcal{F}$ is an ultrafilter on I iff it is a *maximal* proper filter on I, i.e., a proper filter that cannot be extended to a larger proper filter on I (cf. Exercise 2.7(5)).

(7) A collection $\mathcal{H} \subseteq \mathcal{P}(I)$ has the *finite intersection property*, or *fip*, if the intersection of every nonempty finite subcollection of $\mathcal{H}$ is nonempty, i.e.,

$$B_1 \cap \cdots \cap B_n \neq \emptyset$$

for any n and any $B_1, \ldots, B_n \in \mathcal{H}$.

Then the filter $\mathcal{F}^\mathcal{H}$ is proper iff $\mathcal{H}$ has the fip.

(8) If $\mathcal{H}$ has the fip, then for any $A \subseteq I$, at least one of the sets $\mathcal{H} \cup \{A\}$ and $\mathcal{H} \cup \{A^c\}$ has the fip.

2.6 Zorn's Lemma

Fact 2.5(8) suggests a way to construct an ultrafilter: start with a set that has the fip, e.g., $\{I\}$, and go through all the members A of $\mathcal{P}(I)$ in turn, adding whichever of A and A^c preserves the fip. This presupposes that there is such a thing as a listing of the members of $\mathcal{P}(I)$ that could be used to "go through them all in turn".

Now, the assertion that any set can be listed in this way is one of many mathematical statements that are equivalent to the *axiom of choice*, which asserts that for any given collection of sets there exists a function whose range of values selects a member from each set in the collection. The version of the axiom of choice most used in algebra is *Zorn's lemma*:

> *If $(P, \leq)$ is a partially ordered set in which every linearly ordered subset (or "chain") has an upper bound in P, then P contains a $\leq$-maximal element.*

(An element p of a partially ordered set is $\leq$-maximal if there is no element q of P that is *greater than* p in the sense that $p \leq q$ and $p \neq q$.)

Here is an outline of how Zorn's lemma can be proven from the assumption that the axiom of choice is true. Let f be a choice function defined on the collection of all nonempty subsets of P. Thus for each such set X, $f(X) \in X$. Now begin with the element $p_0 = f(P)$. If p_0 is maximal, we have the desired conclusion. Otherwise, we use f to choose an element p_1 that is greater than p_0, i.e., $p_1 = f(X)$, where $X = \{x \in P : p_0 < x\} \neq \emptyset$. If p_1 is maximal, again we are done. Otherwise we can choose p_2 with $p_1 < p_2$. If this process repeats denumerably many times, the p_n's form a chain. By the hypothesis of Zorn's lemma, this chain must then have an upper bound p_ω, giving

$$p_0 < p_1 < \cdots < p_n < \cdots < p_\omega.$$

If p_ω is maximal, we are done; otherwise there exists $p_{\omega+1} > p_\omega$, and so on. Now, this whole construction cannot go on forever, because eventually we will "run out of" elements of P. At some point we must finish with the desired maximal element.

This argument shows what is going on behind the scenes when Zorn's lemma is applied. Of course the part about running out of elements is vague, and to make it precise we would need to introduce the theory of infinite "ordinal" numbers and "well-orderings" in order to show that we can generate a list of all the elements of P. In many applications, appealing directly to Zorn's lemma itself allows us to avoid such machinery. For example:

Theorem 2.6.1 *Any collection of subsets of I that has the finite intersection property can be extended to an ultrafilter on I.*

Proof. If $\mathcal{H}$ has the fip, then the filter $\mathcal{F}^\mathcal{H}$ generated by $\mathcal{F}$ is proper (2.5(7)). Let P be the collection of all proper filters on I that include $\mathcal{F}^\mathcal{H}$, partially ordered by set inclusion $\subseteq$. Then every linearly ordered subset of P has an upper bound in P, since by 2.4(4) the union of this chain is in P. Hence by Zorn's lemma P has a maximal element, which is thereby a maximal proper filter on I and thus an ultrafilter by 2.5(6). □

Corollary 2.6.2 *Any infinite set has a nonprincipal ultrafilter on it.*

Proof. If I is infinite, the cofinite filter $\mathcal{F}^{co}$ is proper and has the finite intersection property, and so is included in an ultrafilter $\mathcal{F}$. But for any $i \in I$ we have $I - \{i\} \in \mathcal{F}^{co} \subseteq \mathcal{F}$, so $\{i\} \notin \mathcal{F}$, whereas $\{i\} \in \mathcal{F}^i$. Hence $\mathcal{F} \neq \mathcal{F}^i$. Thus $\mathcal{F}$ is nonprincipal. □

This result is the key fact we need to begin our construction of the hyperreal number system. We could have simply taken it as an assumption, but there is insight to be gained in showing how it derives from more general principles like Zorn's lemma. In fact, a deeper set-theoretic analysis proves that there are as many nonprincipal ultrafilters on an infinite set I as there possibly could be: an ultrafilter is a member of the double power set $\mathcal{P}(\mathcal{P}(I))$, and there is a one-to-one correspondence between the set of all nonprincipal ultrafilters on I and $\mathcal{P}(\mathcal{P}(I))$ itself.

2.7 Exercises on Filters

(1) If $\emptyset \neq A \subseteq I$, there is an ultrafilter $\mathcal{F}$ on I with $A \in \mathcal{F}$.

(2) There exists a nonprincipal ultrafilter on $\mathbb{N}$ containing the set of even numbers, and another containing the set of odd numbers.

(3) An ultrafilter on a finite set must be principal.

(4) For $\mathcal{H} \subseteq \mathcal{P}(I)$, let $\mathcal{F}^{\mathcal{H}}$ be as defined in Example 2.4(3).

   (i) Show that $\mathcal{F}^{\mathcal{H}}$ is a filter that includes $\mathcal{H}$, i.e., $\mathcal{H} \subseteq \mathcal{F}^{\mathcal{H}}$.

   (ii) Show that $\mathcal{F}^{\mathcal{H}}$ is included in any other filter that includes $\mathcal{H}$.

(5) Let $\mathcal{F}$ be a proper filter on I.

   (i) Show that $\mathcal{F} \cup \{A^c\}$ has the finite intersection property iff $A \notin \mathcal{F}$.

   (ii) Use (i) to deduce that $\mathcal{F}$ is an ultrafilter iff it is a maximal proper filter on I.

3
Ultrapower Construction of the Hyperreals

3.1 The Ring of Real-Valued Sequences

Let $\mathbb{N} = \{1, 2, \dots\}$, and let $\mathbb{R}^{\mathbb{N}}$ be the set of all sequences of real numbers. A typical member of $\mathbb{R}^{\mathbb{N}}$ has the form $r = \langle r_1, r_2, r_3, \dots \rangle$, which may be denoted more briefly as $\langle r_n : n \in \mathbb{N} \rangle$ or just $\langle r_n \rangle$.

For $r = \langle r_n \rangle$ and $s = \langle s_n \rangle$, put

$$r \oplus s = \langle r_n + s_n : n \in \mathbb{N} \rangle,$$
$$r \odot s = \langle r_n \cdot s_n : n \in \mathbb{N} \rangle.$$

Then $\langle \mathbb{R}^{\mathbb{N}}, \oplus, \odot \rangle$ is a commutative ring with zero $\mathbf{0} = \langle 0, 0, 0, \dots \rangle$ and unity $\mathbf{1} = \langle 1, 1, \dots \rangle$, and additive inverses given by

$$-r = \langle -r_n : n \in \mathbb{N} \rangle.$$

It is not, however, a field, since

$$\langle 1, 0, 1, 0, 1, \dots \rangle \odot \langle 0, 1, 0, 1, 0, \dots \rangle = \mathbf{0},$$

so the two sequences on the left of this equation are nonzero elements of $\mathbb{R}^{\mathbb{N}}$ with a zero product; hence neither can have a multiplicative inverse. Indeed, no sequence that has at least one zero term can have such an inverse in $\mathbb{R}^{\mathbb{N}}$.

3.2 Equivalence Modulo an Ultrafilter

Let $\mathcal{F}$ be a fixed nonprincipal ultrafilter on the set $\mathbb{N}$ (such exists by Corollary 2.6.2). $\mathcal{F}$ will be used to construct a quotient ring of $\mathbb{R}^{\mathbb{N}}$.

Define a relation $\equiv$ on $\mathbb{R}^{\mathbb{N}}$ by putting

$$\langle r_n \rangle \equiv \langle s_n \rangle \quad \text{iff} \quad \{n \in \mathbb{N} : r_n = s_n\} \in \mathcal{F}.$$

When this relation holds it may be said that the two sequences *agree on a large set*, or *agree almost everywhere modulo* $\mathcal{F}$, or *agree at almost all* n.

3.3 Exercises on Almost-Everywhere Agreement

(1) $\equiv$ is an equivalence relation on $\mathbb{R}^{\mathbb{N}}$.

(2) $\equiv$ is a congruence on the ring $\langle \mathbb{R}^{\mathbb{N}}, \oplus, \odot \rangle$, which means that if $r \equiv r'$ and $s \equiv s'$, then

$$r \oplus s \equiv r' \oplus s' \text{ and } r \odot s \equiv r' \odot s'.$$

(3) $\langle 1, \frac{1}{2}, \frac{1}{3}, \dots \rangle \not\equiv \langle 0, 0, 0, \dots \rangle$.

3.4 A Suggestive Logical Notation

It is suggestive to denote the agreement set $\{n \in \mathbb{N} : r_n = s_n\}$ by $[\![r = s]\!]$, rather than E_{rs} as in Section 2.2. Thus

$$r \equiv s \quad \text{iff} \quad [\![r = s]\!] \in \mathcal{F}.$$

Then results like 3.3(1) and 3.3(2) can be handled by first proving properties such as those in Section 3.5 below.

The set $[\![r = s]\!]$ may be thought of as the *interpretation*, or *value*, of the statement "$r = s$", or as a measure of the extent to which "$r = s$" is true. Normally we think of a statement as having one of two values: it is either *true* or *false*. Here, instead of assigning truth values, we take the value of a statement to be a subset of $\mathbb{N}$. When $[\![r = s]\!] \in \mathcal{F}$, it is sometimes said that $r = s$ *almost everywhere* (modulo $\mathcal{F}$).

This idea can be applied to other logical assertions, such as inequalities, by defining

$$\begin{aligned} [\![r < s]\!] &= \{n \in \mathbb{N} : r_n < s_n\}, \\ [\![r > s]\!] &= \{n \in \mathbb{N} : r_n > s_n\}, \\ [\![r \leq s]\!] &= \{n \in \mathbb{N} : r_n \leq s_n\}, \end{aligned}$$

and so on.

3.5 Exercises on Statement Values

(1) $[\![r = s]\!] \cap [\![s = t]\!] \subseteq [\![r = t]\!]$.

(2) $[\![r = r']\!] \cap [\![s = s']\!] \subseteq [\![r \oplus s = r' \oplus s']\!] \cap [\![r \odot s = r' \odot s']\!]$.

(3) $[\![r = r']\!] \cap [\![s = s']\!] \cap [\![r < s]\!] \subseteq [\![r' < s']\!]$.

(4) If $r \equiv r'$ and $s \equiv s'$, then $[\![r < s]\!] \in \mathcal{F}$ iff $[\![r' < s']\!] \in \mathcal{F}$.

3.6 The Ultrapower

The equivalence class of a sequence $r \in \mathbb{R}^{\mathbb{N}}$ under $\equiv$ will be denoted by $[r]$. Thus
$$[r] = \{s \in \mathbb{R}^{\mathbb{N}} : r \equiv s\}.$$

The quotient set (set of equivalence classes) of $\mathbb{R}^{\mathbb{N}}$ by $\equiv$ is
$${}^*\mathbb{R} = \{[r] : r \in \mathbb{R}^{\mathbb{N}}\}.$$

Define
$$[r] + [s] = [r \oplus s] = [\langle r_n + s_n \rangle],$$
$$[r] \cdot [s] = [r \odot s] = [\langle r_n \cdot s_n \rangle],$$

and
$$[r] < [s] \quad \text{iff} \quad [\![r < s]\!] \in \mathcal{F} \quad \text{iff} \quad \{n \in \mathbb{N} : r_n < s_n\} \in \mathcal{F}.$$

By 3.3(2) and 3.5(4) these notions are *well-defined*, which means that they are independent of the equivalence class representatives chosen to define them.

A simpler notation, which is attractive but puts some burden on the reader, is to write $[r_n]$ for the equivalence class $[\langle r_n : n \in \mathbb{N} \rangle]$ of the sequence whose nth term is r_n. The definitions of addition and multiplication then read

$$[r_n] + [s_n] = [r_n + s_n],$$
$$[r_n] \cdot [s_n] = [r_n \cdot s_n].$$

Theorem 3.6.1 *The structure $\langle {}^*\mathbb{R}, +, \cdot, < \rangle$ is an ordered field with zero $[0]$ and unity $[1]$.*

Proof. (Sketch) As a quotient ring of $\mathbb{R}^{\mathbb{N}}$, ${}^*\mathbb{R}$ is readily shown to be a commutative ring with zero $[0]$ and unity $[1]$, and additive inverses given by
$$-[\langle r_n : n \in \mathbb{N} \rangle] = [\langle -r_n : n \in \mathbb{N} \rangle],$$

or more briefly, $-[r_n] = [-r_n]$. To show that it has multiplicative inverses, suppose $[r] \neq [\mathbf{0}]$. Then $r \not\equiv \mathbf{0}$, i.e., $\{n \in \mathbb{N} : r_n = 0\} \notin \mathcal{F}$, so as $\mathcal{F}$ is an ultrafilter, $J = \{n \in \mathbb{N} : r_n \neq 0\} \in \mathcal{F}$. Define a sequence s by putting

$$s_n = \begin{cases} \frac{1}{r_n} & \text{if } n \in J, \\ 0 & \text{otherwise.} \end{cases}$$

Then $[\![r \odot s = 1]\!]$ is equal to J, so $[\![r \odot s = 1]\!] \in \mathcal{F}$, giving $r \odot s \equiv 1$ and hence

$$[r] \cdot [s] = [r \odot s] = [\mathbf{1}]$$

in *$\mathbb{R}$. But this means that $[s]$ is the multiplicative inverse $[r]^{-1}$ of $[r]$.

To see that the ordering $<$ on *$\mathbb{R}$ is *linear*, observe that $\mathbb{N}$ is the disjoint union of the three sets

$$[\![r < s]\!], \quad [\![r = s]\!], \quad [\![s < r]\!],$$

so exactly one of the three belongs to $\mathcal{F}$ (by 2.5(4)), and so exactly one of

$$[r] < [s], \quad [r] = [s], \quad [s] < [r]$$

is true. It remains to show that the set $\{[r] : [\mathbf{0}] < [r]\}$ of "positive" elements in *$\mathbb{R}$ is closed under addition and multiplication. This is left as an exercise.

□

In the proof just given we were trying in effect to show that $[r_n]^{-1} = [r_n^{-1}]$, but were constrained by the fact that the real number r_n^{-1} may not exist for some n. The reason why $[r]^{-1}$ nonetheless exists is that r_n^{-1} exists for *almost all n* (i.e., for all n in the set $\{n \in \mathbb{N} : r_n \neq 0\} \in \mathcal{F}$). This relationship between *$\mathbb{R}$ and $\mathbb{R}$ characterises the definitions of the relations $=$, $<$, $>$, etc. in *$\mathbb{R}$, in the sense that

$$\begin{aligned}
[r_n] = [s_n] &\quad \text{iff} \quad r_n = s_n \text{ for almost all } n, \\
[r_n] < [s_n] &\quad \text{iff} \quad r_n < s_n \text{ for almost all } n, \\
[r_n] + [s_n] = [t_n] &\quad \text{iff} \quad r_n + s_n = t_n \text{ for almost all } n, \\
[r_n] \cdot [s_n] = [t_n] &\quad \text{iff} \quad r_n \cdot s_n = t_n \text{ for almost all } n,
\end{aligned}$$

and so on. Let us call this relationship the *almost-all criterion*. As we will see, it holds for many other properties and is the basis of the transfer principle. Theorem 3.6.1 is itself a special case of transfer: *$\mathbb{R}$ is an ordered field because $\mathbb{R}$ is. This is explained further in Section 4.5.

The ring $\mathbb{R}^{\mathbb{N}}$ is an example of what is known in algebra as a *direct power* of $\mathbb{R}$, a special case of the notion of direct product. An *ultrapower* is a quotient of a direct power that arises from the congruence relation defined by an ultrafilter.

3.7 Including the Reals in the Hyperreals

We can identify a real number $r \in \mathbb{R}$ with the constant sequence $\mathbf{r} = \langle r, r, \dots \rangle$ and hence assign to it the *$\mathbb{R}$-element

$$^*r = [\mathbf{r}] = [\langle r, r, \dots \rangle].$$

It can be shown that for $r, s \in \mathbb{R}$, we have

$$\begin{aligned}
^*(r+s) &= {}^*r + {}^*s, \\
^*(r \cdot s) &= {}^*r \cdot {}^*s, \\
^*r < {}^*s &\text{ iff } r < s, \\
^*r = {}^*s &\text{ iff } r = s.
\end{aligned}$$

Hence

Theorem 3.7.1 *The map $r \mapsto {}^*r$ is an order-preserving field isomorphism from $\mathbb{R}$ into *$\mathbb{R}$.* □

This result allows us to identify the real number r with $^*r$ whenever convenient, and hence to regard $\mathbb{R}$ as a subfield of *$\mathbb{R}$. In particular we may identify $[\mathbf{0}]$ with 0 and $[\mathbf{1}]$ with 1.

3.8 Infinitesimals and Unlimited Numbers

Let $\varepsilon = \langle 1, \frac{1}{2}, \frac{1}{3}, \dots \rangle = \langle \frac{1}{n} : n \in \mathbb{N} \rangle$. Then

$$[\![\mathbf{0} < \varepsilon]\!] = \{n \in \mathbb{N} : 0 < \tfrac{1}{n}\} = \mathbb{N} \in \mathcal{F},$$

so $[\mathbf{0}] < [\varepsilon]$ in *$\mathbb{R}$. But if r is any positive real number, then the set

$$[\![\varepsilon < \mathbf{r}]\!] = \{n \in \mathbb{N} : \tfrac{1}{n} < r\}$$

is cofinite (because ε converges to 0 in $\mathbb{R}$!). Now, since $\mathcal{F}$ is nonprincipal, it contains all cofinite sets (2.5(5)), so $[\![\varepsilon < \mathbf{r}]\!] \in \mathcal{F}$ and therefore $[\varepsilon] < {}^*r$ in *$\mathbb{R}$. Thus $[\varepsilon]$ is a *positive infinitesimal*.

Now let $\omega = \langle 1, 2, 3, \dots \rangle$. Then for any $r \in \mathbb{R}$, the set

$$[\![\mathbf{r} < \omega]\!] = \{n \in \mathbb{N} : r < n\}$$

is cofinite (by the Eudoxus–Archimedes principle!) and so belongs to $\mathcal{F}$, showing that $^*r < [\omega]$ in *$\mathbb{R}$. Thus $[\omega]$ is "infinitely large" compared to $\mathbb{R}$ in *$\mathbb{R}$, although we will prefer to use the adjective *unlimited* to describe such entities. In fact $\varepsilon \cdot \omega = \mathbf{1}$, so $[\omega] = [\varepsilon]^{-1}$ and $[\varepsilon] = [\omega]^{-1}$.

The properties observed of $[\varepsilon]$ and $[\omega]$ show that *$\mathbb{R}$ is a proper extension of $\mathbb{R}$, and hence a new structure. Even more directly, for any $r \in \mathbb{R}$, the set $[\![\mathbf{r} = \omega]\!]$ is either $\emptyset$, or equal to $\{r\}$ when $r \in \mathbb{N}$, so cannot belong to $\mathcal{F}$, implying *$r \neq [\omega]$. Thus $[\omega] \in$ *$\mathbb{R} - \mathbb{R}$.

This argument depends crucially on the fact that $\mathcal{F}$ is nonprincipal. If $\mathcal{F}$ were principal, then there would be some fixed $\underline{n} \in \mathbb{N}$ such that

$$\mathcal{F} = \mathcal{F}^{\underline{n}} = \{A \subseteq \mathbb{N} : \underline{n} \in A\}.$$

But then each sequence $s \in \mathbb{R}^{\mathbb{N}}$ would agree almost everywhere with the sequence taking the constant value $s_{\underline{n}}$, and from this it would follow that *$\mathbb{R} = \{{}^*r : r \in \mathbb{R}\}$, and hence *$\mathbb{R}$ would be isomorphic to $\mathbb{R}$. The details of this are left as an exercise: the essential point is that use of a principal ultrafilter to construct *$\mathbb{R}$ does not lead to anything new.

Our discussion of ε and ω shows in fact that if r is any real-valued sequence converging to zero, then $[r]$ is an infinitesimal in *$\mathbb{R}$, while if r diverges to ∞, then $[r]$ is unlimited in *$\mathbb{R}$. Thus we have achieved the objective proposed in Section 2.1 of building a number system with these features.

Now that we have shown that there are infinitesimals in *$\mathbb{R}$, we can begin to apply the field operations to them to construct new numbers. What happens for instance if we multiply or divide an infinitesimal by a positive real number? Or by a negative real number? The general arithmetic of hyperreals will be described in Chapter 5.

Exercise 3.8.1
Use only general properties of ordered fields to deduce from the fact that $[\varepsilon]$ is a positive infinitesimal the conclusion that $[\varepsilon]^{-1}$ is greater than every real number.

3.9 Enlarging Sets

A subset A of $\mathbb{R}$ can be "enlarged" to a subset *A of *$\mathbb{R}$: for each $r \in \mathbb{R}^{\mathbb{N}}$, put

$$[r] \in {}^*A \quad \text{iff} \quad \{n \in \mathbb{N} : r_n \in A\} \in \mathcal{F}.$$

Thus we are declaring, by the almost-all criterion, that $[r_n]$ is in *A iff r_n is in A for almost all n. Again it has to be checked that this is well-defined. Invoking the $[\![\ldots]\!]$ notation, put

$$[\![r \in A]\!] = \{n \in \mathbb{N} : r_n \in A\}.$$

Then

$$[\![r = r']\!] \cap [\![r \in A]\!] \subseteq [\![r' \in A]\!],$$

so

$$r \equiv r' \;\&\; [\![r \in A]\!] \in \mathcal{F} \quad \text{implies} \quad [\![r' \in A]\!] \in \mathcal{F}$$

as required. We have

$$[r] \in {}^*A \quad \text{iff} \quad [\![r \in A]\!] \in \mathcal{F}.$$

Observe that if $s \in A$, then $[\![s \in A]\!] = \mathbb{N} \in \mathcal{F}$ (where $\mathbf{s} = \langle s, s, \dots \rangle$ as usual), so $^*s \in {}^*A$. Identifying s with $^*s$, we may regard $^*A$ as a superset of $A: A \subseteq {}^*A$. Elements of $^*A - A$ may be thought of as new "nonstandard", or "ideal", members of A that live in $^*\mathbb{R}$.

For example, let $A = \mathbb{N}$, and $\omega = \langle 1, 2, 3, \dots \rangle$ as above. Then $[\![\omega \in \mathbb{N}]\!] = \mathbb{N} \in \mathcal{F}$, so $[\omega] \in {}^*\mathbb{N}$. $[\omega]$ is a "nonstandard natural number".

Theorem 3.9.1 *Any infinite subset of $\mathbb{R}$ has nonstandard members.*

Proof. Note first that this result must depend on $\mathcal{F}$ being nonprincipal, because if $\mathcal{F}$ were principal, there would be no nonstandard elements of $^*\mathbb{R}$ at all.

Now, if $A \subseteq \mathbb{R}$ is infinite, then there is a sequence r of elements of A whose terms are all distinct. Then $[\![r \in A]\!] = \mathbb{N} \in \mathcal{F}$, so $[r] \in {}^*A$. But for each $s \in A$, $\{n : r_n = s\}$ is either $\emptyset$ or a singleton, neither of which can belong to $\mathcal{F}$ (2.5(5)), so $[r] \neq {}^*s$. Hence $[r] \in {}^*A - A$. □

The converse of this theorem is also true (cf. the next exercise), so the property of having nonstandard members exactly characterises the infinite sets.

3.10 Exercises on Enlargement

(1) If A is finite, show that $^*A = A$, and hence A has no nonstandard members.

(2) $A \subseteq B$ iff $^*A \subseteq {}^*B$,
    $A = B$ iff $^*A = {}^*B$.

(3) $^*(A \cup B) = {}^*A \cup {}^*B$,
    $^*(A \cap B) = {}^*A \cap {}^*B$,
    $^*(A - B) = {}^*A - {}^*B$,
    $^*\emptyset = \emptyset$.

(4) Is it true that $^*(\bigcup_{n=1}^{\infty} A_n) = \bigcup_{n=1}^{\infty} {}^*A_n$?

(5) Show that if $A \subseteq \mathbb{R}$, then $^*A \cap \mathbb{R} = A$.

(6) For $a, b \in \mathbb{R}$, let $[a, b]$ be the closed interval $\{x \in \mathbb{R} : a \leq x \leq b\}$. Prove that $^*[a, b] = \{x \in {}^*\mathbb{R} : a \leq x \leq b\}$.

(7) $^*\mathbb{Z}$ is a subring of $^*\mathbb{R}$, i.e., $^*\mathbb{Z}$ is closed under $+$, $\cdot$, $-$.

(8) If $\mathbb{R}^+ = \{x \in \mathbb{R} : x > 0\}$, show that $^*(\mathbb{R}^+) = \{x \in {}^*\mathbb{R} : x > 0\}$, i.e., $^*(\mathbb{R}^+) = (^*\mathbb{R})^+$.

3.11 Extending Functions

A function $f : \mathbb{R} \to \mathbb{R}$ extends to $^*f : {}^*\mathbb{R} \to {}^*\mathbb{R}$ as follows. First, for each sequence $r \in \mathbb{R}^\mathbb{N}$, let $f \circ r$ be the sequence $\langle f(r_1), f(r_2), \ldots \rangle$. Then put

$$^*f([r]) = [f \circ r].$$

In other words,

$$^*f([\langle r_1, r_2, \ldots \rangle]) = [\langle f(r_1), f(r_2), \ldots \rangle],$$

or in the simplified notation,

$$^*f([r_n]) = [f(r_n)].$$

Now, in general,

$$[\![r = r']\!] \subseteq [\![f \circ r = f \circ r']\!],$$

and so

$$r \equiv r' \quad \text{implies} \quad f \circ r \equiv f \circ r',$$

ensuring that $^*f$ is well-defined. Observe that $^*f$ obeys the almost-all criterion:

$$^*f([r]) = [s] \quad \text{iff} \quad [\![f \circ r = s]\!] \in \mathcal{F}$$
$$\text{iff} \quad \{n \in \mathbb{N} : f(r_n) = s_n\} \in \mathcal{F}$$
$$\text{iff} \quad f(r_n) = s_n \text{ for almost all } n.$$

For example, the sine function is extended to all of $^*\mathbb{R}$ by

$$^*\sin([r]) = [\langle \sin(r_1), \sin(r_2), \ldots \rangle] = [\sin(r_n)].$$

3.12 Exercises on Extensions

(1) Show that $^*f$ agrees with f on $\mathbb{R}$: if $r \in \mathbb{R}$, then $^*f(r) = f(r)$.

(2) If f is injective, so is $^*f$. What about surjectivity?

(3) For $x \in {}^*\mathbb{R}$, let
$$|x| = \begin{cases} x & \text{if } x > 0, \\ 0 & \text{if } x = 0, \\ -x & \text{if } x < 0 \end{cases}$$
be the usual definition of the absolute value function. Show that this extends the definition of $|\cdot|$ on $\mathbb{R}$: $|[r]| = [\langle |r_1|, |r_2|, \ldots \rangle] = [|r_n|]$.

(4) Let χ_A be the characteristic function of a set $A \subseteq \mathbb{R}$. Show that $^*(\chi_A) = \chi_{*A}$.

(5) Show how to define $^*f$ when f is a function of more than one argument.

3.13 Partial Functions and Hypersequences

Let $f : A \to \mathbb{R}$ be a function whose domain A is a subset of $\mathbb{R}$ (e.g., $f(x) = \tan x$). Then f extends to a function $^*f : {^*A} \to {^*\mathbb{R}}$ whose domain is the enlargement of A, i.e., $\operatorname{dom} {^*f} = {^*(\operatorname{dom} f)}$.

To define this extension, take $r \in \mathbb{R}^\mathbb{N}$ with $[r] \in {^*A}$, so that

$$[\![r \in A]\!] = \{n \in \mathbb{N} : r_n \in A\} \in \mathcal{F}.$$

Let

$$s_n = \begin{cases} f(r_n) & \text{if } n \in [\![r \in A]\!], \\ 0 & \text{if } n \notin [\![r \in A]\!] \end{cases}$$

(it is enough to define s_n for almost all n). Then put

$$^*f([r]) = [s].$$

Essentially, we have defined

$$^*f([r_n]) = [f(r_n)]$$

as in Section 3.11, but with a modification to cater for the complication that $f(r_n)$ may not be defined for some n. The construction works because $f(r_n)$ exists for almost all n modulo $\mathcal{F}$.

It is readily shown that if $r \in A$, then $^*f(^*r) = {^*(f(r))}$, or identifying $^*r$ with r etc., we have $^*f(r) = f(r)$, so $^*f$ extends f. Therefore it would do no harm to drop the $*$ symbol and just use f for the extension as well, and **we will do so most of the time**. It is a particularly natural practice for the more common mathematical functions. For instance, the function $\sin x$ is now defined for all hyperreals $x \in {^*\mathbb{R}}$.

An important case of this construction concerns *sequences*. A real-valued sequence is just a function $s : \mathbb{N} \to \mathbb{R}$, and so the construction extends this to a *hypersequence* $s : {^*\mathbb{N}} \to {^*\mathbb{R}}$. Hence the term s_n is now defined even when $n \in {^*\mathbb{N}} - \mathbb{N}$.

3.14 Enlarging Relations

Let P be a k-ary relation on $\mathbb{R}$. Thus P is a set of k-tuples: a subset of $\mathbb{R}^k$. For given sequences $r^1, \ldots, r^k \in \mathbb{R}^\mathbb{N}$, define

$$[\![P(r^1, \ldots, r^k)]\!] = \{n \in \mathbb{N} : P(r_n^1, \ldots, r_n^k)\}.$$

Now P can be enlarged to a k-ary relation $*P$ on $*\mathbb{R}$, i.e., a subset of $(*\mathbb{R})^k$. For this we use the notation $*P([r^1],\ldots,[r^k])$ to mean that the k-tuple $\langle[r^1],\ldots,[r^k]\rangle$ belongs to $*P$. The definition is:

$$*P([r^1],\ldots,[r^k]) \quad \text{iff} \quad [\![P(r^1,\ldots,r^k)]\!] \in \mathcal{F}$$
$$\text{iff} \quad P(r_n^1,\ldots,r_n^k) \text{ for almost all } n.$$

As always with a definition involving equivalence classes named by particular elements, it must be shown that the notion is well-defined. In this case we can prove

$$[\![r^1 = s^1]\!] \cap \cdots \cap [\![r^k = s^k]\!] \cap [\![P(r^1,\ldots,r^k)]\!] \subseteq [\![P(s^1,\ldots,s^k)]\!],$$

so that *if $r^1 \equiv s^1$ and ... and $r^k \equiv s^k$ and $[\![P(r^1,\ldots,r^k)]\!] \in \mathcal{F}$, then $[\![P(s^1,\ldots,s^k)]\!] \in \mathcal{F}$.*

When $r^1,\ldots,r^k$ are real numbers,

$$P(r^1,\ldots,r^k) \quad \text{iff} \quad *P(*r^1,\ldots,*r^k),$$

showing that $*P$ is an extension of P.

This definition of k-ary $*P$ encompasses the work of Sections 3.9–3.13 on extensions of sets and functions. A subset A of $\mathbb{R}$ is just a *unary* relation ($k=1$), so the definition of $*A$ is a special case of that of $*P$. When P is any of the relations $=, <, >, \leq$ on $\mathbb{R}$, then $*P$ is the corresponding relation that we defined on $*\mathbb{R}$, because

$$[r] = [s] \quad \text{iff} \quad [\![r = s]\!] \in \mathcal{F},$$
$$[r] < [s] \quad \text{iff} \quad [\![r < s]\!] \in \mathcal{F},$$

and so on.

An m-ary function $f : \mathbb{R}^m \to \mathbb{R}$ can be identified with its $(m+1)$-ary *graph*

$$\text{Graph } f = \{\langle r^1,\ldots,r^m,s\rangle : f(r^1,\ldots,r^m) = s\}.$$

Then the extension of Graph f to $*\mathbb{R}$ is just the graph of the extension $*f : *\mathbb{R}^m \to *\mathbb{R}$ of f (Exercise 3.12(5)), i.e.,

$$*(\text{Graph } f) = \text{Graph}\,(*f).$$

Moreover, Graph f is defined even when f is a *partial* function, and so that case is covered as well.

3.15 Exercises on Enlarged Relations

(1) If $A_1,\ldots,A_k$ are subsets of $\mathbb{R}$, put $P = A_1 \times \cdots \times A_k$ and apply the definition of $*P$ to show that

$$*(A_1 \times \cdots \times A_k) = *A_1 \times \cdots \times *A_k.$$

In particular, explain why $*(\mathbb{R}^k) = (*\mathbb{R})^k$, so that it is okay to write $*\mathbb{R}^k$.

(2) Let dom P denote the domain of a binary relation. If $P \subseteq \mathbb{R}^2$, show that $*(\text{dom } P) = \text{dom } *P$.

(3) Generalise Exercise (2) to k-ary relations. In particular, show that if f is a partial m-ary function, then the domain of $*f$ is given by

$$\text{dom } *f = *(\text{dom } f) \subseteq *\mathbb{R}^m.$$

3.16 Is the Hyperreal System Unique?

The construction of $*\mathbb{R}$ as a quotient ring of $\mathbb{R}^\mathbb{N}$ depends on the choice of the nonprincipal ultrafilter $\mathcal{F}$ that determines the congruence $\equiv$. But there are many such ultrafilters on $\mathbb{N}$, as many as there are subsets of $\mathcal{P}(\mathbb{N})$—the set of all nonprincipal ultrafilters on $\mathbb{N}$ is in bijective correspondence with $\mathcal{P}(\mathcal{P}(\mathbb{N}))$.

Now, it has been shown that under a certain set-theoretic assumption called the *continuum hypothesis* the choice of $\mathcal{F}$ is irrelevant: all quotients of $\mathbb{R}^\mathbb{N}$ with respect to nonprincipal ultrafilters on $\mathbb{N}$ are isomorphic as ordered fields. To explain this assumption, let us say that a set A is *smaller than* set B, and that B is *larger than* A, if there exists an injective function from A to B, but none from B to A. A famous result of Cantor is that $\mathbb{R}$ is bigger than $\mathbb{N}$ (and more generally that a set A is always smaller than its power set $\mathcal{P}(A)$). The continuum hypothesis asserts that there is no subset of $\mathbb{R}$ that is smaller than $\mathbb{R}$ but bigger than $\mathbb{N}$. This implies that $\mathbb{R}$ represents the least "infinite size" greater than the size of $\mathbb{N}$.

The continuum hypothesis is neither provable nor disprovable from the generally accepted axioms of set theory, including the axiom of choice. Thus we can say that if we take the continuum hypothesis as an axiom, then our construction of $*\mathbb{R}$ produces a unique result. Without this assumption the situation is undetermined.

4
The Transfer Principle

What properties are preserved in passing from $\mathbb{R}$ to *$\mathbb{R}$? We have seen a number of examples, and will now consider some more in order to illustrate the powerful logical transfer principle that underlies them. To formulate this principle we will need to develop a precise language in which to describe transferable properties. Ultimately this will allow us to abandon the ultrapower description of *$\mathbb{R}$ and ultrafilter calculations, in the same way that the Dedekind completeness principle allows us to abandon the view of real numbers as cuts or equivalence classes of Cauchy sequence of rationals.

Later it will be seen that the strength of nonstandard analysis lies in the ability to transfer properties back from *$\mathbb{R}$ to $\mathbb{R}$, providing a new technique for exploring real analysis.

4.1 Transforming Statements

1. *The Eudoxus–Archimedes Principle.* The statement
$$\forall x\, \exists m\, (x < m \text{ and } m \in \mathbb{N})$$
is true when the variable x ranges over $\mathbb{R}$, but is no longer true when x ranges over *$\mathbb{R}$ (e.g., let $x = [\langle 1, 2, 3, \ldots \rangle]$). But if $\mathbb{N}$ is replaced by its "*-transform" *$\mathbb{N}$, the result is the statement
$$\forall x\, \exists m\, (x < m \text{ and } m \in {}^*\mathbb{N}),$$
which *is* true when x ranges over all of *$\mathbb{R}$.

This example shows that in order to determine the truth value of a sentence, we need to explain what values a quantified variable is allowed to take. We can achieve this by using *bounded* quantifiers, a notational device that displays the range of quantification explicitly. Thus the first sentence can be conveniently written as

$$\forall x \in \mathbb{R}\, \exists m \in \mathbb{N}\, (x < m),$$

which is simply *true*. Its ∗-transform

$$\forall x \in {}^*\mathbb{R}\, \exists m \in {}^*\mathbb{N}\, (x < m)$$

is also true. On the other hand,

$$\forall x \in {}^*\mathbb{R}\, \exists m \in \mathbb{N}\, (x < m)$$

is false.

2. *Density of the Rationals.* This is expressed by the true statement

$$\forall x, y \in \mathbb{R}\, (x < y \ \text{ implies }\ \exists q \in \mathbb{Q}\, (x < q < y)).$$

The ∗-transform

$$\forall x, y \in {}^*\mathbb{R}\, (x < y \ \text{ implies }\ \exists q \in {}^*\mathbb{Q}\, (x < q < y))$$

is also true.

3. *Finiteness.* Let $A = \{r_1, \ldots, r_k\}$ be a finite subset of $\mathbb{R}$. Then the statement

$$\forall x \in A\, (x = r_1 \text{ or } x = r_2 \text{ or } \cdots \text{ or } x = r_k)$$

is true, and so is its ∗-transform

$$\forall x \in {}^*A\, (x = {}^*r_1 \text{ or } x = {}^*r_2 \text{ or } \cdots \text{ or } x = {}^*r_k).$$

Since we identify r_i with ${}^*r_i$ in regarding $\mathbb{R}$ as a subset of ${}^*\mathbb{R}$, this implies that ${}^*A = A$. Hence finite sets of standard numbers have no nonstandard elements (Ex. 3.10(1)).

Question: why does this argument not work for infinite sets (Theorem 3.9.1)?

4. *Finitary Set Operations.* If $A, B \subseteq \mathbb{R}$, then the statement

$$\forall x \in \mathbb{R}\, (x \in A \cup B \ \text{ iff }\ x \in A \text{ or } x \in B)$$

transforms to the true statement

$$\forall x \in {}^*\mathbb{R}\, (x \in {}^*(A \cup B) \ \text{ iff }\ x \in {}^*A \text{ or } x \in {}^*B),$$

which shows that $^*(A \cup B) = {^*A} \cup {^*B}$. Similarly for the other results of Exercise 3.10(3).

Question: why does the argument not work for unions of infinitely many sets (3.10(4))?

5. *Discreteness of* $\mathbb{N}$. If $n \in \mathbb{N}$, then the statement

$$\forall x \in \mathbb{N} \, (n \leq x \leq n+1 \quad \text{implies} \quad x = n \text{ or } x = n+1)$$

transforms to

$$\forall x \in {^*\mathbb{N}} \, ({^*n} \leq x \leq {^*(n+1)} \quad \text{implies} \quad x = {^*n} \text{ or } x = {^*(n+1)}),$$

which again is true. Since $n = {^*n}$ and likewise $^*(n+1) = n+1$, this shows that there are no nonstandard members of $^*\mathbb{N}$ occurring *between* any standard natural numbers. Also, there are no members of $^*\mathbb{N}$ smaller than 1, i.e.,

$$\forall x \in {^*\mathbb{N}} \, (x \geq 1);$$

hence any member of $^*\mathbb{N} - \mathbb{N}$ must be *greater than* all members of $\mathbb{N}$, and so is unlimited, i.e., infinitely large (Section 3.8).

6. *Unbounded Sets of Reals.* If we assume that there is an unlimited $N \in {^*\mathbb{N}}$, then we can *deduce* the Eudoxus–Archimedes principle in the following way. If r is any real number, then $r < N$, since N is unlimited, and so the statement

$$\exists n \in {^*\mathbb{N}} \, (r < n)$$

is true. This is the $*$-transform of the statement

$$\exists n \in \mathbb{N} \, (r < n),$$

and as we shall see, a statement must be true if its $*$-transform is. This shows that there is a positive integer greater than r.

More generally, this argument can be used to show that if the enlargement $^*A$ of a set of reals has an unlimited member, then A itself must be unbounded in $\mathbb{R}$ in the sense that for any real r there is a member of A that is greater than r. In brief: if A has an unlimited nonstandard member, then it has arbitrarily large standard members.

It appears from these examples that the $*$-transform of a statement arises by attaching the "$*$" prefix to symbols that name particular entities, but not attaching it to *variable* symbols. The precise definition of $*$-transform will be laid out in Section 4.4.

Exercise 4.1.1
Verify the truth of the $*$-transforms given in 1–5 above.

4.2 Relational Structures

The examples just given used a semiformal logical symbolism to express statements that were asserted to be true or false of the structures $\mathbb{R}$ and $^*\mathbb{R}$. This symbolism will now be explicitly described.

A *relational structure* is a system of the form

$$\mathcal{S} = \langle S, Rel_\mathcal{S}, Fun_\mathcal{S}\rangle,$$

where S is a nonempty set, $Rel_\mathcal{S}$ is a collection of finitary relations on S, and $Fun_\mathcal{S}$ is a collection of finitary functions on S (possibly including partial functions). For instance, associated with any set S is the *full* structure

$$\langle S, Rel_S, Fun_S\rangle$$

based on S, where Rel_S consists of *all* the finitary relations on S, and Fun_S consists of *all* the finitary functions on S. Since sets are unary relations, a full structure includes all subsets of S in Rel_S.

The full structure based on $\mathbb{R}$ will be denoted by $\mathfrak{R}$. Associated with it is the structure

$$^*\mathfrak{R} = \langle ^*\mathbb{R}, \{^*P : P \in Rel_\mathbb{R}\}, \{^*f : f \in Fun_\mathbb{R}\}\rangle.$$

Thus $^*\mathfrak{R}$ consists of the extensions $^*P$ and $^*f$ of all relations and functions on $\mathbb{R}$, as defined in Sections 3.9, 3.11, and 3.14. $^*\mathfrak{R}$ is *not*, however, a full structure, since there are relations on $^*\mathbb{R}$ that are not of the form $^*P$ for any $P \in Rel_\mathfrak{R}$.

Exercise 4.2.1
Show that none of the sets $\mathbb{N}, \mathbb{Z}, \mathbb{Q}, \mathbb{R}$, and indeed no infinite subset of $\mathbb{R}$, can belong to $Rel_{^*\mathfrak{R}}$.

4.3 The Language of a Relational Structure

Associated with each relational structure $\mathcal{S}$ is a language $\mathcal{L}_\mathcal{S}$ based on the following alphabet:

- **Logical Connectives:**

$\wedge$	and
$\vee$	or
$\neg$	not
$\rightarrow$	implies
$\leftrightarrow$	if and only if

- **Quantifier Symbols:**

$\forall$	for all
$\exists$	there exists

4.3 The Language of a Relational Structure

- **Parentheses:** $(,),[,]$

- **Variables:** A countable collection of symbols, for which we use letters like x, y, z, x_1, x', etc.

Terms of $\mathcal{L}_S$

These are strings of symbols defined inductively by the following rules:

- Each variable is an $\mathcal{L}_S$-term.

- Each element s of S is an $\mathcal{L}_S$-term, called a ***constant***.

- If $f \in Fun_S$ is an m-ary function, and $\tau_1, \ldots, \tau_m$ are $\mathcal{L}_S$-terms, then $f(\tau_1, \ldots, \tau_m)$ is an $\mathcal{L}_S$-term.

We will adopt the customary conventions of notation that depart from this formal definition. For instance, we continue to use the usual "infix" notation for binary operations, writing $\tau_1 + \tau_2$ or $\tau_1 \cdot \tau_2$ for $f(\tau_1, \tau_2)$ when f is addition or multiplication, etc. We will also retain such standard notations as $1/x$, $\frac{1}{x}$, x^2, $|x|$, e^x, etc.

What Does a Term Name?

A ***closed*** term is one that has no variables and therefore is made up of constants and function symbols. Such a term is intended to name a particular element of the structure S. But there are many opportunities in mathematics to write down symbolic expressions that have no meaning because the element they purport to name does not exist, as in

$$\tan(\pi/2).$$

(In ordinary language there is the similar phenomenon of syntactically well-formed expressions that do not denote anything, such as Chomsky's famous "green ideas".)

A closed term is *undefined* if it does not name anything. Here are the rules that determine when, and what, a closed term names:

- The constant s names itself.

- If $\tau_1, \ldots, \tau_m$ name the elements $s_1, \ldots, s_m$, respectively, and the m-tuple $\langle s_1, \ldots, s_m \rangle$ is in the domain of f, then $f(\tau_1, \ldots, \tau_m)$ names the element $f(s_1, \ldots, s_m)$.

- $f(\tau_1, \ldots, \tau_m)$ is undefined if one of $\tau_1, \ldots, \tau_m$ is undefined, or if they are all defined but name an m-tuple that is not in the domain of f.

Atomic Formulae of $\mathcal{L}_\mathcal{S}$

These are strings of the form

$$P(\tau_1, \ldots, \tau_k)$$

where $P \in Rel_\mathcal{S}$ is k-ary, and the τ_i are $\mathcal{L}_\mathcal{S}$-terms. Such strings assert basic relationships between elements of $\mathcal{S}$ and serve as the building blocks for more complex expressions.

We also use conventional notation for atomic formulae where appropriate. For binary relations ($k = 2$) there is the usual infix notation: $P(\tau_1, \tau_2)$ is written

$$\tau_1 = \tau_2$$

when P is the identity relation $\{\langle a, b \rangle \in S \times S : a = b\}$, and as

$$\tau_1 < \tau_2$$

when $P = \{\langle a, b \rangle : a < b\}$. Similarly for $\tau_1 > \tau_2$, $\tau_1 \leq \tau_2$, $\tau_1 \geq \tau_2$.

When $k = 1$ we have *unary*, or *monadic*, atomic formulae of the form $P(\tau)$, with P being a subset of S. Such a formula expresses membership of P and so will usually be written in the form

$$\tau \in P.$$

Formulae

- Each atomic $\mathcal{L}_\mathcal{S}$-formula is an $\mathcal{L}_\mathcal{S}$-formula.

- If φ and ψ are $\mathcal{L}_\mathcal{S}$-formulae, then so are $\varphi \wedge \psi$, $\varphi \vee \psi$, $\neg \varphi$, $\varphi \to \psi$, $\varphi \leftrightarrow \psi$.

- If φ is an $\mathcal{L}_\mathcal{S}$-formula, x is any variable symbol, and $P \in Rel_\mathcal{S}$ is unary, i.e., P is a subset of S, then

$$(\forall x \in P)\,\varphi, \qquad (\exists x \in P)\,\varphi$$

are $\mathcal{L}_\mathcal{S}$-formulae. Here P is the *bound* of the quantifier in question.

A formula is said to be *defined* if and only if all of its closed terms are defined.

Parentheses will be inserted or deleted in formulae where convenient to aid legibility. Various abbreviations and informalities will be used, such as writing

$$x \leq y \leq z$$

for the formula $(x \leq y) \wedge (y \leq z)$, or collapsing a string of similar quantifiers with the same bound like

$$(\forall x \in P)\,(\forall y \in P)\,(\forall z \in P)$$

to the form $(\forall x, y, z \in P)$.

Sentences

An occurrence of the variable x within a formula ψ is called *bound* if it is located within a formula of the form $(\forall x \in P)\,\varphi$ or $(\exists x \in P)\,\varphi$ that is part of ψ. An occurrence that is not bound is *free*. Thus in

$$(x < 1) \wedge (\forall x \in \mathbb{N})\,(x > y),$$

the first occurrence of x is free, while the others are bound, and the only occurrence of y is free.

If a formula contains free variables, then it has no particular meaning until we assign some values to those free variables. Thus the above formula makes a true assertion if $x = y = 0$, but if $x = 2$, then it cannot be true whatever the value of y is.

A *sentence* is a formula in which all variables are bound. The role of each symbol in a sentence is determined. There are no free variables that need to be assigned a value, and if the closed terms of the sentence are all defined then it has a fixed meaning and makes a definite assertion. A defined sentence is either ***true*** or ***false***.

An *atomic sentence* is just an atomic formula $P(\tau_1, \ldots, \tau_k)$ that is a sentence. This means that the terms $\tau_1, \ldots, \tau_k$ are all closed, i.e., the formula has no variables at all.

Truth and Quantification

Suppose that there is only one variable, say x, that has any free occurrence in a certain formula φ. Then we write $\varphi(s)$ for the sentence that is obtained by substituting the constant s in place of all free occurrences of x in φ. For example, if φ is

$$\tan(-x) = -\tan(x),$$

then $\varphi(\pi/2)$ is the (undefined) atomic sentence

$$\tan(-\pi/2) = -\tan(\pi/2).$$

Now consider the truth of a defined sentence of the form $(\forall x \in P)\,\varphi$. Here only the variable x can have any free occurrence in φ, so we can form sentences of the type $\varphi(s)$. Intuitively, $(\forall x \in P)\,\varphi$ asserts that whatever φ "says about x" is true of each member of P, provided that this is defined, and so it asserts that the sentence $\varphi(s)$ is true for every element s of P for which it is defined. Thus

- $(\forall x \in P)\,\varphi$ *is true if and only if for all s in P, if the sentence $\varphi(s)$ is defined, then it is true.*

For example, the following sentence is true:

$$(\forall x \in \mathbb{R})\,[\tan(-x) = -\tan(x)].$$

The corresponding analysis of the existential quantifier is

- $(\exists x \in P)\varphi$ is true if and only if there is some s in P for which $\varphi(s)$ is (defined and) true.

The standard meanings of the symbolic connectives $\land, \lor, \neg, \to, \leftrightarrow$ are given by the rules:

- $\varphi \land \psi$ is true if and only if φ is true and ψ is true.
- $\varphi \lor \psi$ is true if and only if φ is true or ψ is true.
- $\neg \varphi$ is true if and only if φ is not true (i.e., is false).
- $\varphi \to \psi$ is true if and only if the truth of φ implies that of ψ (i.e., either φ is false or else ψ is true).
- $\varphi \leftrightarrow \psi$ is true if and only if $\varphi \to \psi$ and $\psi \to \varphi$ are true (i.e., φ and ψ are either both true or both false).

With all these rules, calculation of the *truth value* of a sentence is reduced to the determination of the truth value of atomic sentences. For them we have

- $P(\tau_1, \ldots, \tau_k)$ is true if and only if the closed terms $\tau_1, \ldots, \tau_k$ are all defined and the k-tuple of elements they name belongs to P.

This analysis of the meaning of "true" may appear to be making the obvious seem complex and convoluted. But as was said in Section 1.2, it is precisely this exact formulation of the syntax of mathematical statements, with an associated account of their truth conditions, that makes the theory of infinitesimals possible. We are able to distinguish exactly which properties are transferable between $\mathbb{R}$ and *$\mathbb{R}$ because we can give an explicit description of the sentences that express such properties.

4.4 *-Transforms

A formula in the language $\mathcal{L}_\mathfrak{R}$ of the real-number structure $\mathfrak{R}$ has symbols P, f for relations and functions of $\mathfrak{R}$. It can be turned into a formula of the language $\mathcal{L}_{*\mathfrak{R}}$ of the hyperreal structure *$\mathfrak{R}$ by replacing P by *P, and f by *f. Any constant r naming a real number is left as is, since we identify r in $\mathfrak{R}$ with *r in fs.

More precisely, we first define the *-transform *τ of an $\mathcal{L}_\mathfrak{R}$-term τ. This is obtained by replacing each function symbol f occurring in τ by *f, leaving the variables and constants of τ alone. Even more formally, we can give the definition by induction on the formation of τ, using the following rules:

- If τ is a variable or an $\mathcal{L}_\mathfrak{R}$-constant, then *τ is just τ.

- If τ is $f(\tau_1,\ldots,\tau_m)$, then *τ is *$f(*\tau_1,\ldots,*\tau_m)$.

The *-transform *φ of an $\mathcal{L}_\Re$-formula φ is obtained by

- replacing each term τ occurring in φ by *τ;

- replacing the relation symbol P of any atomic formula occurring in φ by *P; and

- replacing the "bound" P of any quantifier $(\forall x \in P)$ or $(\exists x \in P)$ occurring in φ by *P.

Again we can spell this out by induction on the formation of φ:

$$
\begin{aligned}
*(P(\tau_1,\ldots,\tau_k)) &:= *P(*\tau_1,\ldots,*\tau_k) \\
*(\varphi \wedge \psi) &:= *\varphi \wedge *\psi \\
*(\varphi \vee \psi) &:= *\varphi \vee *\psi \\
(\neg \varphi) &:= \neg(\varphi) \\
*(\varphi \to \psi) &:= *\varphi \to *\psi \\
*(\varphi \leftrightarrow \psi) &:= *\varphi \leftrightarrow *\psi \\
*(\forall x \in P)\varphi &:= (\forall x \in *P)\,*\varphi \\
*(\exists x \in P)\varphi &:= (\exists x \in *P)\,*\varphi.
\end{aligned}
$$

We tend to drop the * symbol when referring to the transforms of some of the more well-known relations like $=, \neq, <, \geq$, etc., and well-known mathematical functions like sin, cos, log, e^x, etc. For instance,

$$
\begin{aligned}
*(\pi < f(x+1)) &:= (\pi < *f(x+1)), \\
*(\sin e^x \in \mathbb{Q}) &:= (\sin e^x \in *\mathbb{Q}),
\end{aligned}
$$

and so on. Even further, we noted in Section 3.13 that it would do no harm to drop the * symbol in referring to the extension *f of *any* function f. If this practice is adopted systematically, then the transform *τ of each term τ will just be τ itself. Then atomic formulae like

$$\tau_1 = \tau_2, \quad \tau_1 \neq \tau_2, \quad \tau_1 < \tau_2, \quad \tau_1 \geq \tau_2,$$

etc. that express basic equalities and inequalities will be left alone under *-transformation, while a *membership formula* $\tau \in P$ becomes $\tau \in *P$.

With all these conventions in place, the general procedure for "adding the stars" reduces simply to replacing

$$
\begin{aligned}
P(\tau_1,\ldots,\tau_k) &\quad \text{by} \quad *P(\tau_1,\ldots,\tau_k), \\
\forall x \in P &\quad \text{by} \quad \forall x \in *P, \\
\exists x \in P &\quad \text{by} \quad \exists x \in *P.
\end{aligned}
$$

To summarise all of this in words; the essence of *-transformation is to

(i) replace the bound P of any quantifier by its enlargement $^*P$; and

(ii) replace relations appearing in atomic formulae by their enlargements, but only in the (unary) case of a membership formula ($\tau \in P$), or for relations of arity greater than one other than the common relations $=, \neq, <, \geq$, etc.

Exercise 4.4.1
Review the examples of Section 4.1, formalising them precisely in $\mathcal{L}_\Re$, and verify that they conform to our definition of $*$-transform.

4.5 The Transfer Principle

The notion of an $\mathcal{L}_\Re$ sentence and its $*$-transform provides an explanation of the notion of an "appropriately formulated statement" as discussed in Section 1.2, and hence provides a first answer to the question as to which properties are subject to transfer between $\mathbb{R}$ and $^*\mathbb{R}$: any property expressible by an $\mathcal{L}_\Re$-sentence is transferable. Formally, the *transfer principle* is stated as follows:

- A defined $\mathcal{L}_\Re$-sentence φ is true if and only if $^*\varphi$ is true.

As a first illustration of this, beyond the examples of Section 4.1, consider the proof that $^*\mathbb{R}$ is an ordered field (Theorem 3.6.1). Now, the fact that $\mathbb{R}$ is an ordered field can be expressed in a finite number of $\mathcal{L}_\Re$-sentences, like

$$(\forall x, y \in \mathbb{R})\,(x + y = y + x),$$
$$(\forall x \in \mathbb{R})\,(x \cdot 1 = x),$$
$$(\forall x, y \in \mathbb{R})\,(x < y \lor x = y \lor y < x),$$

and so on. By transfer we can immediately conclude that the $*$-transforms of these sentences are true, showing that $^*\mathbb{R}$ is an ordered field. In particular, to show that multiplicative inverses exist in $^*\mathbb{R}$, instead of making an ultrapower construction of the inverses as in the proof of Theorem 3.6.1 we simply observe that it is true that

$$(\forall x \in \mathbb{R})\,[x \neq 0 \to (\exists y \in \mathbb{R})\, x \cdot y = 1]$$

and conclude by transfer that

$$(\forall x \in {}^*\mathbb{R})\,[x \neq 0 \to (\exists y \in {}^*\mathbb{R})\, x \cdot y = 1].$$

For another example, consider the closed interval

$$[a, b] = \{x \in \mathbb{R} : a \leq x \leq b\}$$

4.5 The Transfer Principle

in the real line defined by points $a, b \in \mathbb{R}$. Then it is true that

$$(\forall x \in \mathbb{R})\,(x \in [a,b] \leftrightarrow a \leq x \leq b),$$

so by transfer we see that the enlargement of $[a,b]$ is the hyperreal interval defined by a and b (Exercise 3.10(6)):

$${}^*[a,b] = \{x \in {}^*\mathbb{R} : a \leq x \leq b\}.$$

Similarly, we can transfer to $^*\mathbb{R}$ many familiar facts about standard mathematical functions. Thus the following are true:

$$(\forall x \in {}^*\mathbb{R})\, \sin(\pi - x) = \sin x,$$
$$(\forall x \in {}^*\mathbb{R})\, \cosh x + \sinh x = e^x,$$
$$(\forall x, y \in {}^*\mathbb{R}^+)\, \log xy = \log x + \log y.$$

All of the above examples involve taking a *universally quantified* $\mathcal{L}_\mathfrak{R}$-sentence of the form $(\forall x, y, \ldots \in \mathbb{R})\,\varphi$ and transforming it to an $\mathcal{L}_{*\mathfrak{R}}$ sentence $(\forall x, y, \ldots \in {}^*\mathbb{R})\,{}^*\varphi$. They are instances of the following general principle.

- **Universal Transfer:** *if a property holds for all real numbers, then it holds for all hyperreal numbers.*

Of course the meaning of "property" has to be explained here, and that is what the formal language $\mathcal{L}_\mathfrak{R}$ was introduced for. **To use nonstandard analysis we need to develop the ability to show that a given property can be expressed in a transferable form.**

Dual to universal transfer is

- **Existential Transfer:** *if there exists a hyperreal number satisfying a certain property, then there exists a real number with this property.*

For example, take a real-valued sequence $s : \mathbb{N} \to \mathbb{R}$ for which we can show (by some means) that the extended hypersequence $^*s : {}^*\mathbb{N} \to {}^*\mathbb{R}$ never takes infinitely large values. Then existential transfer can be used to conclude that the original sequence must be bounded in $\mathbb{R}$. To see this, let ω be a member of $^*\mathbb{N} - \mathbb{N}$. By hypothesis it is true that

$$(\forall n \in {}^*\mathbb{N})\,(|{}^*s(n)| < \omega).$$

Now, this sentence is not the $*$-transform of an $\mathcal{L}_\mathfrak{R}$-sentence, because it contains the constant ω. But the constant can be removed in favour of an existentially quantified variable, by observing that the sentence implies

$$(\exists y \in {}^*\mathbb{R})(\forall n \in {}^*\mathbb{N})\,(|{}^*s(n)| < y),$$

which is now "appropriately formulated". Existential transfer then yields that
$$(\exists y \in \mathbb{R})(\forall n \in \mathbb{N})\,(|s(n)| < y),$$
which is the desired conclusion. Put informally, from the existence of a hyperreal bound on *s we infer the existence of a real bound on s.

Typically, in order to show that a real number of a certain type exists, it may be easier to show that a hyperreal of this type exists and then apply existential transfer.

Exercise 4.5.1
Which of Exercises 3.10 can be proven using transfer?

4.6 Justifying Transfer

In constructing the ordered field *$\mathbb{R}$ we repeatedly used the criterion that a particular property was to hold of hyperreals $[r], [s], \ldots$ iff the corresponding property held of the real numbers $r_n, s_n, \ldots$ for almost all n. In fact, this *almost-all criterion* works for any property expressible by an $\mathcal{L}_{\mathfrak{R}}$-formula, and that ultimately is why the transfer principle holds.

To spell this out some further technical notation is needed. For a formula φ we write
$$\varphi(x_1, \ldots, x_p)$$
to indicate that the list $x_1, \ldots, x_p$ includes all the variables that occur *free* in the formula φ. Then
$$\varphi(s_1, \ldots, s_p)$$
is the *sentence* obtained by replacing each *free* occurrence of x_i in φ by the constant s_i. For example, if $\varphi(x_1, x_2)$ is the formula
$$(\exists y \in \mathbb{Q})\,(x_1^2 + x_2^2 < y),$$
then $\varphi(\pi, \sqrt{2})$ is the *sentence*
$$(\exists y \in \mathbb{Q})\,(\pi^2 + (\sqrt{2})^2 < y).$$
Now, if $\varphi(x_1, \ldots, x_p)$ is a formula of $\mathcal{L}_{\mathfrak{R}}$, and $r^1, \ldots, r^p \in \mathbb{R}^{\mathbb{N}}$, put
$$[\![\varphi(r^1, \ldots, r^p)]\!] = \{n \in \mathbb{N} : \varphi(r_n^1, \ldots, r_n^p) \text{ is true}\}.$$
This extends the definitions of $[\![r = s]\!]$, $[\![r < s]\!]$, etc. to $\mathcal{L}_{\mathfrak{R}}$-formulae in general. Then such statements as

$$\begin{aligned}
[r] = [s] &\quad \text{iff} \quad [\![r = s]\!] \in \mathcal{F}, \\
[r] < [s] &\quad \text{iff} \quad [\![r < s]\!] \in \mathcal{F}, \\
[r] \in {}^*\!A &\quad \text{iff} \quad [\![r \in A]\!] \in \mathcal{F} \\
{}^*P([r^1], \ldots, [r^k]) &\quad \text{iff} \quad [\![P(r^1, \ldots, r^k)]\!] \in \mathcal{F}
\end{aligned}$$

(cf. Sections 3.6, 3.9, 3.14) are seen to be cases of the following fundamental result.

> For any $\mathcal{L}_\Re$-formula $\varphi(x_1,\ldots,x_p)$ and any $r^1,\ldots,r^p \in \mathbb{R}^\mathbb{N}$, the sentence $^*\varphi([r^1],\ldots,[r^p])$ is true if and only if $\varphi(r^1_n,\ldots,r^p_n)$ is true for almost all $n \in \mathbb{N}$.

In other words,

$$^*\varphi([r^1],\ldots,[r^p]) \text{ is true} \quad \text{iff} \quad [\![\varphi(r^1,\ldots,r^p)]\!] \in \mathcal{F}.$$

This result is known as *Łoś's theorem*, after the Polish mathematician who first proved it in the early 1950s. It includes transfer as a special case, because if φ is a sentence, then it has no free variables, so that $\varphi(s_1,\ldots,s_p)$ is just φ and likewise for $^*\varphi$. Hence $[\![\varphi(r^1,\ldots,r^p)]\!]$ is $\mathbb{N}$ if φ is true and $\emptyset$ otherwise, independently of the sequences r^j. Since $\emptyset \notin \mathcal{F}$, Łoś's theorem in this case simply says

$$^*\varphi \text{ is true} \quad \text{iff} \quad \varphi \text{ is true,}$$

which is the transfer principle!

A proof of Łoś's theorem would proceed by induction on the formation of the formula φ, considering first atomic formulae and then dealing with inductive cases for the logical connectives and quantifiers. We will not enter into those details here, but rely on the examples already discussed to lend plausibility to the assertion of Łoś's theorem, and hence to transfer.

4.7 Extending Transfer

We defined general relational structures S and their languages $\mathcal{L}_S$, but applied these ideas only to the language $\mathcal{L}_\Re$ in describing the transfer principle. In fact, it is possible to use the ultrapower construction to build an "enlargement" of any structure S and obtain a transfer principle for it. For instance, by replacing $\mathbb{R}$ by $\mathbb{C}$ this would give us a way of embarking on the nonstandard study of complex analysis.

It is important also to realise that the language $\mathcal{L}_\Re$ is limited by the fact that its quantifiable variables can range only over elements of $\mathbb{R}$, and not over more complicated entities like subsets of $\mathbb{R}$, sequences, real-valued functions, etc. For example, the Dedekind completeness principle,

> *every subset of $\mathbb{R}$ that is nonempty and bounded above has a least upper bound,*

cannot be formulated in $\mathcal{L}_\Re$ because the language does not allow quantifiers of the type

$$\forall x \in \mathcal{P}(\mathbb{R})$$

that apply to a variable (x) whose range of values is the set of all subsets of $\mathbb{R}$.

Later on (Chapter 13), a language will be introduced that does have such "higher-order" quantifiers and for which an appropriate transfer principle exists. Before then we will see that $\mathcal{L}_\mathfrak{R}$ is still powerful enough to develop a great deal of the standard theory of $\mathbb{R}$, including the convergence of sequences and series, differential and integral calculus, and the basic topology of the real line. Indeed, for the next half-dozen chapters we will forget about the ultrapower construction and explore all these topics using only the fact that *$\mathbb{R}$ is an ordered field that

- has $\mathbb{R}$ as a subfield;

- includes unlimited numbers $N \in {}^*\mathbb{N} - \mathbb{N}$, hence infinitesimals (such as $\frac{1}{N}$); and

- satisfies the transfer principle.

5
Hyperreals Great and Small

Members of *ℝ are called *hyperreal numbers*, while members of ℝ are *real* and sometimes called *standard*. *ℚ consists of *hyperrationals*, *ℤ of *hyperintegers*, and *ℕ of *hypernaturals*. That *ℚ consists precisely of quotients m/n of hyperintegers $m, n \in$ *ℤ follows by transfer of the sentence

$$\forall x \in \mathbb{R}\, [x \in \mathbb{Q} \leftrightarrow \exists y, z \in \mathbb{Z}\, (z \neq 0 \wedge x = y/z)].$$

It is now time to examine the basic arithmetical and algebraic structure of *ℝ, particularly in its relation to the structure of ℝ.

5.1 (Un)limited, Infinitesimal, and Appreciable Numbers

A hyperreal number b is:

- *limited* if $r < b < s$ for some $r, s \in \mathbb{R}$;
- *positive unlimited* if $r < b$ for all $r \in \mathbb{R}$;
- *negative unlimited* if $b < r$ for all $r \in \mathbb{R}$;
- *unlimited* if it is positive or negative unlimited;
- *positive infinitesimal* if $0 < b < r$ for all positive $r \in \mathbb{R}$;
- *negative infinitesimal* if $r < b < 0$ for all negative $r \in \mathbb{R}$;

- *infinitesimal* if it is positive infinitesimal, negative infinitesimal, or 0.
- *appreciable* if it is limited but not infinitesimal, i.e., $r < |b| < s$ for some $r, s \in \mathbb{R}^+$.

Thus all real numbers, and all infinitesimals, are limited. The only infinitesimal real is 0: all other reals are appreciable. An appreciable number is one that is neither infinitely small nor infinitely big. Observe that b is

- limited iff $|b| < n$ for some $n \in \mathbb{N}$;
- unlimited iff $|b| > n$ for all $n \in \mathbb{N}$;
- infinitesimal iff $|b| < \frac{1}{n}$ for all $n \in \mathbb{N}$;
- appreciable iff $\frac{1}{n} < |b| < n$ for some $n \in \mathbb{N}$.

We denote the set $^*\mathbb{N} - \mathbb{N}$ of unlimited hypernaturals by $^*\mathbb{N}_\infty$. Similarly, $^*\mathbb{R}_\infty^+$ denotes the set of positive unlimited hyperreals, and $^*\mathbb{R}_\infty^-$ the set of negative unlimited numbers. This notation may be adapted to an arbitary subset X of $^*\mathbb{R}$, putting $X_\infty = \{x \in X : x \text{ is unlimited}\}$, $X^+ = \{x \in X : x > 0\}$, etc.

We will also use $\mathbb{L}$ for the set of all limited numbers, and $\mathbb{I}$ for the set of infinitesimals.

5.2 Arithmetic of Hyperreals

Let ε, δ be infinitesimal, b, c appreciable, and H, K unlimited. Then

- **Sums:**

    $\varepsilon + \delta$ is infinitesimal

    $b + \varepsilon$ is appreciable

    $b + c$ is limited (possibly infinitesimal)

    $H + \varepsilon$ and $H + b$ are unlimited

- **Opposites:**

    $-\varepsilon$ is infinitesimal

    $-b$ is appreciable

    $-H$ is unlimited

- **Products:**

    $\varepsilon \cdot \delta$ and $\varepsilon \cdot b$ are infinitesimal

    $b \cdot c$ is appreciable

    $b \cdot H$ and $H \cdot K$ are unlimited

- **Reciprocals:**
  $\frac{1}{\varepsilon}$ is unlimited if $\varepsilon \neq 0$
  $\frac{1}{b}$ is appreciable
  $\frac{1}{H}$ is infinitesimal

- **Quotients:**
  $\frac{\varepsilon}{b}, \frac{\varepsilon}{H}$, and $\frac{b}{H}$ are infinitesimal
  $\frac{b}{c}$ is appreciable (if $c \neq 0$)
  $\frac{b}{\varepsilon}, \frac{H}{\varepsilon}$, and $\frac{H}{b}$ are unlimited ($\varepsilon, b \neq 0$)

- **Roots:**
  If $\varepsilon > 0$, $\sqrt[n]{\varepsilon}$ is infinitesimal
  If $b > 0$, $\sqrt[n]{b}$ is appreciable
  If $H > 0$, $\sqrt[n]{H}$ is unlimited

- **Indeterminate Forms:**
  $\frac{\varepsilon}{\delta}, \frac{H}{K}, \varepsilon \cdot H, H + K$

It follows from these rules that the set $\mathbb{L}$ of limited numbers and the set $\mathbb{I}$ of infinitesimals are each a subring of *$\mathbb{R}$. Also, the infinitesimals form an ideal in the ring of limited numbers. What then is the associated quotient ring $\mathbb{L}/\mathbb{I}$? Read on to Theorem 5.6.3.

With regard to nth roots, for fixed $n \in \mathbb{N}$ the function $x \mapsto \sqrt[n]{x}$ is defined for all positive reals, so extends to a function defined for all positive hyperreals. But we could also consider nth roots for unlimited n. The statement

$$(\forall n \in \mathbb{N})\,(\forall x \in \mathbb{R}^+)\,(\exists y \in \mathbb{R})\,(y^n = x)$$

asserts that any positive real has a real nth root for all $n \in \mathbb{N}$. Its transform asserts that every hyperreal has a hyperreal nth root for all $n \in $ *$\mathbb{N}$.

Exercise 5.2.1
For any positive hyperreal a, explain why the function $x \mapsto a^x$ is defined for all $x \in $ *$\mathbb{R}$. Use transfer to explore its properties.

5.3 On the Use of "Finite" and "Infinite"

The words "finite" and "infinite" are sometimes used for "limited" and "unlimited", but this does not accord well with the philosophy of our subject. A set is regarded as being finite if it has n elements for some $n \in \mathbb{N}$, and therefore is in bijective correspondence with the set

$$\{1, 2, \ldots, n\} = \{k \in \mathbb{N} : k \leq n\}.$$

However, if N is an unlimited hypernatural, then the collection
$$\{1, 2, \ldots, N\} = \{k \in {}^*\mathbb{N} : k \leq N\}$$
is set-theoretically infinite but, by transfer, has many properties enjoyed by finite sets. Collections of this type are called *hyperfinite*, and will be examined fully later. They are fundamental to the methodology of hyperreal analysis.

There is also potential conflict with other traditional uses of the word "infinite" in mathematics, such as in describing a series or an integral as being infinite when it it is divergent or undefined, or in referring to the area or volume or some more general measure of a set as being infinite when this has nothing to do with unlimited hyperreals.

5.4 Halos, Galaxies, and Real Comparisons

- Hyperreal b is *infinitely close* to hyperreal c, denoted by $b \simeq c$, if $b - c$ is infinitesimal. This defines an equivalence relation on ${}^*\mathbb{R}$, and the *halo* of b is the $\simeq$-equivalence class
$$\mathrm{hal}(b) = \{c \in {}^*\mathbb{R} : b \simeq c\}.$$

- Hyperreals b, c are *of limited distance apart*, denoted by $b \sim c$, if $b - c$ is limited. The *galaxy* of b is the $\sim$-equivalence class
$$\mathrm{gal}(b) = \{c \in {}^*\mathbb{R} : b \sim c\}.$$

So, b is infinitesimal iff $b \simeq 0$, and limited iff $b \sim 0$. Thus $\mathrm{hal}(0) = \mathbb{I}$, the set of infinitesimals, while $\mathrm{gal}(0) = \mathbb{L}$, the set of limited hyperreals.

Abraham Robinson called $\mathrm{hal}(b)$ the "monad" of b and used the notation $\mu(b)$, which is quite common in the literature. The more evocative name "halo" has been popularised by a French school of nonstandard analysis, founded by George Reeb, which is also responsible for "shadow" (see Section 5.6). The work of this school is described in the book listed as item 10 in the bibliography of Chapter 20.

5.5 Exercises on Halos and Galaxies

(1) Verify that $\simeq$ and $\sim$ are equivalence relations.

(2) If $b \simeq x \leq y \simeq c$ with b and c real, show that $b \leq c$. What if b and/or c are not real?

(3) $\mathrm{hal}(b) = \{b + \varepsilon : \varepsilon \in \mathrm{hal}(0)\}$.

(4) $\text{gal}(b) = \{b + c : c \in \text{gal}(0)\}$.

(5) If $x \simeq y$ and b is limited, prove that $b \cdot x \simeq b \cdot y$. Show that the result can fail for unlimited b.

(6) Show that any galaxy contains members of *$\mathbb{Z}$, of *$\mathbb{Q}$ − *$\mathbb{Z}$, and of *$\mathbb{R}$ − *$\mathbb{Q}$.

Real Comparisons

Exercise (2) above embodies an important general principle, which we will often use, about comparing the sizes of two *real* numbers b, c. If $b > c$, then the halos of the two numbers are disjoint, with everything in $\text{hal}(b)$ greater than everything in $\text{hal}(c)$. Thus to show that $b \leq c$ it is enough to show that something in $\text{hal}(b)$ is less than or equal to something in $\text{hal}(c)$. In particular, this will hold if there is some x with either $b \simeq x \leq c$ or $b \leq x \simeq c$.

5.6 Shadows

Theorem 5.6.1 *Every limited hyperreal b is infinitely close to exactly one real number, called the **shadow** of b, denoted by $\text{sh}(b)$.*

Proof. Let $A = \{r \in \mathbb{R} : r < b\}$. Since b is limited, there exist real r, s with $r < b < s$, so A is nonempty and bounded above in $\mathbb{R}$ by s. By the completeness of $\mathbb{R}$, it follows that A has a least upper bound $c \in \mathbb{R}$.

To show $b \simeq c$, take any positive real $\varepsilon \in \mathbb{R}$. Since c is an upper bound of A, we cannot have $c + \varepsilon \in A$; hence $b \leq c + \varepsilon$. Also, if $b \leq c - \varepsilon$, then $c - \varepsilon$ would be an upper bound of A, contrary to the fact that c is the smallest such upper bound. Hence $b \not\leq c - \varepsilon$. Altogether then, $c - \varepsilon < b \leq c + \varepsilon$, so $|b - c| \leq \varepsilon$. Since this holds for all positive real ε, b is infinitely close to c.

Finally, for uniqueness, if $b \simeq c' \in \mathbb{R}$, then as $b \simeq c$, we get $c \simeq c'$, and so $c = c'$, since both are real. □

Theorem 5.6.2 *If b and c are limited and $n \in \mathbb{N}$, then*

(1) $\text{sh}(b \pm c) = \text{sh}(b) \pm \text{sh}(c)$,

(2) $\text{sh}(b \cdot c) = \text{sh}(b) \cdot \text{sh}(c)$,

(3) $\text{sh}(b/c) = \text{sh}(b)/\text{sh}(c)$ *if* $\text{sh}(c) \neq 0$ (*i.e., if c is appreciable*),

(4) $\text{sh}(b^n) = \text{sh}(b)^n$,

(5) $\text{sh}(|b|) = |\text{sh}(b)|$,

(6) $\text{sh}(\sqrt[n]{b}) = \sqrt[n]{\text{sh}(b)}$ *if* $b \geq 0$,

(7) *if $b \leq c$ then $\operatorname{sh}(b) \leq \operatorname{sh}(c)$.*

Proof. Exercise. □

We see from these last facts that the **shadow map** $\operatorname{sh}: b \mapsto \operatorname{sh}(b)$ is an order-preserving homomorphism from the ring $\mathbb{L}$ of limited numbers onto $\mathbb{R}$. The kernel of this homomorphism is the set $\{b \in \mathbb{L} : \operatorname{sh}(b) \simeq 0\}$ of infinitesimals, and the cosets of the kernel are the halos $\operatorname{hal}(b)$ for limited b (cf. Exercise 5.5(3)). Thus we have an answer to our question about the quotient of $\mathbb{L}$ by $\mathbb{I}$:

Theorem 5.6.3 *The quotient ring $\mathbb{L}/\mathbb{I}$ is isomorphic to the real number field $\mathbb{R}$ by the correspondence $\operatorname{hal}(b) \mapsto \operatorname{sh}(b)$. Hence $\mathbb{I}$ is a maximal ideal of the ring $\mathbb{L}$.* □

The shadow $\operatorname{sh}(b)$ is often called the *standard part of b*.

5.7 Exercises on Infinite Closeness

(1) Show that if b, c are limited and $b \simeq b'$, $c \simeq c'$, then $b \pm c \simeq b' \pm c'$, $b \cdot c \simeq b' \cdot c'$, and $b/c \simeq b'/c'$ if $c \not\simeq 0$. Show that the last result can fail when $c \simeq 0$.

(2) If ε is infinitesimal, show that

$$\sin \varepsilon \simeq 0,$$
$$\cos \varepsilon \simeq 1,$$
$$\tan \varepsilon \simeq 0,$$
$$\sin \varepsilon / \varepsilon \simeq 1,$$
$$(\cos \varepsilon - 1)/\varepsilon \simeq 0$$

(use transfer of standard properties of trigonometric functions).

(3) Show that every hyperreal is infinitely close to some hyperrational number.

(4) Show that $\mathbb{R}$ is isomorphic to the ring of limited hyperrationals $^*\mathbb{Q} \cap \mathbb{L}$ factored by its ideal $^*\mathbb{Q} \cap \mathbb{I}$ of hyperrational infinitesimals.

5.8 Shadows and Completeness

We saw in the proof of Theorem 5.6.1 that the existence of shadows of limited numbers follows from the Dedekind completeness of $\mathbb{R}$. In fact, their existence turns out to be an alternative formulation of completeness, as the next result shows.

Theorem 5.8.1 *The assertion "every limited hyperreal is infinitely close to a real number" implies the completeness of $\mathbb{R}$.*

Proof. Let $s : \mathbb{N} \to \mathbb{R}$ be a Cauchy sequence. Recall that this means that its terms get arbitrarily close to each other as we move along the sequence. In particular, there exists a $k \in \mathbb{N}$ such that all terms of s beyond s_k are within a distance of 1 of each other, i.e.,

$$\forall m, n \in \mathbb{N}\, (m, n \geq k \to |s_m - s_n| < 1)$$

is true. Hence the $*$-transform of this sentence is also true, and applies to the extended hypersequence $\langle s_n : n \in {}^*\mathbb{N}\rangle$ as defined in Section 3.13. In particular, if we take N to be an unlimited member of $^*\mathbb{N}$, then $k, N \geq k$, so

$$|s_k - s_N| < 1,$$

and therefore s_N is limited. By the assertion quoted in the statement of the theorem, it follows that $s_N \simeq L$ for some $L \in \mathbb{R}$. We will show that the original sequence s converges to the real number L.

If ε is any positive real number, then again, since s is Cauchy, there exists $j_\varepsilon \in \mathbb{N}$ such that beyond s_{j_ε} all terms are within ε of each other:

$$\forall m, n \in \mathbb{N}\, (m, n \geq j_\varepsilon \to |s_m - s_n| < \varepsilon).$$

But now we can show that beyond s_{j_ε} all terms are within ε of L. The essential reason is that all such terms are within ε of s_N, which is itself infinitely close to L. For if $m \in \mathbb{N}$ with $m \geq j_\varepsilon$, we have $m, N \geq j_\varepsilon$, so by transfer of the last sentence we get that s_m is within ε of s_N:

$$|s_m - s_N| < \varepsilon.$$

Since s_N is infinitely close to L, this forces s_m to be within ε of L. Indeed,

$$|s_m - L| \leq |s_m - s_N| + |s_N - L| < \varepsilon + \text{infinitesimal},$$

so as $s_m - L$ and ε are real, $|s_m - L| \leq \varepsilon$.

This establishes that all the terms $s_j, s_{j+1}, s_{j+2}, \ldots$ are within ε of L, which is enough to prove that the sequence s converges to the real number L. All told, we have demonstrated that every real Cauchy sequence is convergent in $\mathbb{R}$, a property that is equivalent to Dedekind completeness (cf. Exercise 5.9 below). □

This result will be revisited in the next chapter (cf. Theorem 6.5.2 and the remarks following it).

5.9 Exercise on Dedekind Completeness

For Theorem 5.8.1, instead of showing that Cauchy sequences converge we can develop a direct proof that any subset $A \subseteq \mathbb{R}$ with a real upper bound has a least real upper bound. First, for each $n \in \mathbb{N}$, let s_n be the least $k \in \mathbb{Z}$ such that k/n is an upper bound of A. Then take an unlimited $N \in {}^*\mathbb{N}_\infty$ and let $L \in \mathbb{R}$ be infinitely close to s_N/N.

(a) Verify that s_n exists as defined for $n \in \mathbb{N}$.

(b) Show that s_N/N is limited, so that such a real L exists under the hypothesis of Theorem 5.8.1.

(c) Prove that L is a least upper bound of A in $\mathbb{R}$.

Hint: consider $(s_N - 1)/N$.

5.10 The Hypernaturals

We now develop a more detailed description of $^*\mathbb{N}$. First, by transfer, $^*\mathbb{N}$ is seen to be closed under addition and multiplication. Next observe that the only *limited* hypernaturals are the members of $\mathbb{N}$. For if $k \in {}^*\mathbb{N}$ is limited, then $k \le n$ for some $n \in \mathbb{N}$. But then by transfer of the sentence

$$\forall x \in \mathbb{N}\,(x \le n \to x = 1 \lor x = 2 \lor \cdots \lor x = n)$$

it follows that $k \in \{1, 2, \ldots, n\}$, so $k \in \mathbb{N}$.

Thus all members of $^*\mathbb{N} - \mathbb{N}$ are unlimited, and hence greater than all members of $\mathbb{N}$. Fixing $K \in {}^*\mathbb{N} - \mathbb{N}$, put

$$\gamma(K) = \{K\} \cup \{K \pm n : n \in \mathbb{N}\}.$$

Then all members of $\gamma(K)$ are unlimited, and together form a "copy of $\mathbb{Z}$" under the ordering $<$. Moreover, it may be seen that

$$\gamma(K) = \{H \in {}^*\mathbb{N} : K \sim H\} = \text{gal}(K) \cap {}^*\mathbb{N},$$

the restriction to $^*\mathbb{N}$ of the galaxy of K. The set $\gamma(K)$ will be called a *$^*\mathbb{N}$-galaxy*. We can also view $\mathbb{N}$ itself as a $^*\mathbb{N}$-galaxy, since $\mathbb{N} = \text{gal}(1) \cap {}^*\mathbb{N}$. Thus we define $\gamma(K) = \mathbb{N}$ when $K \in \mathbb{N}$. Then in general,

$$\gamma(K) = \gamma(H) \quad \text{iff} \quad K \sim H,$$

and the $^*\mathbb{N}$-galaxies may be ordered by putting

$$\gamma(K) < \gamma(H) \quad \text{iff} \quad K < H$$

whenever $K \not\sim H$ (i.e., whenever $|K - H|$ is unlimited).

There is no greatest $^*\mathbb{N}$-galaxy, since $\gamma(K) < \gamma(2K)$. Also, there is no smallest *unlimited* one: since one of K and $K+1$ is even (by transfer) and $\gamma(K) = \gamma(K+1)$, we can assume that K is even and note that $K/2 \in {}^*\mathbb{N}$, with $\gamma(K/2) < \gamma(K)$ and $K/2$ unlimited when K is. Finally, between any two *N-galaxies there is a third, for if $\gamma(K) < \gamma(H)$, with K, H both even, then

$$\gamma(K) < \gamma((H+K)/2) < \gamma(H).$$

To sum up: the ordering $<$ of *N consists of N followed by a densely ordered set of *N-galaxies (copies of Z) with no first or last such galaxy.

5.11 Exercises on Hyperintegers and Primes

(1) Provide an analogous description of the order structure of the hyperintegers *Z.

(2) Show that for any $M \in$ *N there is an $N \in$ *N that is divisible in *N by all members of $\{1, 2, \ldots, M\}$. Hence show that there exists a hypernatural number N that is divisible by every standard positive integer.

(3) Develop a theory of prime factors in *N: if Π is the set of standard prime numbers, with enlargement *$\Pi \subseteq$ *N, prove the following.

   (a) *Π consists precisely of those hypernaturals > 1 that have no nontrivial factors in *N.

   (b) Every hypernatural number > 1 has a "hyperprime" factor, i.e., is divisible by some member of *Π.

   (c) Two hypernaturals are equal if they have exactly the same factors of the form p^n with $p \in$ *Π and $n \in$ *N.

   (d) A hypernatural number is divisible by every standard positive integer iff it is divisible by p^n for every standard prime p and every $n \in$ N.

5.12 On the Existence of Infinitely Many Primes

The set Π of all standard prime numbers is infinite, a fact whose proof is attributed to Euclid. Therefore, the enlargement *Π has nonstandard members (by Theorem 3.9.1), so there are unlimited hypernatural numbers that are prime in the sense of having no nontrivial factors in *N.

But by using ideas suggested in the above exercises, together with a nonstandard adaptation of Euclid's own argument, we can show directly that *Π has nonstandard members, thereby giving an alternative proof that Π must be infinite, since if it were finite it would be equal to *Π.

Let N be a hypernatural number that is divisible by every member of N (Exercise 5.11(2)), and let q be a member of *Π that divides $N+1$ (Exercise

5.11.(3b)). Then q is our desired nonstandard prime. For if $q \in \Pi$, then q would divide N by assumption on N. But since it divides $N+1$, it would then divide the difference $N+1-N = 1$, which is false for a prime number. Hence q cannot be standard.

Part II

Basic Analysis

6
Convergence of Sequences and Series

A real-valued sequence $\langle s_n : n \in \mathbb{N} \rangle$ is a function $s : \mathbb{N} \to \mathbb{R}$, and so extends to a hypersequence $s : {}^*\mathbb{N} \to {}^*\mathbb{R}$ by the construction of Section 3.13. Hence the term s_n becomes defined for unlimited hypernaturals $n \in {}^*\mathbb{N}_\infty$ (a fact that was already used in Theorem 5.8.1), and in this case we say that s_n is an *extended term* of the sequence. The collection

$$\{s_n : n \in {}^*\mathbb{N}_\infty\}$$

of extended terms is the *extended tail* of s.

6.1 Convergence

In real analysis, $\langle s_n : n \in \mathbb{N} \rangle$ *converges to the limit* $L \in \mathbb{R}$ when each open interval $(L - \varepsilon, L + \varepsilon)$ around L in $\mathbb{R}$ contains some standard tail of the sequence, i.e., contains all the terms

$$s_n, s_{n+1}, s_{n+2}, \ldots$$

from some point on (with this point depending on ε). Formally, this is expressed by the statement

$$(\forall \varepsilon \in \mathbb{R}^+) \, (\exists m_\varepsilon \in \mathbb{N}) \, (\forall n \in \mathbb{N}) \, (n > m_\varepsilon \to |s_n - L| < \varepsilon),$$

which is intended to capture the idea that we can approximate L as closely as we like by moving far enough along the sequence. It turns out that this is equivalent to the requirement that if we go "infinitely far" along the sequence, then we become infinitely close to L:

Theorem 6.1.1 *A real-valued sequence $\langle s_n : n \in \mathbb{N}\rangle$ converges to $L \in \mathbb{R}$ if and only if $s_n \simeq L$ for all unlimited n.*

Proof. Suppose $\langle s_n : n \in \mathbb{N}\rangle$ converges to L, and fix an $N \in {}^*\mathbb{N}_\infty$. In order to show that $s_N \simeq L$ we have to show that $|s_N - L| < \varepsilon$ for any positive real ε. But given such an ε, the standard convergence condition implies that there is an $m_\varepsilon \in \mathbb{N}$ such that the standard tail beyond s_{m_ε} is within ε of L:
$$(\forall n \in \mathbb{N})\,(n > m_\varepsilon \to |s_n - L| < \varepsilon).$$
Then by (universal) transfer this holds for the extended tail as well:
$$(\forall n \in {}^*\mathbb{N})\,(n > m_\varepsilon \to |s_n - L| < \varepsilon).$$
But in fact, $N > m_\varepsilon$ because N is unlimited and m_ε is limited, and so this last sentence implies $|s_N - L| < \varepsilon$ as desired.

For the converse, suppose $s_n \simeq L$ for all unlimited n. We have to show that any given interval $(L - \varepsilon, L + \varepsilon)$ in $\mathbb{R}$ contains some standard tail of the sequence. The essence of the argument is to invoke the fact that the *extended* tail is infinitely close to L, hence contained in ${}^*(L-\varepsilon, L+\varepsilon)$, and then apply transfer.

To spell this out, fix an unlimited $N \in {}^*\mathbb{N}_\infty$. Then for any $n \in {}^*\mathbb{N}$, if $n > N$, it follows that n is also unlimited, so $s_n \simeq L$ and therefore $|s_n - L| < \varepsilon$. This shows that
$$(\forall n \in {}^*\mathbb{N})\,(n > N \to |s_n - L| < \varepsilon).$$
Hence the sentence
$$(\exists z \in {}^*\mathbb{N})\,(\forall n \in {}^*\mathbb{N})\,(n > z \to |s_n - L| < \varepsilon)$$
is true. But this is the $*$-transform of
$$(\exists z \in \mathbb{N})\,(\forall n \in \mathbb{N})\,(n > z \to |s_n - L| < \varepsilon),$$
so by (existential) transfer the latter holds true, giving the desired conclusion. □

Thus convergence to L amounts to the requirement that the extended tail of the sequence is contained in the halo of L. In this characterisation the role of the standard tails is taken over by the extended tail, while the standard open neighbourhoods $(L-\varepsilon, L+\varepsilon)$ are replaced by the "infinitesimal neighbourhood" $\mathrm{hal}(L)$.

6.2 Monotone Convergence

As a first application of this infinitesimal approach to convergence, here is an interesting alternative proof of a fundamental result about the behaviour of monotonic sequences.

Theorem 6.2.1 *A real-valued sequence $\langle s_n : n \in \mathbb{N} \rangle$ converges in $\mathbb{R}$ if either*

(1) *it is bounded above in $\mathbb{R}$ and nondecreasing: $s_1 \leq s_2 \leq \cdots$; or*

(2) *it is bounded below in $\mathbb{R}$ and nonincreasing: $s_1 \geq s_2 \geq \cdots$.*

Proof. Consider case (1). Let s_N be an extended term. We will show that s_N has a shadow, and that this shadow is a least upper bound of the set $\{s_n : n \in \mathbb{N}\}$ in $\mathbb{R}$. Since a set can have only one least upper bound, this implies that all extended terms have the same shadow, and so by Theorem 6.1.1 the original sequence converges to this shadow in $\mathbb{R}$.

Now, by hypothesis there is a real number b that is an upper bound for $\{s_n : n \in \mathbb{N}\}$. Then the statement $s_1 \leq s_n \leq b$ holds for all $n \in \mathbb{N}$, so it holds for all $n \in {}^*\mathbb{N}$ by universal transfer. In particular, $s_1 \leq s_N \leq b$, showing that s_N is limited, so indeed has a shadow L.

Next we show that L is an upper bound of the real sequence. Since this sequence is nondecreasing, by universal transfer we have

$$n \leq m \to s_n \leq s_m$$

for all $n, m \in {}^*\mathbb{N}$. In particular, if $n \in \mathbb{N}$, then $n \leq N$, so $s_n \leq s_N \simeq L$, giving $s_n \leq L$, as both numbers are real.

Finally, we show that L is the *least* upper bound in $\mathbb{R}$. For if r is any real upper bound of $\{s_n : n \in \mathbb{N}\}$, then by transfer, $s_n \leq r$ for all $n \in {}^*\mathbb{N}$, so $L \simeq s_N \leq r$, giving $L \leq r$, as both are real. □

One significant use of this result in standard analysis is to prove that if c is a real number between 0 and 1, then

$$\lim_{n \to \infty} c^n = 0.$$

To show this from the nonstandard perspective, note that if $0 < c < 1$, then the sequence $\langle c^n : n \in \mathbb{N} \rangle$ is nonincreasing and bounded below, and hence by Theorem 6.2.1 converges to some real number L. Thus if N is unlimited, then both $c^N \simeq L$ and $c^{N+1} \simeq L$. But then

$$\begin{aligned} L &\simeq c^{N+1} = c \cdot c^N && \text{(by transfer of } (\forall n \in \mathbb{N})\, c^{n+1} = c \cdot c^n) \\ &\simeq c \cdot L && \text{(by Exercise 5.5(5), as } c \text{ is real),} \end{aligned}$$

so we must have $L \simeq c \cdot L$. Hence $L = c \cdot L$, as both numbers are real, so as $c \neq 1$, it follows that $L = 0$ as desired.

6.3 Limits

It follows readily from Theorem 6.1.1 that a real-valued sequence has at most one limit. For if $\langle s_n \rangle$ converges to L and M in $\mathbb{R}$, then taking an unlimited n, we have $s_n \simeq L$ and $s_n \simeq M$, so $L \simeq M$, and therefore $L = M$ because L and M are real.

Theorem 6.3.1 *If* $\lim_{n\to\infty} s_n = L$ *and* $\lim_{n\to\infty} t_n = M$ *in* $\mathbb{R}$, *then*

(1) $\lim_{n\to\infty}(s_n + t_n) = L + M$,

(2) $\lim_{n\to\infty}(cs_n) = cL$, *for any* $c \in \mathbb{R}$,

(3) $\lim_{n\to\infty}(s_n t_n) = LM$,

(4) $\lim_{n\to\infty}(s_n / t_n) = L/M$, *if* $M \neq 0$.

Proof. Use Exercise 5.7(1). □

6.4 Boundedness and Divergence

Theorem 6.4.1 *A real-valued sequence* $\langle s_n \rangle$ *is bounded in* $\mathbb{R}$ *if and only if its extended terms are all limited.*

Proof. To say that $\langle s_n : n \in \mathbb{N} \rangle$ is bounded in $\mathbb{R}$ means that it is contained within some real interval $[-b, b]$, or equivalently that its absolute values $|s_n|$ have some real upper bound b:

$$(\forall n \in \mathbb{N}) \, |s_n| < b.$$

Then by universal transfer the extended sequence is contained in $^*[-b, b]$, i.e., $|s_n| < b$ for all $n \in {}^*\mathbb{N}$; hence s_n is limited in general.

For the converse, if s_n is limited for all unlimited $n \in {}^*\mathbb{N}_\infty$, then it is limited for all $n \in {}^*\mathbb{N}$. Hence if $r \in {}^*\mathbb{R}_\infty^+$ is any positive unlimited hyperreal, we observe that the entire extended sequence lies in the interval $\{x \in {}^*\mathbb{R} : -r < x < r\}$ and apply transfer. More formally, we have $|s_n| < r$ for all $n \in {}^*\mathbb{N}$, so the sentence

$$(\exists y \in {}^*\mathbb{R}) \, (\forall n \in {}^*\mathbb{N}) \, |s_n| < y$$

is true. But then by existential transfer it follows that there is some real number that is an upper bound to $|s_n|$ for all $n \in \mathbb{N}$. □

This proof can be refined to show the following:

- the real-valued sequence $\langle s_n \rangle$ is bounded *above* in $\mathbb{R}$, i.e., there is a real upper bound to $\{s_n : n \in \mathbb{N}\}$, if and only if it has no positive unlimited extended terms;

- $\langle s_n \rangle$ is bounded *below* in $\mathbb{R}$, i.e., there is a real lower bound to $\{s_n : n \in \mathbb{N}\}$, if and only if it has no negative unlimited terms.

We say that $\langle s_n \rangle$ *diverges to infinity* if for each real r there is some $n \in \mathbb{N}$ such that all terms of the standard tail $s_n, s_{n+1}, s_{n+2}, \ldots$ are greater than r. Correspondingly, $\langle s_n \rangle$ *diverges to minus infinity* if for each real r there is some $n \in \mathbb{N}$ such that $s_m < r$ for all $m \geq n$.

Theorem 6.4.2 *A real-valued sequence*

(1) *diverges to infinity if and only if all of its extended terms are positive unlimited; and*

(2) *diverges to minus infinity if and only if all of its extended terms are negative unlimited.*

Proof. Exercise. □

6.5 Cauchy Sequences

The standard definition of a Cauchy sequence is one that satisfies
$$\lim_{m,n \to \infty} |s_n - s_m| = 0,$$
meaning that the terms get arbitrarily close to each other as we move along the sequence. Formally this is rendered by the sentence
$$(\forall \varepsilon \in \mathbb{R}^+)(\exists j \in \mathbb{N})(\forall m, n \in \mathbb{N})(m, n \geq j \to |s_m - s_n| < \varepsilon).$$

Theorem 6.5.1 *A real-valued sequence $\langle s_n \rangle$ is Cauchy in $\mathbb{R}$ if and only if all its extended terms are infinitely close to each other, i.e., iff $s_m \simeq s_n$ for all $m, n \in {}^*\mathbb{N}_\infty$.*

Proof. Exercise. □

Theorem 6.5.2 (Cauchy's Convergence Criterion). *A real-valued sequence converges in $\mathbb{R}$ if and only if it is Cauchy.*

Proof. If $\langle s_n : n \in \mathbb{N} \rangle$ is Cauchy, then it is bounded (standard result—why is it true?). Thus taking an unlimited number $m \in {}^*\mathbb{N}_\infty$, we have that s_m is limited (Theorem 6.4.1) and so it has a shadow $L \in \mathbb{R}$. But all extended terms of the sequence are infinitely close to each other (Theorem 6.5.1), hence are infinitely close to s_m, and therefore are infinitely close to L as $s_m \simeq L$. This shows that the extended tail of the sequence is contained in the halo of L, implying by Theorem 6.1.1 that $\langle s_n \rangle$ converges to $L \in \mathbb{R}$.
Converse: exercise. □

Note that the assertion that Cauchy sequences converge is often taken as an "axiom" for the real number system, and is equivalent to the Dedekind completeness assertion that sets that are bounded above have least upper bounds in $\mathbb{R}$. We used Dedekind completeness to prove the existence of shadows (Theorem 5.6.1), which were then applied in Theorem 6.5.2 above. But we saw also in Theorem 5.8.1 that the existence of shadows in turn implies convergence of Cauchy sequences (and existence of least upper bounds in Exercise 5.9). The constructions in the proofs of Theorems 5.8.1 and 6.5.2 are essentially the same.

6.6 Cluster Points

A real number L is a *cluster point* of the real-valued sequence $\langle s_n : n \in \mathbb{N}\rangle$ if each open interval $(L - \varepsilon, L + \varepsilon)$ in $\mathbb{R}$ contains infinitely many terms of the sequence. This is expressed by the sentence

$$(\forall \varepsilon \in \mathbb{R}^+)(\forall m \in \mathbb{N})(\exists n \in \mathbb{N})(n > m \land |s_n - L| < \varepsilon). \tag{i}$$

From this it can be shown that the original sequence has a subsequence converging to L. Cluster points are also known as *limit points* of the sequence.

Theorem 6.6.1 *$L \in \mathbb{R}$ is a cluster point of the real-valued sequence $\langle s_n : n \in \mathbb{N}\rangle$ if and only if the sequence has an extended term infinitely close to L, i.e., iff $s_N \simeq L$ for some unlimited N.*

Proof. Assume that (i) holds. Let ε be a positive infinitesimal and $m \in {}^*\mathbb{N}_\infty$. Then by transfer of (i), there is some $n \in {}^*\mathbb{N}$ with $n > m$, and hence n is unlimited, and

$$|s_n - L| < \varepsilon \simeq 0.$$

Thus s_n is an extended term infinitely close to L. (Indeed, the argument shows that any interval of infinitesimal width around L contains terms arbitrarily far along the extended tail.)

Conversely, suppose there is an unlimited N with $s_N \simeq L$. To prove (i), take any positive $\varepsilon \in \mathbb{R}$ and $m \in \mathbb{N}$. Then $N > m$ and $|s_N - L| < \varepsilon$. This shows that

$$(\exists n \in {}^*\mathbb{N})(n > m \land |s_n - L| < \varepsilon).$$

Thus by existential transfer, $|s_n - L| < \varepsilon$ for some $n \in \mathbb{N}$, with $n > m$. □

This characterisation shows that a shadow of an extended term is a cluster point of a real sequence, and indeed that the cluster points are precisely the shadows of those extended terms that have them, i.e., are limited. But if the sequence is bounded, then *all* of its extended terms are limited and so have shadows that must be cluster points. In particular, this gives a very direct proof of a famous result:

Theorem 6.6.2 (Bolzano–Weierstrass) *Every bounded sequence of real numbers has a cluster point in $\mathbb{R}$.* □

6.7 Exercises on Limits and Cluster Points

(1) Let $\langle s_n\rangle$ and $\langle t_n\rangle$ be real-valued sequences with limits L, M respectively. Show that if $s_n \leq t_n$ for $n \in \mathbb{N}$, then $L \leq M$.

(2) If $r_n \leq s_n \leq t_n$ in $\mathbb{R}$ for all $\in \mathbb{N}$, and $\lim_{n\to\infty} r_n = \lim_{n\to\infty} t_n$, show that $\langle s_n \rangle$ converges to this same limit.

(3) If a sequence converges in $\mathbb{R}$, show that it has exactly one cluster point.

(4) Suppose that a real-valued sequence has a single cluster point. If the sequence is bounded, must it be convergent? What if it is unbounded?

6.8 Limits Superior and Inferior

Let $s = \langle s_n : n \in \mathbb{N} \rangle$ be a bounded real-valued sequence. Even if the sequence does not converge, its behaviour can be analysed into patterns of regularity: the Bolzano–Weierstrass theorem guarantees that it has a cluster point, and it may have many such cluster points with subsequences converging to each of them. If C_s is the set of all cluster points of s, then by the characterisation of Theorem 6.6.1 it can also be described as the set of all shadows of the extended tail:

$$C_s = \{\operatorname{sh}(s_n) : n \text{ is unlimited}\}.$$

Now, any real upper bound of the original sequence is an upper bound of C_s, for if $s_n \leq b$ for all $n \in \mathbb{N}$ (with $b \in \mathbb{R}$), then for an unlimited n we get $\operatorname{sh}(s_n) \simeq s_n \leq b$ and hence $\operatorname{sh}(s_n) \leq b$, since both are real. Similarly, any real lower bound of s is a lower bound of C_s. Since s is a bounded sequence, it follows that the set C_s is bounded above and below in $\mathbb{R}$, and so has a real least upper bound, known as the *limit superior* of s, and a real greatest lower bound, known as the *limit inferior*. The notations

$$\limsup_{n\to\infty} s_n \quad \text{and} \quad \liminf_{n\to\infty} s_n$$

are used for these two numbers.

Writing "sup" for the least upper bound (supremum) and "inf" for the greatest lower bound (infimum), we have

$$\limsup_{n\to\infty} s_n = \sup\{\operatorname{sh}(s_n) : n \in {}^*\mathbb{N}_\infty\},$$
$$\liminf_{n\to\infty} s_n = \inf\{\operatorname{sh}(s_n) : n \in {}^*\mathbb{N}_\infty\}.$$

The symbols $\overline{\lim}$ and $\underline{\lim}$ are also used for lim sup and lim inf.

Exercise 6.8.1
Prove, by nonstandard reasoning, that both the limit superior and the limit inferior are cluster points of the sequence s. □

This exercise shows that $\overline{\lim}\, s$ and $\underline{\lim}\, s$ both belong to C_s, and hence are the *maximum* and *minimum* elements of C_s respectively.

Theorem 6.8.2 *A real number L is equal to $\overline{\lim}\, s$ if and only if*

(1) $s_n < L$ *or* $s_n \simeq L$ *for all unlimited n; and*

(2) $s_n \simeq L$ *for at least one unlimited n.*

Proof. The condition "$s_n < L$ or $s_n \simeq L$" holds iff $\text{sh}(s_n) \leq L$. Thus (1) is equivalent to the assertion that L is an upper bound of C_s. But (2) asserts that L is a cluster point, so (1) and (2) together assert precisely that L is the maximum element of C_s, i.e., that $L = \overline{\lim}\, s$. □

Formulation of the nonstandard characterisation of $\underline{\lim}\, s$ corresponding to Exercise 6.8.1 is left as a further exercise.

Theorem 6.8.3 *A bounded real-valued sequence s converges to $L \in \mathbb{R}$ if and only if*

$$\limsup_{n \to \infty} s_n = \liminf_{n \to \infty} s_n = L.$$

Proof. Since $\overline{\lim}\, s$ and $\underline{\lim}\, s$ are the maximum and minimum elements of C_s, requiring that they both be equal to L amounts to requiring that $C_s = \{L\}$. But that just means that the shadow of every extended term is equal to L, which is equivalent to having s converge to L by Theorem 6.1.1. □

Theorem 6.8.4 *If s is a bounded real-valued sequence with limit superior $\overline{\lim}$, then for any positive real ε:*

(1) *some standard tail of s has all its terms smaller than $\overline{\lim} + \varepsilon$, i.e., $s_n < \overline{\lim} + \varepsilon$ for all but finitely many $n \in \mathbb{N}$;*

(2) $\overline{\lim} - \varepsilon < s_n$ *for infinitely many $n \in \mathbb{N}$.*

Proof.

(1) If $m \in {}^*\mathbb{N}$ is unlimited, then $\text{sh}(s_m) \leq \overline{\lim}$, so

$$s_m \simeq \text{sh}(s_m) < \overline{\lim} + \varepsilon,$$

showing that $s_m < \overline{\lim} + \varepsilon$ because $\text{sh}(s_m)$ and $\overline{\lim} + \varepsilon$ are both real. Thus all extended terms are smaller than $\overline{\lim} + \varepsilon$, and in particular, this holds for all terms after s_N for any fixed unlimited N:

$$(\forall m \in {}^*\mathbb{N})\, (m \geq N \to s_m < \overline{\lim} + \varepsilon).$$

Existential transfer then provides an $n \in \mathbb{N}$ such that all of

$$s_n, s_{n+1}, s_{n+2}, \ldots$$

are smaller than $\overline{\lim} + \varepsilon$.

(2) $\overline{\lim}$ is the shadow of some extended term s_N with N unlimited. Then $\overline{\lim} - \varepsilon < \overline{\lim} \simeq s_N$, so $\overline{\lim} - \varepsilon < s_N$. But now for any limited $m \in \mathbb{N}$ we have $m < N$ and $\overline{\lim} - \varepsilon < s_N$. Existential transfer then ensures that there is a limited n with $m < n$ and $\overline{\lim} - \varepsilon < s_n$. This shows that $\overline{\lim} - \varepsilon < s_n$ for arbitrarily large $n \in \mathbb{N}$, giving the desired conclusion. □

The two parts of the proof just given illustrate two fundamental principles of nonstandard reasoning:

(a) If a property (in the above case $s_n < \overline{\lim} + \varepsilon$) holds throughout the extended tail of a sequence, then it holds throughout some standard tail. More generally, if a certain property holds for all unlimited hypernatural n, then it holds for all but finitely many (limited) $n \in \mathbb{N}$.

(b) If a certain property (in the above case $\overline{\lim} - \varepsilon < s_n$) holds for some unlimited n, then it holds for arbitrarily large limited n.

Principle (a) was also at work in the proof of Theorem 6.1.1, where we saw that if the extended tail lies between $L - \varepsilon$ and $L + \varepsilon$, then so does some standard tail. The principle itself is an instance of "underflow" (Section 11.8).

Principle (b) is manifest in the situation of Section 5.12: the existence of a prime number bigger than some unlimited n implies the existence of arbitrarily large standard primes.

Now, if a real sequence s is bounded in $\mathbb{R}$, any standard tail $s_n, s_{n+1}, \ldots$ is also bounded and hence has a least upper bound, which we denote by S_n. Thus we are putting

$$S_n = \sup_{m \geq n} s_m.$$

In general, $S_n \geq S_{n+1}$, so the sequence $S = \langle S_n : n \in \mathbb{N} \rangle$ is nonincreasing. Moreover, it is bounded below, for if $b \in \mathbb{R}$ is a lower bound for s, then in general $b \leq s_n \leq S_n$. Thus by monotone convergence (Theorem 6.2.1), S converges in $\mathbb{R}$, and indeed its limit is the same as its greatest lower bound. We will now see that this limit is also the limit superior.

Theorem 6.8.5 *For any bounded real-valued sequence s,*

$$\limsup_{n \to \infty} s_n = \lim_{n \to \infty} S_n = \lim_{n \to \infty} \left(\sup_{m \geq n} s_m \right).$$

Proof. First we show that

$$\overline{\lim} \leq S_m \quad \text{for all } m \in \mathbb{N}. \tag{ii}$$

To see this, take an extended term s_N whose shadow is infinitely close to the cluster point $\overline{\lim}$. Then if $m \in \mathbb{N}$, we have $s_n \leq S_m$ for all limited

70   6. Convergence of Sequences and Series

$n \geq m$, and hence for all hypernatural $n \geq m$ by transfer. In particular, $s_N \leq S_m$, so as $\overline{\lim} \simeq s_N$, this forces $\overline{\lim} \leq S_m$ as required for (ii).

Now let $L = \lim_{n\to\infty} S_n$. Then L is the greatest lower bound of the sequence S, and by (ii), $\overline{\lim}$ is a lower bound for this sequence, so $\overline{\lim} \leq L$. But if $\overline{\lim} < L$, we can choose a positive real ε with $\overline{\lim} + \varepsilon < L$, and then by Theorem 6.8.4(1) there is some $n \in \mathbb{N}$ such that the standard tail $s_n, s_{n+1}, s_{n+2}, \ldots$ is bounded above by $\overline{\lim} + \varepsilon$. This implies that the least upper bound S_n of this tail is no bigger than $\overline{\lim} + \varepsilon$. However, that gives

$$S_n \leq \overline{\lim} + \varepsilon < L,$$

which contradicts the fact that L is a lower bound of S, and so $L \leq S_n$.

We are left with the conclusion that $\overline{\lim} = L$, as desired. □

The notion of limit superior can be defined for any real-valued sequence $s = \langle s_n \rangle$, bounded or not, by a consideration of cases in the following way.

(1) If s is not bounded above, put $\limsup_{n\to\infty} s_n = +\infty$. In this case s has at least one positive unlimited extended term (Section 6.4).

(2) If s is bounded above, hence has no positive unlimited extended terms, then there are two subcases:

  (i) s diverges to minus infinity. Then put $\limsup_{n\to\infty} s_n = -\infty$. In this case all extended terms are negative unlimited (Theorem 6.4.2), so there are no limited extended terms and therefore no cluster points.

  (ii) s does not diverge to minus infinity. Then there is at least one extended term that is not negative unlimited, and hence is limited because there are no positive unlimited terms (as s is bounded above). The shadow of this term is a cluster point of s. Thus the set C_s of cluster points is nonempty and bounded above (by any upper bound for s). Then we define $\limsup_{n\to\infty} s_n$ to be the least upper bound of C_s as previously.

6.9   Exercises on lim sup and lim inf

(1) Formulate the definition of the limit inferior of an arbitrary real-valued sequence.

(2) Formulate and prove theorems about the limit inferior of a bounded sequence that correspond to Theorems 6.8.4 and 6.8.5.

(3) If s is a bounded sequence, show that for each $\varepsilon \in \mathbb{R}^+$ there exists an $n \in \mathbb{N}$ such that the standard tail $s_n, s_{n+1}, s_{n+2}, \ldots$ is contained in the interval $(\underline{\lim} - \varepsilon, \overline{\lim} + \varepsilon)$.

6.10 Series

A real infinite series $\sum_1^\infty a_i$ is convergent iff the sequence $s = \langle s_n : n \in \mathbb{N} \rangle$ of *partial sums*
$$s_n = a_1 + \cdots + a_n$$
is convergent. We write $\sum_1^n a_i$ for s_n, and $\sum_m^n a_i$ for $s_n - s_{m-1}$ when $n \geq m$. Extending s to a hypersequence $\langle s_n : n \in {}^*\mathbb{N} \rangle$, we get that s_n and s_{m-1} are defined for all hyperintegers n, m, so the expressions $\sum_1^n a_i$ and $\sum_m^n a_i$ become meaningful for all $n, m \in {}^*\mathbb{N}$, and may be thought of as *hyperfinite sums* when n is unlimited.

Applying our results on convergence of sequences to the sequence of partial sums, we have:

- $\sum_1^\infty a_i = L$ in $\mathbb{R}$ iff $\sum_1^n a_i \simeq L$ for all unlimited n.

- $\sum_1^\infty a_i$ converges in $\mathbb{R}$ iff $\sum_m^n a_i \simeq 0$ for all unlimited m, n with $m \leq n$.

The second of these is given by the Cauchy convergence criterion (Theorem 6.5.2), since $\sum_m^n a_i \simeq 0$ iff $s_n \simeq s_{m-1}$ for unlimited m, n. Taking the case $m = n$ here, we get that if the series $\sum_1^\infty a_i$ converges, then $a_n \simeq 0$ whenever n is unlimited. This shows, by Theorem 6.1.1, that

- if $\sum_1^\infty a_i$ converges, then $\lim_{i \to \infty} a_i = 0$.

Observe that for a convergent real series we have
$$\sum_1^\infty a_i = \mathrm{sh}\left(\sum_1^n a_i\right)$$
for any unlimited n.

A series $\sum_1^\infty a_i$ always converges if it is *absolutely convergent*, which means that the series $\sum_1^\infty |a_i|$ of absolute values converges. The standard proof of this uses the comparison test, which is itself illuminated by non-standard ideas (see Exercise (4) below).

6.11 Exercises on Convergence of Series

(1) Give an example of a series that diverges but has a_n infinitesimal for all unlimited n.

(2) Give nonstandard proofs of the usual rules for arithmetically combining convergent series:
$$\sum_1^\infty a_i + \sum_1^\infty b_i = \sum_1^\infty (a_i + b_i),$$
$$\sum_1^\infty a_i - \sum_1^\infty b_i = \sum_1^\infty (a_i - b_i),$$
$$\sum_1^\infty c a_i = c \left(\sum_1^\infty a_i\right).$$

(3) Suppose that $a_i \geq 0$ for all $i \in \mathbb{N}$. Prove that $\sum_1^\infty a_i$ converges iff $\sum_1^n a_i$ is limited for all unlimited n, and that this holds iff $\sum_1^n a_i$ is limited for *some* unlimited n.

(4) (Comparison Test) Let $\sum_1^\infty a_i$ and $\sum_1^\infty b_i$ be two real series of non-negative terms, with $\sum_1^\infty b_i$ convergent. If $a_i \leq b_i$ for all $i \in \mathbb{N}$, use result (3) to show that $\sum_1^\infty a_i$ is convergent.

(5) Show that the comparison test holds under the weaker assumption that $a_n \leq b_n$ for all unlimited n (hint: use the Cauchy convergence criterion). Show that this weaker assumption is equivalent to requiring that there be some limited $k \in \mathbb{N}$ with $a_n \leq b_n$ for all $n \geq k$.

(6) Let $\sum_1^\infty a_i$ and $\sum_1^\infty b_i$ be two series of positive terms such that the sequence $\langle a_i/b_i : i \in \mathbb{N} \rangle$ is convergent in $\mathbb{R}$. Show that for unlimited m and n, $\sum_m^n a_i$ is infinitesimal if and only if $\sum_m^n b_i$ is infinitesimal. Deduce that either both series converge, or both diverge.

(7) Let $c \in \mathbb{R}$. Recall the identity
$$1 + c + c^2 + \cdots + c^n = \frac{1 - c^{n+1}}{1 - c}.$$

(a) Considering the case of unlimited n, show that the series $\sum_1^\infty c^i$ converges if $|c| < 1$.

(b) Show that $\sum_m^n c^i$ is infinitesimal when m and n are unlimited, either by applying result (a) or by making further use of the above identity.

(8) (Ratio Test: Convergence) Suppose that
$$\limsup_{i \to \infty} \frac{|a_{i+1}|}{|a_i|} < 1$$
in $\mathbb{R}$ (i.e., the limit superior of the sequence of ratios is a real number smaller than 1). Prove that the series $\sum_1^\infty a_i$ is absolutely convergent, by the following reasoning.

(a) Show that there exists a positive real $c < 1$ with $|a_{n+1}| < c|a_n|$ for all unlimited n.

(b) Hence show that there is some limited $k \in \mathbb{N}$ such that $|a_{n+1}| < c|a_n|$ for all $n \geq k$.

(c) Deduce from (b) that in general,
$$|a_{k+n}| < c^n |a_k|,$$
and hence
$$\sum_{k+m}^{k+n} |a_i| \leq \left(\sum_m^n c^i \right) |a_k|.$$

(d) Use the last inequality and result 7(b) to conclude that $\sum_1^\infty a_i$ converges absolutely.

(9) (Ratio Test: Divergence) Suppose that

$$\liminf_{i \to \infty} \frac{|a_{i+1}|}{|a_i|} > 1$$

in $\mathbb{R}$ (i.e., the lim inf is a real number greater than 1). Prove that the series $\sum_1^\infty a_i$ diverges, as follows.

(a) Show that $|a_{n+1}| > |a_n|$ for all unlimited n. Hence prove that there is some limited $k \in \mathbb{N}$ such that $|a_n| > |a_k| > 0$ for all $n > k$.

(b) Deduce, by considering unlimited n, that $\sum_1^\infty a_i$ cannot converge.

(10) Apply the ratio test with $a_i = (x^i/i!)$ to show that for any real number x, the hyperreal $x^n/n!$ is infinitesimal when n is unlimited.

(11) (Leibniz's Alternating Series Test) Suppose $\langle a_i : i \in \mathbb{N}\rangle$ is a real sequence that is nonincreasing (i.e., $a_i \geq a_{i+1}$) and converges to 0. Prove that the alternating series

$$\sum_1^\infty (-1)^{i+1} a_i = a_1 - a_2 + a_3 - a_4 + \cdots$$

converges by showing that in general

$$\left|\sum_m^n (-1)^{i+1} a_i\right| \leq |a_m|$$

and then considering the case of unlimited m.

7
Continuous Functions

Let f be an $\mathbb{R}$-valued function defined on an open interval (a,b) of $\mathbb{R}$. In passing to *$\mathbb{R}$, we may regard f as being defined for all hyperreal x between a and b, since *$(a,b) = \{x \in {}^*\mathbb{R} : a < x < b\}$.

7.1 Cauchy's Account of Continuity

Informally, we describe the assertion

> f is continuous at a point c in the interval (a,b)

as meaning that $f(x)$ stays "close to" $f(c)$ whenever x is "close to" c. The way Cauchy put it in 1821 was that

> the function $f(x)$ is continuous with respect to x between the given limits if between these limits an infinitely small increase in the variable always produces an infinitely small increase in the function.

From the enlarged perspective of *$\mathbb{R}$, this account can be made precise:

Theorem 7.1.1 *f is continuous at the real point c if and only if $f(x) \simeq f(c)$ for all $x \in {}^*\mathbb{R}$ such that $x \simeq c$, i.e., iff*

$$f(\mathrm{hal}(c)) \subseteq \mathrm{hal}(f(c)).$$

Proof. The standard definition is that f is continuous at c iff for each open interval $(f(c) - \varepsilon, f(c) + \varepsilon)$ around $f(c)$ in $\mathbb{R}$ there is a corresponding open interval $(c-\delta, c+\delta)$ around c that is mapped into $(f(c)-\varepsilon, f(c)+\varepsilon)$ by f. Since $a < c < b$, the number δ can be chosen small enough so that the interval $(c-\delta, c+\delta)$ is contained with (a,b), ensuring that f is indeed defined at all points that are within δ of c.

Continuity at c is thus formally expressed by the sentence

$$(\forall \varepsilon \in \mathbb{R}^+)\,(\exists \delta \in \mathbb{R}^+)\,(\forall x \in \mathbb{R})\,(\,|x-c|<\delta \to |f(x)-f(c)|<\varepsilon).\quad \text{(i)}$$

Now suppose $x \simeq c$ implies $f(x) \simeq f(c)$. To show that (i) holds, let ε be a positive real number. Then we have to find a real δ small enough to fulfill (i). First we show that this can be achieved if "small enough" is replaced by "infinitely small", and then apply transfer. For if d is any positive infinitesimal, then for any $x \in {}^*\mathbb{R}$, if $|x-c|<d$, we have $x \simeq c$, hence $f(x) \simeq f(c)$ by assumption, so $|f(x)-f(c)|<\varepsilon$, as ε is real. Replacing d by an existentially quantified variable, this shows that the sentence

$$(\exists \delta \in {}^*\mathbb{R}^+)\,(\forall x \in {}^*\mathbb{R})\,(\,|x-c|<\delta \to |f(x)-f(c)|<\varepsilon)$$

is true. By existential transfer we then infer

$$(\exists \delta \in \mathbb{R}^+)\,(\forall x \in \mathbb{R})\,(\,|x-c|<\delta \to |f(x)-f(c)|<\varepsilon),$$

which is enough to complete the demonstration of (i).

Conversely, assume that (i) holds. Let ε be any positive real. Then by (i) there is a positive $\delta \in \mathbb{R}$ such that the sentence

$$(\forall x \in \mathbb{R})\,(\,|x-c|<\delta \to |f(x)-f(c)|<\varepsilon)$$

is true, and hence by universal transfer we have

$$(\forall x \in {}^*\mathbb{R})\,(\,|x-c|<\delta \to |f(x)-f(c)|<\varepsilon).$$

But now if $x \simeq c$ in ${}^*\mathbb{R}$, then $|x-c|<\delta$, and so by this last sentence $|f(x)-f(c)|<\varepsilon$. Since this holds for arbitrary $\varepsilon \in \mathbb{R}^+$, it follows that $f(x) \simeq f(c)$.

In other words, the halo hal(c) is mapped by f into the interval $(f(c)-\varepsilon, f(c)+\varepsilon)$ for any positive real ε, and hence is mapped into the halo hal$(f(c))$. $\square$

A close inspection of the first part of this proof reveals that in order to establish the standard criterion for continuity at c it suffices to know that $f(x) \simeq f(c)$ for all x that are within some positive infinitesimal distance d of c. Thus we have this stronger conclusion:

Corollary 7.1.2 *The following are equivalent.*

(1) *f is continuous at $c \in \mathbb{R}$.*

(2) *$f(x) \simeq f(c)$ whenever $x \simeq c$.*

(3) *There is some positive $d \simeq 0$ such that $f(x) \simeq f(c)$ whenever $|x-c| < d$.* □

If A is a subset of the domain of function f, then f is *continuous on the set A* if it is continuous at all points c that belong to A. Sometimes we would like A to be something other than an open interval (a,b), such as a half-open or closed interval $(a,b]$, $[a,b)$, or $[a,b]$, or a union of such sets. In this case the definition of continuity is modified to specify that for each positive ε there is a corresponding δ such that $f(x)$ belongs to $(f(c)-\varepsilon, f(c)+\varepsilon)$ whenever x is a point of A that belongs to $(c-\delta, c+\delta)$. In other words, the bounded quantification of x in sentence (i) is restricted to the set A: f is continuous at all points c in A if

$$(\forall c \in A)\,(\forall \varepsilon \in \mathbb{R}^+)\,(\exists \delta \in \mathbb{R}^+)\,(\forall x \in A)\,(\,|x-c| < \delta \to |f(x)-f(c)| < \varepsilon).$$

Reworking the proofs of Theorem 7.1.1 and Corollary 7.1.2, we obtain a hyperreal characterisation of this refinement:

Theorem 7.1.3 *The following are equivalent.*

(1) *f is continuous at c in A.*

(2) *$f(x) \simeq f(c)$ for all $x \in {}^*A$ with $x \simeq c$.*

(3) *There is some positive $d \simeq 0$ such that $f(x) \simeq f(c)$ for all $x \in {}^*A$ with $|x-c| < d$.*

It would be natural at this point to ask whether continuity of f on A entails that the condition $f(\mathrm{hal}(c)) \subseteq \mathrm{hal}(f(c))$ must hold for all points $c \in {}^*A$ and not just the real ones. It turns out that this need not be so: it is a stronger requirement, which, remarkably, is equivalent to the standard notion of *uniform* continuity. We take this up in Section 7.7.

7.2 Continuity of the Sine Function

To illustrate the use of Theorem 7.1.1, let c be real and $x \simeq c$. Then $x = c+\varepsilon$ for an infinitesimal ε, and

$$\begin{aligned}
\sin x - \sin c &= \sin(c+\varepsilon) - \sin c \\
&= \sin c \cos \varepsilon + \cos c \sin \varepsilon - \sin c \\
&= \sin c\,(\cos \varepsilon - 1) + \cos c \sin \varepsilon \\
&= \text{an infinitesimal},
\end{aligned}$$

since $\cos \varepsilon \simeq 1$ and $\sin \varepsilon \simeq 0$ (Exercise 5.7(2)), while $\sin c$ and $\cos c$ are real. Hence $\sin x \simeq \sin c$. This proves that the sine function is continuous at all $c \in \mathbb{R}$.

Note that in this proof we used the addition formula

$$\sin(c+\varepsilon) = \sin c \cos \varepsilon + \cos c \sin \varepsilon.$$

This holds for all real numbers, and hence by transfer it holds for all hyperreals.

7.3 Limits of Functions

Continuity is often defined in terms of limits of functions. By similar arguments to those of Section 7.1, one can establish that if $c, L \in \mathbb{R}$ and f is defined on $A \subseteq \mathbb{R}$, then

- $\lim_{x \to c^+} f(x) = L$ iff $f(x) \simeq L$ for all $x \in {}^*A$ with $x \simeq c$ and $x > c$.
- $\lim_{x \to c^-} f(x) = L$ iff $f(x) \simeq L$ for all $x \in {}^*A$ with $x \simeq c$ and $x < c$.
- $\lim_{x \to c} f(x) = L$ iff $f(x) \simeq L$ for all $x \in {}^*A$ with $x \simeq c$ and $x \neq c$.
- $\lim_{x \to c} f(x) = +\infty$ iff $f(x) \in {}^*\mathbb{R}_\infty^+$ for all $x \in {}^*A$ with $x \simeq c$ and $x \neq c$.
- $\lim_{x \to c} f(x) = -\infty$ iff $f(x) \in {}^*\mathbb{R}_\infty^-$ for all $x \in {}^*A$ with $x \simeq c$ and $x \neq c$.
- $\lim_{x \to +\infty} f(x) = L$ iff $f(x) \simeq L$ for all positive unlimited $x \in {}^*A$ (and such x exist).
- $\lim_{x \to -\infty} f(x) = L$ iff $f(x) \simeq L$ for all negative unlimited $x \in {}^*A$ (and such x exist).

7.4 Exercises on Limits

(1) Review the standard definitions of limits of functions, and derive the above characterisations.

(2) Use these results to show the following standard facts:

- $\lim_{x \to c} f(x) = L$ iff $\lim_{x \to c^+} f(x) = L$ and $\lim_{x \to c^-} f(x) = L$.
- If $\lim_{x \to c} f(x)$ and $\lim_{x \to c} g(x)$ exist, then

$$\lim_{x \to c}[f(x) + g(x)] = \lim_{x \to c} f(x) + \lim_{x \to c} g(x),$$
$$\lim_{x \to c}[f(x)g(x)] = \lim_{x \to c} f(x) \cdot \lim_{x \to c} g(x),$$
$$\lim_{x \to c}[f(x)/g(x)] = \lim_{x \to c} f(x) / \lim_{x \to c} g(x), \quad \text{if } \lim_{x \to c} g(x) \neq 0.$$

- f is continuous at c iff $\lim_{x \to c} f(x) = f(c)$.

(3) Use infinitesimals to discuss the continuity of the following functions:

$$f_1(x) = \begin{cases} \sin \frac{1}{x} & \text{if } x \neq 0, \\ 0 & \text{if } x = 0. \end{cases}$$

$$f_2(x) = \begin{cases} x \sin \frac{1}{x} & \text{if } x \neq 0, \\ 0 & \text{if } x = 0. \end{cases}$$

$$f_3(x) = \begin{cases} 1 & \text{if } x \text{ is rational,} \\ 0 & \text{if } x \text{ is irrational.} \end{cases}$$

$$f_4(x) = \begin{cases} x & \text{if } x \text{ is rational,} \\ -x & \text{if } x \text{ is irrational.} \end{cases}$$

$$f_5(x) = \begin{cases} 0 & \text{if } x \text{ is irrational,} \\ \frac{1}{n} & \text{if } x = \frac{m}{n} \in \mathbb{Q} \text{ in simplest form with } n \geq 1. \end{cases}$$

7.5 The Intermediate Value Theorem

This fundamental result of standard real analysis states that

if the real function f is continuous on the closed interval $[a, b]$ in $\mathbb{R}$, then for every real number d strictly between $f(a)$ and $f(b)$ there exists a real $c \in (a, b)$ such that $f(c) = d$.

There is an intuitively appealing proof of this using infinitesimals. The basic idea is to partition the interval $[a, b]$ into subintervals of equal infinitesimal width, and locate a subinterval whose end points have f-values on either side of d. Then c will be the common shadow of these end points. In this way we "pin down" the point at which the f-values pass through d.

We deal with the case $f(a) < f(b)$, so that $f(a) < d < f(b)$. First, for each (limited) $n \in \mathbb{N}$, partition $[a, b]$ into n equal subintervals of width $(b - a)/n$. Thus these intervals have end points $p_k = a + k(b - a)/n$ for $0 \leq k \leq n$. Then let s_n be the greatest partition point whose f-value is less than d. Indeed, the set

$$\{p_k : f(p_k) < d\}$$

is finite and nonempty (it contains $p_0 = a$ but not $p_n = b$). Hence s_n exists as the maximum of this set, and is given by some p_k with $k < n$.

Now, for all $n \in \mathbb{N}$ we have

$$a \leq s_n < b \quad \text{and} \quad f(s_n) < d \leq f(s_n + (b - a)/n),$$

and so by transfer, these conditions hold for all $n \in {}^*\mathbb{N}$.

To obtain an infinitesimal-width partition, choose an unlimited hypernatural N. Then s_N is limited, as $a \leq s_N < b$, so has a shadow $c = \text{sh}(s_N) \in \mathbb{R}$

(by transfer, s_N is a number of the form $a + K(b-a)/N$ for some $K \in {}^*\mathbb{N}$). But $(b-a)/N$ is infinitesimal, so s_N and $s_N + (b-a)/N$ are both infinitely close to c. Since f is continuous at c and c is real, it follows (Theorem 7.1.1) that $f(s_N)$ and $f(s_N + (b-a)/N)$ are both infinitely close to $f(c)$. But

$$f(s_N) < d \leq f(s_N + (b-a)/N),$$

so d is also infinitely close to $f(c)$. Since $f(c)$ and d are both real, they must then be equal. □

7.6 The Extreme Value Theorem

If the real function f is continuous on the closed interval $[a, b]$ in $\mathbb{R}$, then f attains an absolute maximum and an absolute minimum on $[a, b]$, i.e., there exist real $c, d \in [a, b]$ such that $f(c) \leq f(x) \leq f(d)$ for all $x \in [a, b]$.

Proof. To obtain the asserted maximum we construct an infinitesimal-width partition of $[a, b]$, and show that there is a particular partition point whose f-value is as big as any of the others. Then d will be the shadow of this particular partition point. As with the intermediate value theorem, the construction is first approximated by finite partitions with subintervals of limited width $\frac{1}{n}$. In these cases there is always a partition point with maximum f-value. Then transfer is applied.

For each limited $n \in \mathbb{N}$, partition $[a, b]$ into n equal subintervals, with end points $a + k(b-a)/n$ for $0 \leq k \leq n$. Then let $s_n \in [a, b]$ be a partition point at which f takes its largest value. In other words, for all integers k such that $0 \leq k \leq n$,

$$a \leq s_n \leq b \quad \text{and} \quad f(a + k(b-a)/n) \leq f(s_n). \tag{ii}$$

By transfer, (ii) holds for all $n \in {}^*\mathbb{N}$ and all hyperintegers k such that $0 \leq k \leq n$.

Similarly to the intermediate value theorem, choose an unlimited hypernatural N and put $d = \mathrm{sh}(s_N) \in \mathbb{R}$. Then by continuity

$$f(s_N) \simeq f(d). \tag{iii}$$

Now the "infinitesimal-width partition"

$$P = \{a + k(b-a)/N : k \in {}^*\mathbb{N} \text{ and } 0 \leq k \leq N\}$$

has the important property that it provides infinitely close approximations to all real numbers between a and b: the halo of each $x \in [a, b]$ contains points from this partition. To show this, observe that if x is an arbitrary real number in $[a, b]$, then for each $n \in \mathbb{N}$ there exists an integer $k < n$ with

$$a + k(b-a)/n \leq x \leq a + (k+1)(b-a)/n.$$

Hence by transfer there exists a hyperinteger $K < N$ such that x lies in the interval
$$[a + K(b-a)/N, a + (K+1)(b-a)/N]$$
of infinitesimal width $(b-a)/N$. Therefore $x \simeq a + K(b-a)/N$, so x is indeed infinitely close to a member of P. It follows by continuity of f at x that
$$f(x) \simeq f(a + K(b-a)/N). \qquad \text{(iv)}$$
But the values of f on P are dominated by $f(s_N)$, as (ii) holds for all $n \in {}^*\mathbb{N}$, so
$$f(a + K(b-a)/N) \leq f(s_N). \qquad \text{(v)}$$
Putting (iii), (iv), and (v) together gives
$$f(x) \simeq f(a + K(b-a)/N) \leq f(s_N) \simeq f(d),$$
which implies $f(x) \leq f(d)$, since $f(x)$ and $f(d)$ are real (Exercise 5.5(2)). Thus f attains its maximum value at d.

The proof that f attains a minimum is similar. □

7.7 Uniform Continuity

If $A \subseteq \mathbb{R}$ and $f : A \to \mathbb{R}$, then f is *uniformly continuous on A* if the following sentence is true:

$$(\forall \varepsilon \in \mathbb{R}^+)(\exists \delta \in \mathbb{R}^+)(\forall x, y \in A)(|x-y| < \delta \to |f(x) - f(y)| < \varepsilon)$$

(compare this to the formal sentence just prior to Theorem 7.1.3). Essentially, this says that for a given ε, the same δ for the continuity condition works at all points of A.

Theorem 7.7.1 *f is uniformly continuous on A if and only if $x \simeq y$ implies $f(x) \simeq f(y)$ for all hyperreals $x, y \in {}^*A$.*

Proof. Exercise. □

Theorem 7.7.1 displays the distinction between uniform and ordinary continuity in a more intuitive and readily comprehensible way than the standard definitions do. For by Theorem 7.1.1, f is continuous on $A \subseteq \mathbb{R}$ iff $x \simeq y$ implies $f(x) \simeq f(y)$ for all $x, y \in {}^*A$ with y *standard*. Thus uniform continuity amounts to preservation of the "infinite closeness" relation $\simeq$ at all hyperreal points in the enlargement ${}^*A$ of A, while continuity only requires preservation of this relation at the real points.

Of course for some sets, these two requirements come to the same thing:

Theorem 7.7.2 *If the real function f is continuous on the closed interval $[a,b]$ in $\mathbb{R}$, then f is uniformly continuous on $[a,b]$.*

Proof. Take hyperreals $x, y \in {}^*[a,b]$ with $x \simeq y$. Let $c = \text{sh}(x)$. Then since $a \leq x \leq b$ and $x \simeq c$, we have $c \in [a,b]$, and so f is continuous at c. Applying Theorem 7.1.1, we get $f(x) \simeq f(c)$ and $f(y) \simeq f(c)$, whence $f(x) \simeq f(y)$. Hence f is uniformly continuous by Theorem 7.7.1. □

7.8 Exercises on Uniform Continuity

(1) Explain why the argument just given fails for intervals (a, b), $(a, b]$, $(a, +\infty)$, $(-\infty, b)$, etc. that are not closed.

(2) Show that $f(x) = 1/x$ is not uniformly continuous on $(0, 1)$.

(3) If f is uniformly continuous on $\mathbb{R}$ and $\langle s_n : n \in \mathbb{N} \rangle$ is a Cauchy sequence, show that $\langle f(s_n) : n \in \mathbb{N} \rangle$ is a Cauchy sequence.

(4) Let the real function f be monotonic on $[a, b]$, and suppose that for all real r between $f(a)$ and $f(b)$ there exists a real $c \in [a, b]$ such that $f(c) = r$. Prove that f is continuous on $[a, b]$.

The property of closed intervals that makes Theorem 7.7.2 work will be examined further in Section 10.4 when we study *compactness* from the hyperreal perspective.

7.9 Contraction Mappings and Fixed Points

A function $f : \mathbb{R} \to \mathbb{R}$ is said to satisfy a *Lipschitz condition* if there is a positive real constant c such that

$$|f(x) - f(y)| \leq c\,|x - y| \qquad \text{(vi)}$$

for all $x, y \in \mathbb{R}$. Such a function is always continuous, indeed *uniformly* continuous, as is readily explained by infinitesimal reasoning. First observe that (vi) holds for all hyperreal x, y by transfer. But then if $x \simeq y$, we have that $c|x - y|$ is infinitesimal, since c is real and $|x - y| \simeq 0$, so by (vi) $|f(x) - f(y)|$ is infinitesimal, making $f(x) \simeq f(y)$. Hence f is uniformly continuous by the characterisation of Theorem 7.7.1.

A *contraction mapping* is a Lipschitz function with constant c less than 1. Such a function acts on any two points to move them closer to each other. It turns out that a contraction mapping has a *fixed point*: a point x satisfying $f(x) = x$. It certainly cannot have two such points, for if $f(x) = x$ and $f(y) = y$, then

$$|x - y| = |f(x) - f(y)| \leq c\,|x - y|,$$

7.9 Contraction Mappings and Fixed Points

and since $c < 1$, this is possible only if $|x - y| = 0$ and hence $x = y$.

Consider for example the contraction mapping f defined by

$$f(x) = \frac{x}{2} + \frac{1}{2}$$

(what is the constant c here?). Its fixed point is the unique solution to $\frac{x}{2} + \frac{1}{2} = x$, namely $x = 1$. Moreover, this fixed point can be approached by starting at any real number x and repeatedly applying f to generate the sequence

$$x, \quad f(x) = \frac{x}{2} + \frac{1}{2}, \quad f(f(x)) = \frac{x}{4} + \frac{3}{4}, \quad f(f(f(x))) = \frac{x}{8} + \frac{7}{8}, \quad \ldots$$

The nth term of this sequence is

$$\frac{x}{2^n} + 1 - \frac{1}{2^n},$$

so the sequence does indeed converge to 1 regardless of what x is (this can also be effectively demonstrated visually by plotting a graph of the function and the terms of the sequence).

Theorem 7.9.1 *Any contraction mapping* $f : \mathbb{R} \to \mathbb{R}$ *has a (unique) fixed point.*

Proof. Let c be the Lipschitz constant for f. Take any $x \in \mathbb{R}$, put $s_0 = x$, and inductively define

$$s_{n+1} = f(s_n). \tag{vii}$$

Observe that

$$\begin{aligned}
|s_1 - s_2| &\leq c|s_0 - s_1|, \\
|s_2 - s_3| &\leq c|s_1 - s_2| \leq c^2|s_0 - s_1|, \\
|s_3 - s_4| &\leq c|s_2 - s_3| \leq c^3|s_0 - s_1|, \\
&\vdots
\end{aligned}$$

and so on. In general, for $n \in \mathbb{N}$ we get

$$|s_n - s_{n+1}| \leq c^n |s_0 - s_1|. \tag{viii}$$

Hence

$$\begin{aligned}
|s_0 - s_n| &\leq |s_0 - s_1| + |s_1 - s_2| + |s_2 - s_3| + \cdots + |s_{n-1} - s_n| \\
&\leq |s_0 - s_1| + c|s_0 - s_1| + c^2|s_0 - s_1| + \cdots + c^{n-1}|s_0 - s_1| \\
&= |s_0 - s_1|(1 + c + c^2 + \cdots + c^{n-1}) \\
&= \frac{1 - c^n}{1 - c}|s_0 - s_1|,
\end{aligned}$$

and therefore
$$|s_0 - s_n| \le \frac{1}{1-c}|s_0 - s_1| \qquad \text{(ix)}$$
for all $n \in \mathbb{N}$.

The standard proof of this theorem uses the more general formula
$$|s_m - s_n| \le \frac{c^m}{1-c}|s_0 - s_1|$$
to prove that the sequence $\langle s_n : n \in \mathbb{N}\rangle$ is Cauchy, hence convergent, and that its limit is a fixed point for f. Here we will instead extend to the hypersequence $\langle s_n :\in {}^*\mathbb{N}\rangle$ and take the shadow of any term in the extended tail.

Thus if $n \in {}^*\mathbb{N}$ is unlimited, then by transfer $|s_0 - s_n|$ is bounded by the real right side of (ix), so s_n is limited and has a shadow $L \in \mathbb{R}$. Then $s_n \simeq L$, and so as f is continuous, $f(s_n) \simeq f(L)$. But $f(s_n) = s_{n+1}$ by transfer of (vii), and $s_{n+1} \simeq s_n$ by transfer of (viii), since c^n is infinitesimal when $c < 1$ and n is unlimited, making $|s_n - s_{n+1}|$ infinitesimal. Altogether then we have
$$f(L) \simeq f(s_n) = s_{n+1} \simeq s_n \simeq L,$$
giving $f(L) \simeq L$. Since $f(L)$ and L are real, it follows that they are equal, so L is the desired fixed point. □

Notice that the fact that the sequence $x, f(x), f(f(x)), \ldots$ converges to a fixed point of f now becomes a consequence of our proof, rather than being part of the proof as in the standard argument. For we have shown that for any unlimited n, the shadow $\mathrm{sh}(s_n)$ exists and is a fixed point. Since there can be only one fixed point, it follows that all extended terms have the same shadow, and hence (Theorem 6.1.1) that the original sequence converges to this shadow.

Theorem 7.9.1 is an instance of the *Banach fixed point theorem*, which asserts the existence of a fixed point for a contraction mapping on any "complete metric space". Essentially the same nonstandard analysis can be used for the proof in that more general setting.

7.10 A First Look at Permanence

One of the distinctive features of nonstandard analysis is the presence of so-called *permanence principles*, which assert that certain functions must exist, or be defined, on a larger domain than that which is originally used to define them. For instance, any real function $f : A \to \mathbb{R}$ automatically extends to the enlargement ${}^*A$ of its real domain A.

In discussing continuity of a real function f at a real point c, we may want (the extension of) f to be defined at points infinitely close to c. For this it suffices that f be defined on some real neighbourhood $(c-\varepsilon, c+\varepsilon)$ in $\mathbb{R}$, for then the domain of the extension of f includes the enlarged interval *$(c-\varepsilon, c+\varepsilon)$, which contains the halo hal(c) of c. But the converse of this is also true: if the extension of f is defined on hal(c), then f must be defined on some real interval of the form $(c-\varepsilon, c+\varepsilon)$, and hence on *$(c-\varepsilon, c+\varepsilon)$.

In fact, for this last conclusion it can be shown that it suffices that f be defined on some hyperreal interval $(c-d, c+d)$ of *infinitesimal* radius d. This is our first example of a permanence statement that is sometimes called *Cauchy's principle*. It asserts that if a property holds for all points within some infinitesimal distance of c, then it must actually hold for all points within some real (hence appreciable) distance of c. At present we can show this for the transforms of properties expressible in the formal language $\mathcal{L}_\mathfrak{R}$. If $\varphi(x)$ is a formula of this language for which there is some positive $d \simeq 0$ such that

$$^*\varphi(x) \text{ is true for all hyperreal } x \text{ with } c - d < x < c + d,$$

then the sentence

$$(\exists y \in {}^*\mathbb{R}^+)(\forall x \in {}^*\mathbb{R})\,(\,|x - c| < y \rightarrow {}^*\varphi)$$

is seen to be true by interpreting y as d. But then by existential transfer there is some real $\varepsilon > 0$ such that

$$(\forall x \in \mathbb{R})\,(\,|x - c| < \varepsilon \rightarrow \varphi),$$

so that φ is true throughout $(c-\varepsilon, c+\varepsilon)$ in $\mathbb{R}$. Hence by universal transfer back to *$\mathbb{R}$,

$$(\forall x \in {}^*\mathbb{R})\,(\,|x - c| < \varepsilon \rightarrow {}^*\varphi),$$

showing that

$$^*\varphi(x) \text{ is true for all hyperreal } x \text{ with } c - \varepsilon < x < c + \varepsilon.$$

Note that in this argument c is a real number. Later it will be shown that permanence works for any hyperreal number in place of c, and applies to a much broader class of properties than those expressible in the language $\mathcal{L}_\mathfrak{R}$ (cf. Theorems 11.9.1 and 15.1.1).

7.11 Exercises on Permanence of Functions

(1) If f is a real function and $c \in \mathbb{R}$, verify in detail that $f(x)$ is defined for all $x \simeq c$ if and only if $f(x)$ is defined for all real x in some open interval $(c-\varepsilon, c+\varepsilon)$ with real radius $\varepsilon > 0$.

(2) Let f be a real function that is defined on some open neighbourhood of $c \in \mathbb{R}$. Show that if f is constant on hal(c), then it is constant on some interval $(c-\varepsilon, c+\varepsilon) \subseteq \mathbb{R}$.

(3) Let f be a real function that is continuous on some interval $A \subseteq \mathbb{R}$. If $f(x)$ is *real* for all $x \in {}^*A$, show with the help of the previous exercise that f is constant on A.

7.12 Sequences of Functions

Let $\langle f_n : n \in \mathbb{N} \rangle$ be a sequence of functions $f_n : A \to \mathbb{R}$ defined on some subset A of $\mathbb{R}$. The sequence is said to converge *pointwise* to the function $f : A \to \mathbb{R}$ if for each $x \in A$ the $\mathbb{R}$-valued sequence $\langle f_n(x) : n \in \mathbb{N} \rangle$ converges to the number $f(x)$. Symbolically, this asserts that

$$(\forall x \in A) \lim_{n \to \infty} f_n(x) = f(x),$$

which is rendered in full by the sentence

$$(\forall x \in A)\, (\forall \varepsilon \in \mathbb{R}^+)\, (\exists m \in \mathbb{N})\, (\forall n \in \mathbb{N})\, (n > m \to |f_n(x) - f(x)| < \varepsilon).$$

In this statement, the integer m that is asserted to exist depends on the choice of $x \in A$ as well as on ε. More strongly, we say that $\langle f_n : n \in \mathbb{N} \rangle$ converges *uniformly* to the function f if m depends only on ε in the sense that for a given ε, the same m works for all $x \in A$:

$$(\forall \varepsilon \in \mathbb{R}^+)\, (\exists m \in \mathbb{N})\, (\forall x \in A)\, (\forall n \in \mathbb{N})\, (n > m \to |f_n(x) - f(x)| < \varepsilon).$$

Now, we know how to extend a sequence of *numbers* to a hypersequence (Section 3.13), but at this point we would like to do the same for a sequence of functions. For $n \in \mathbb{N}$, the function f_n extends to a function with domain ${}^*A$, but we would like to define $f_n : {}^*A \to {}^*\mathbb{R}$ also for unlimited n. To achieve this we first identify the original sequence $\langle f_n : n \in \mathbb{N} \rangle$ of functions with the single function

$$F : \mathbb{N} \times A \to \mathbb{R}$$

defined by putting $F(n, x) = f_n(x)$ for all $n \in \mathbb{N}$ and $x \in A$. This function F has an extension

$${}^*F : {}^*\mathbb{N} \times {}^*A \to {}^*\mathbb{R},$$

which can then be used to *define* $f_n : {}^*A \to {}^*\mathbb{R}$ by putting $f_n(x) = {}^*F(n, x)$. Thus we now have a hypersequence of functions $\langle f_n : n \in {}^*\mathbb{N} \rangle$ as desired.

For each standard integer $n \in \mathbb{N}$, the new construction of f_n just reproduces the extension of the original function f_n, as defined in Section 3.13. This follows by transfer of

$$(\forall x \in A)\, (f_n(x) = F(n, x)).$$

Moreover, for each $x \in A$, the real-number sequence $s = \langle f_n(x) : n \in \mathbb{N} \rangle$ has as its extension the hypersequence $\langle f_n(x) : n \in {}^*\mathbb{N} \rangle$. This follows by transfer of

$$(\forall n \in \mathbb{N})\, (s(n) = F(n, x)).$$

In view of the characterisation of converging number sequences given by Theorem 6.1.1, we can thus immediately infer

Theorem 7.12.1 *The sequence $\langle f_n : n \in \mathbb{N}\rangle$ of real-valued functions defined on $A \subseteq \mathbb{R}$ converges pointwise to the function $f : A \to \mathbb{R}$ if and only if for each $x \in A$ and each unlimited $n \in {}^*\mathbb{N}$, $f_n(x) \simeq f(x)$.* □

On the other hand, as the discussion in Section 7.7 might suggest, we have:

Theorem 7.12.2 $\langle f_n : n \in \mathbb{N}\rangle$ *converges uniformly to the function $f : A \to \mathbb{R}$ if and only if for each $x \in {}^*A$ and each unlimited $n \in {}^*\mathbb{N}$, $f_n(x) \simeq f(x)$.*

Proof. Exercise. □

The ideas underlying this characterisation are well illustrated by the behaviour of the sequence $\langle f_n : n \in \mathbb{N}\rangle$ given by $f_n(x) = x^n$ on $A = [0, 1]$. This converges pointwise to the function f that is constantly zero on $[0, 1)$ and has $f(1) = 1$. Thus when $x < 1$, the sequence $\langle x^n : n \in \mathbb{N}\rangle$ converges to 0, but as x moves towards 1 the rate of convergence slows down, in the sense that for a fixed $\varepsilon \in \mathbb{R}^+$, as x approaches 1 we have to move further and further along the sequence of powers of x before reaching a point where the terms are less than ε. Ultimately, when x becomes infinitely close to 1 (but still less than 1), it takes "infinitely long" for x^n to become infinitely close to 0. Indeed, by transferring the statement of pointwise convergence and taking ε to be a positive infinitesimal, it follows that there will be some $M \in {}^*\mathbb{N}$ such that for $n > M$ we have $x^n < \varepsilon$ and hence $x^n \simeq 0$. Now, this M will be unlimited, because when n is limited, $x \simeq 1$ implies $x^n \simeq 1$. Hence $\{x^n : n \in \mathbb{N}\}$ is contained entirely within the halo of 1. But there is a permanence principle that concludes from this that there is some *unlimited* N such that $\{x^n : n \leq N\}$ is contained in the halo of 1 (cf. Robinson's sequential lemma in Section 15.2). In particular, $x^N \not\simeq 0$, i.e., $f_N(x) \not\simeq f(x)$, showing that the condition of Theorem 7.12.2 is violated, and therefore that the original standard sequence $\langle f_n : n \in \mathbb{N}\rangle$ is not *uniformly* convergent to f.

7.13 Continuity of a Uniform Limit

A sequence $\langle f_n : n \in \mathbb{N}\rangle$ of *continuous* functions can converge pointwise to a discontinuous function. We have just discussed the standard example: take $f_n(x) = x^n$ on $A = [0, 1]$. Under uniform convergence this phenomenon cannot occur. Here is a hyperreal approach to this classical result:

Theorem 7.13.1 *If the functions $\langle f_n : n \in \mathbb{N}\rangle$ are all continuous on $A \subseteq \mathbb{R}$, and the sequence converges uniformly to the function $f : A \to \mathbb{R}$, then f is continuous on A.*

Proof. Let c belong to A. To prove that f is continuous at c, we invoke Theorem 7.1.3(2). If $x \in {}^*A$ with $x \simeq c$, we want $f(x) \simeq f(c)$, i.e., $|f(x) - f(c)| < \varepsilon$ for any positive real ε. The key to this is to analyse the inequality

$$|f(x) - f(c)| \leq |f(x) - f_n(x)| + |f_n(x) - f_n(c)| + |f_n(c) - f(c)|. \quad (\text{x})$$

On the right side, the middle term $|f_n(x) - f_n(c)|$ will be infinitesimal for any $n \in \mathbb{N}$ because $x \simeq c$ and f_n is continuous at c. By taking a large enough n, the first and last terms on the right can be made small enough that the sum of the three terms is less than ε.

To see how this works in detail, for a given $\varepsilon \in \mathbb{R}^+$ we apply the definition of uniform convergence to the number $\varepsilon/4$ to get that there is some integer $m \in \mathbb{N}$ such that

$$n > m \quad \text{implies} \quad |f_n(x) - f(x)| < \varepsilon/4$$

for all $n \in \mathbb{N}$ and all $x \in A$, and hence for all $n \in {}^*\mathbb{N}$ and all $x \in {}^*A$ by universal transfer.

Now fix n as a standard integer, say by putting $n = m + 1$. Then for any $x \in {}^*A$ with $x \simeq c$ it follows, since $x, c \in {}^*A$, that

$$|f_n(x) - f(x)|, \; |f_n(c) - f(c)| < \varepsilon/4,$$

and so in (x) we get

$$|f(x) - f(c)| < \varepsilon/4 + \text{infinitesimal} + \varepsilon/4 < \varepsilon$$

as desired. $\square$

Note that this proof is a mixture of standard and nonstandard arguments: it uses the hyperreal characterisation of continuity of f_n and f, but the standard definition of uniform convergence of $\langle f_n : n \in \mathbb{N} \rangle$ rather than the characterisation given by Theorem 7.12.2.

In fact, we could invoke Theorem 7.12.2 to make the first and last terms on the right of the inequality (x) become infinitesimal (instead of less than $\varepsilon/4$), by choosing n to be unlimited. But then what happens to the middle term $|f_n(x) - f_n(c)|$ when n is unlimited? Can we constrain it to still be infinitesimal? We will take this up first as a separate question.

7.14 Continuity in the Extended Hypersequence

Given a sequence $\langle f_n : n \in \mathbb{N} \rangle$ of functions that are continuous on a set $A \subseteq \mathbb{R}$, it is natural to wonder about the continuity properties of the extended terms f_n of the associated hypersequence $\langle f_n : n \in {}^*\mathbb{N} \rangle$.

Now, when n is unlimited, then f_n is a function from ${}^*A$ to ${}^*\mathbb{R}$, and even when restricted to A it may take values that are not real, e.g., when

$f_n(x) = x^n$. Thus f_n may not be the extension of any $\mathbb{R}$-valued function, and so we cannot apply the transfer argument of Theorem 7.1.1 directly. However, we can demonstrate a continuity property of f_n that corresponds to that of 7.1.3(3). Thus even though f_n may not map the whole of the halo of a point y into the halo of $f_n(y)$, it will do this to some sufficiently small infinitesimal neighbourhood of y:

Theorem 7.14.1 *If the functions $\langle f_n : n \in \mathbb{N} \rangle$ are all continuous on $A \subseteq \mathbb{R}$, then for any $n \in {}^*\mathbb{N}$ and any $y \in {}^*A$ there is a positive infinitesimal d such that $f_n(x) \simeq f_n(y)$ for all $x \in {}^*A$ with $|x - y| < d$.*

Proof. The fact that f_n is continuous on A for all $n \in \mathbb{N}$ is expressed by the sentence

$(\forall n \in \mathbb{N})\,(\forall y \in A)$
$\quad (\forall \varepsilon \in \mathbb{R}^+)\,(\exists \delta \in \mathbb{R}^+)\,(\forall x \in A)(\,|x - y| < \delta \to |f_n(x) - f_n(y)| < \varepsilon),$

which states that "for all $n \in \mathbb{N}$ and all $y \in A$, f_n is continuous at y in A". Transfer this, take $n \in {}^*\mathbb{N}$ and $y \in {}^*A$, and let ε be a positive infinitesimal. Then from the transferred sentence we get that there is some hyperreal $\delta \in {}^*\mathbb{R}^+$ such that $|x - y| < \delta$ implies $|f_n(x) - f_n(y)| < \varepsilon$ for all $x \in {}^*A$. Hence for all such x,

$$|x - y| < \delta \text{ implies } f_n(x) \simeq f_n(y)$$

because ε is infinitesimal. Now replace δ by any positive infinitesimal $d < \delta$, and the desired conclusion follows. □

This last result can now be used to give the suggested alternative proof that uniform convergence preserves continuity (Theorem 7.13.1), a proof that does use the hyperreal characterisation of uniform convergence given by Theorem 7.12.2. As explained, the idea was to take the inequality

$$|f(x) - f(c)| \leq |f(x) - f_n(x)| + |f_n(x) - f_n(c)| + |f_n(c) - f(c)| \quad (\text{x})$$

used in the proof of Theorem 7.13.1 and make the terms $|f(x) - f_n(x)|$ and $|f_n(c) - f(c)|$ become infinitesimal, thereby forcing $|f(x) - f(c)|$ to be infinitesimal. To achieve this we must take n to be unlimited. In fact, the inequality itself can be dispensed with in favour of a direct examination of the infinitely close proximity of the terms involved.

To review this argument, let f_n be continuous on $A \subseteq \mathbb{R}$ for all $n \in \mathbb{N}$, and suppose $\langle f_n : n \in \mathbb{N} \rangle$ converges uniformly to $f : A \to \mathbb{R}$. Take $c \in \mathbb{R}$, with the object of showing that f is continuous at c. Choose an unlimited $n \in {}^*\mathbb{N}$. Then by our new result, Theorem 7.14.1, there is some infinitesimal $d > 0$ such that

$$|x - c| < d \text{ implies } f_n(x) \simeq f_n(c)$$

whenever $x \in {}^*A$. But now if $x \in {}^*A$ and $|x - c| < d$, we have $f_n(x) \simeq f(x)$ and $f_n(c) \simeq f(c)$ by uniform convergence (Theorem 7.12.2) because n is unlimited and $x, c \in {}^*A$. Then

$$f(x) \simeq f_n(x) \simeq f_n(c) \simeq f(c),$$

making $f(x) \simeq f(c)$.

All told, we have established that there is a positive $d \simeq 0$ such that for all $x \in {}^*A$,

$$|x - c| < d \text{ implies } f(x) \simeq f(c).$$

By Corollary 7.1.2(3) this guarantees that f is continuous at c.

7.15 Was Cauchy Right?

The ideas discussed in this chapter are of central importance in analysis, and have caused difficulty and controversy in the past. There is a famous "theorem" of Cauchy (1821) stating:

> *If the different terms of the series*
>
> $$u_0 + u_1 + \cdots + u_n + u_{n+1} + \cdots$$
>
> *are functions of the same variable x, continuous with respect to this variable in the neighbourhood of a particular value for which the series is convergent, the sum s of the series is also, in the neighbourhood of this particular value, a continuous function of x.*

It has been widely held that this statement is in error, because it leaves out the hypothesis of *uniform* convergence. But the development of Robinson's nonstandard analysis has caused a reassessment of Cauchy's ideas, producing the view that his theorem is correct if it is understood that he intended to assert continuity of $u_n(x)$ even for infinitely large n, and that "neighbourhood of a particular value" refers to points infinitely close to that value. The following articles explore this issue in depth:

> JOHN P. CLEAVE. Cauchy, Convergence, and Continuity. *British J. Phil. Sci.*, **22** (1971), 27–37.

> IMRE LAKATOS. Cauchy and the Continuum: The Significance of Non-Standard Analysis for the History and Philosophy of Mathematics. In *Mathematics, Science and Epistomology*, Philosophical Papers volume 2, edited by John Worrall and Gregory Currie. Cambridge University Press, 1978, 43–60.

8
Differentiation

We come now to an examination—from the *modern* infinitesimal perspective—of the cornerstone concept of the calculus.

8.1 The Derivative

Newton called the derivative the *fluxion* $\dot{y}$ of a fluent quantity y, thinking of it as the "speed" with which the quantity flows. In more modern parlance, the derivative of a function f at a real number x is the real number $f'(x)$ that represents the rate of change of the function as it varies near x. Alternatively, it is the slope of the tangent to the graph of f at x. Formally it is defined as the number

$$\lim_{h \to 0} \frac{f(x+h) - f(x)}{h}.$$

Theorem 8.1.1 *If f is defined at $x \in \mathbb{R}$, then the real number $L \in \mathbb{R}$ is the derivative of f at x if and only if for every nonzero infinitesimal ε, $f(x+\varepsilon)$ is defined and*

$$\frac{f(x+\varepsilon) - f(x)}{\varepsilon} \simeq L . \qquad (i)$$

Proof. Let $g(h) = \frac{f(x+h)-f(x)}{h}$ and apply the characterisation of

$$\text{``} \lim_{h \to 0} g(h) = L\text{''}$$

given in Section 7.3. □

Thus when f is differentiable (i.e., has a derivative) at x, we have

$$f'(x) = \text{sh}\left(\frac{f(x+\varepsilon) - f(x)}{\varepsilon}\right)$$

for all infinitesimal $\varepsilon \neq 0$.

If (i) holds only for all *positive* infinitesimal ε, then L is the *right-hand derivative* of f at x, defined classically as

$$\lim_{h \to 0^+} \frac{f(x+h) - f(x)}{h}.$$

Similarly, if (i) holds for all negative $\varepsilon \simeq 0$, then L is the left-hand derivative given by the limit as $h \to 0^-$.

Exercise 8.1.2

Use the characterisation of Theorem 8.1.1 to prove that the derivative of $\sin x$ is $\cos x$ at real x (cf. Section 7.2 and Exercise 5.7(2)).

8.2 Increments and Differentials

Let Δx denote an arbitrary nonzero infinitesimal representing a change or *increment* in the value of variable x. The corresponding *increment in the value of the function* f at x is

$$\Delta f = f(x + \Delta x) - f(x).$$

To be quite explicit we should denote this increment by $\Delta f(x, \Delta x)$, since its value depends both on the value of x and the choice of the infinitesimal Δx. The more abbreviated notation is, however, convenient and suggestive.

If f is differentiable at $x \in \mathbb{R}$, Theorem 8.1.1 implies that

$$\frac{\Delta f}{\Delta x} \simeq f'(x),$$

so the *Newton quotient* $\frac{\Delta f}{\Delta x}$ is limited. Hence as

$$\Delta f = \frac{\Delta f}{\Delta x}\Delta x,$$

it follows that the increment Δf in f is *infinitesimal*. Thus $f(x + \Delta x) \simeq f(x)$ for all infinitesimal Δx, and this proves

Theorem 8.2.1 *If f is differentiable at $x \in \mathbb{R}$, then f is continuous at x.* □

The *differential of f at x corresponding to* Δx is defined to be

$$df = f'(x)\Delta x.$$

Thus whereas Δf represents the increment of the "y-coordinate" along the graph of f at x, df represents the increment along the tangent line to this graph at x. Writing dx for Δx, the definition of df yields

$$\frac{df}{dx} = f'(x).$$

Now, since $f'(x)$ is limited and Δx is infinitesimal, it follows that df is infinitesimal. Hence df and Δf are infinitely close to each other. In fact, their difference is infinitely smaller than Δx, for if

$$\varepsilon = \frac{\Delta f}{\Delta x} - f'(x),$$

then ε is infinitesimal, because $\frac{\Delta f}{\Delta x} \simeq f'(x)$, and

$$\Delta f - df = \Delta f - f'(x)\Delta x = \varepsilon \Delta x,$$

which is also infinitesimal (being a product of infinitesimals). But

$$\frac{\Delta f - df}{\Delta x} = \frac{\varepsilon \Delta x}{\Delta x} = \varepsilon \simeq 0,$$

and in this sense $\Delta f - df$ is infinitesimal compared to Δx (equivalently, $\frac{\Delta x}{\Delta f - df}$ is unlimited). These relationships are summarised in

Theorem 8.2.2 (Incremental Equation) *If $f'(x)$ exists at real x and $\Delta x = dx$ is infinitesimal, then Δf and df are infinitesimal, and there is an infinitesimal ε, dependent on x and Δx, such that*

$$\Delta f = f'(x)\Delta x + \varepsilon \Delta x = df + \varepsilon dx,$$

and so

$$f(x+\Delta x) = f(x) + f'(x)\Delta x + \varepsilon \Delta x.$$

□

This last equation elucidates the role of the derivative function f' as the *best linear approximation* to the function f at x. For the graph of the linear function

$$l(\Delta x) = f(x) + f'(x)\Delta x$$

gives the tangent to f at x when the origin is translated to the point $(x,0)$, and $l(\Delta x)$ differs from $f(x+\Delta x)$ by the amount $\varepsilon \Delta x$, which we saw above is itself infinitely smaller than Δx when Δx is infinitesimal, and in that sense is "negligible".

8.3 Rules for Derivatives

If f and g are differentiable at $x \in \mathbb{R}$, then so to are $f+g$, fg, and f/g, provided that $g(x) \neq 0$. Moreover,

(1) $(f+g)'(x) = f'(x) + g'(x)$,

(2) $(fg)'(x) = f'(x)g(x) + f(x)g'(x)$,

(3) $(f/g)'(x) = \dfrac{f'(x)g(x) - f(x)g'(x)}{g(x)^2}$.

Proof. We prove Leibniz's rule (2), and leave the others as exercises.

If $\Delta x \neq 0$ is infinitesimal, then, by Theorem 8.1.1, $f(x + \Delta x)$ and $g(x + \Delta x)$ are both defined, and hence so is

$$(fg)(x + \Delta x) = f(x + \Delta x)g(x + \Delta x).$$

Then the increment of fg at x corresponding to Δx is

$$\begin{aligned}\Delta(fg) &= f(x+\Delta x)g(x+\Delta x) - f(x)g(x) \\ &= (f(x) + \Delta f)(g(x) + \Delta g) - f(x)g(x) \\ &= (\Delta f)g(x) + f(x)\Delta g + \Delta f \Delta g\end{aligned}$$

(compare this to Leibniz's reasoning as discussed in Section 1.2). It follows that

$$\begin{aligned}\frac{\Delta(fg)}{\Delta x} &= \frac{\Delta f}{\Delta x}g(x) + f(x)\frac{\Delta g}{\Delta x} + \Delta f \frac{\Delta g}{\Delta x} \\ &\simeq f'(x)g(x) + f(x)g'(x) + 0,\end{aligned}$$

since $\frac{\Delta f}{\Delta x} \simeq f'(x)$, $\frac{\Delta g}{\Delta x} \simeq g'(x)$, $\Delta f \simeq 0$ and all quantities involved are limited.

Hence by Theorem 8.1.1, $f'(x)g(x) + f(x)g'(x)$ is the derivative of fg at x. □

8.4 Chain Rule

If f is differentiable at $x \in \mathbb{R}$, and g is differentiable at $f(x)$, then $g \circ f$ is differentiable at x with derivative $g'(f(x))f'(x)$.

Proof. Let Δx be a nonzero infinitesimal. Then $f(x + \Delta x)$ is defined and $f(x + \Delta x) \simeq f(x)$, as we saw in Section 8.2. But g is defined at all points infinitely close to $f(x)$, since $g'(f(x))$ exists, so $(g \circ f)(x + \Delta x) = g(f(x + \Delta x))$ is defined.

Now let
$$\Delta f = f(x + \Delta x) - f(x),$$
$$\Delta(g \circ f) = g(f(x + \Delta x)) - g(f(x))$$
be the increments of f and $g \circ f$ at x corresponding to Δx. Then Δf is infinitesimal, and
$$\Delta(g \circ f) = g(f(x) + \Delta f) - g(f(x)),$$
which shows, crucially, that

$\Delta(g \circ f)$ *is also the increment of g at $f(x)$ corresponding to Δf.*

In the full incremental notation, this reads
$$\Delta(g \circ f)(x, \Delta x) = \Delta g(f(x), \Delta f).$$
By the incremental equation (Theorem 8.2.2) for g, it then follows that there exists an infinitesimal ε such that
$$\Delta(g \circ f) = g'(f(x))\Delta f + \varepsilon \Delta f.$$
Hence
$$\frac{\Delta(g \circ f)}{\Delta x} = g'(f(x))\frac{\Delta f}{\Delta x} + \varepsilon \frac{\Delta f}{\Delta x}$$
$$\simeq g'(f(x))f'(x) + 0,$$
establishing that $g'(f(x))f'(x)$ is the derivative of $g \circ f$ at x. □

8.5 Critical Point Theorem

Let f have a maximum or a minimum at x on some real interval (a, b). If f is differentiable at x, then $f'(x) = 0$.

Proof. Suppose f has a maximum at x. By transfer,
$$f(x + \Delta x) \leq f(x)$$
for all infinitesimal Δx. Hence if ε is positive infinitesimal and δ is negative infinitesimal,
$$f'(x) \simeq \frac{f(x + \varepsilon) - f(x)}{\varepsilon} \leq 0 \leq \frac{f(x + \delta) - f(x)}{\delta} \simeq f'(x),$$
and so as $f'(x)$ is real, it must be equal to 0.

The case of f having a minimum at x is similar. □

Using the critical point and extreme value theorems, the following results can be successively derived about a function f that is continuous on $[a, b] \subseteq \mathbb{R}$ and differentiable on (a, b). The proofs do not require any further reasoning about infinitesimals or limits.

- **Rolle's Theorem**: if $f(a) = f(b) = 0$, then $f'(x) = 0$ for some $x \in (a, b)$.

- **Mean Value Theorem**: for some $x \in (a, b)$,
$$f'(x) = \frac{f(b) - f(a)}{b - a}.$$

- If f' is zero/positive/negative on (a, b), then f is constant/increasing/decreasing on $[a, b]$.

8.6 Inverse Function Theorem

Let f be continuous and strictly monotone (increasing or decreasing) on (a, b), and suppose g is the inverse function of f. If f is differentiable at x in (a, b), with $f'(x) \neq 0$, then g is differentiable at $y = f(x)$, with $g'(y) = 1/f'(x)$.

Proof. Using the intermediate value theorem and monotonicity of f it can be shown that g is defined on some real open interval around y. The result $g'(f(x)) = 1/f'(x)$ would follow easily by the chain rule applied to the equation $g(f(x)) = x$ *if* we knew that g was differentiable at $f(x)$. But that is what we have to prove!

Now let Δy be a nonzero infinitesimal. We need to show that
$$\frac{g(y + \Delta y) - g(y)}{\Delta y} \simeq \frac{1}{f'(x)}.$$

Now, if $g(y + \Delta y)$ were not infinitely close to $g(y)$, then there would be a real number r strictly between them. But then, by monotonicity of f, $f(r)$ would be a real number strictly between $y + \Delta y$ and y. Since y is real, this would mean that $y + \Delta y$ and y were an appreciable distance apart, which is not so. Hence
$$\Delta x = g(y + \Delta y) - g(y)$$
is infinitesimal and is nonzero. (Thus the argument so far establishes that g is continuous at y.) Observe that Δx is, by definition, the increment $\Delta g(y, \Delta y)$ of g at y corresponding to Δy.

Since $g(y) = x$, the last equation gives $g(y + \Delta y) = x + \Delta x$, so
$$f(x + \Delta x) = f(g(y + \Delta y)) = y + \Delta y.$$

Hence
$$\begin{aligned}\Delta y &= f(x + \Delta x) - f(x) \\ &= \Delta f, \text{ the increment of } f \text{ at } x \text{ corresponding to } \Delta x.\end{aligned}$$

Altogether we have
$$\frac{\Delta f(x, \Delta x)}{\Delta x} = \frac{\Delta y}{\Delta x}$$
and
$$\frac{\Delta g(y, \Delta y)}{\Delta y} = \frac{\Delta x}{\Delta y} = \frac{\Delta x}{\Delta f}.$$
Put more briefly, we have shown that
$$\frac{\Delta g}{\Delta y} = \frac{1}{\Delta f / \Delta x}.$$
To derive from this the conclusion $g'(y) = 1/f'(x)$ we invoke the hypothesis that $f'(x) \neq 0$ (which is essential: consider what happens at $x = 0$ when $f(x) = x^3$). Since $\text{sh}(\Delta f / \Delta x) = f'(x)$, it follows that $\Delta f / \Delta x$ is appreciable. But then
$$\text{sh}\left(\frac{\Delta x}{\Delta f}\right) = \text{sh}\left(\frac{1}{\Delta f / \Delta x}\right)$$
$$= \frac{1}{\text{sh}(\Delta f / \Delta x)} \quad \text{by 5.6.2(3)}$$
$$= \frac{1}{f'(x)}.$$
Therefore,
$$\frac{\Delta g(y, \Delta y)}{\Delta y} = \frac{\Delta x}{\Delta y} \simeq \frac{1}{f'(x)}.$$
Because Δy is an arbitrary nonzero infinitesimal, this establishes that the real number $1/f'(x)$ is the derivative of g at y, as desired. □

8.7 Partial Derivatives

Let $z = f(x, y)$ be a real-valued function of two variables, with partial derivatives denoted by f_x and f_y. At a real point (a, b), $f_x(a, b)$ is the derivative of the function $x \mapsto f(x, b)$ at a, while $f_y(a, b)$ is the derivative of $y \mapsto f(a, y)$ at b. Thus for nonzero infinitesimals $\Delta x, \Delta y$,
$$f_x(a, b) \simeq \frac{f(a + \Delta x, b) - f(a, b)}{\Delta x},$$
$$f_y(a, b) \simeq \frac{f(a, b + \Delta y) - f(a, b)}{\Delta y}.$$
Points (x_1, y_1) and (x_2, y_2) in the hyperreal plane $^*\mathbb{R}^2$ are *infinitely close* if both $x_1 \simeq x_2$ and $y_1 \simeq y_2$, which is equivalent to requiring that their Euclidean distance apart,
$$\sqrt{(x_1 - x_2)^2 + (y_1 - y_2)^2},$$

be infinitesimal.

The function f is *continuous* at the real point (a,b) if $(x,y) \simeq (a,b)$ implies $f(x,y) \simeq f(a,b)$ for all hyperreal x, y. For this to hold it is necessary that f be defined on some open disk about (a,b) in the real plane.

We say that f is *smooth* at (a,b) if f_x and f_y both exist and are continuous at (a,b).

The *increment* of f at a point (a,b) corresponding to $\Delta x, \Delta y$ is defined to be
$$\Delta f = f(a + \Delta x, b + \Delta y) - f(a,b),$$
while the *total differential* is
$$df = f_x(a,b)\Delta x + f_y(a,b)\Delta y.$$

The graph of $z = f(x,y)$ is a surface in three-dimensional space, and Δf is the change in z-value on this surface in moving from the point (a,b) to the point $(a + \Delta x, b + \Delta y)$. The total differential df is the corresponding change on the tangent plane to the surface at (a,b).

Theorem 8.7.1 (Incremental Equation for Two Variables) *If f is smooth at the real point (a,b) and Δx and Δy are infinitesimal, then*
$$\Delta f = df + \varepsilon \Delta x + \delta \Delta y$$
for some infinitesimals ε and δ.

Proof. The increment of f at (a,b) corresponding to $\Delta x, \Delta y$ can be written as
$$\Delta f = [f(a + \Delta x, b + \Delta y) - f(a + \Delta x, b)] + [f(a + \Delta x, b) - f(a,b)]. \quad \text{(ii)}$$

The second main summand of (ii) is the increment at a corresponding to Δx of the one-variable function $x \mapsto f(x,b)$, whose derivative $f_x(a,b)$ is assumed to exist. Applying the one-variable incremental equation (Theorem 8.2.2) thus gives
$$f(a + \Delta x, b) - f(a,b) = f_x(a,b)\Delta x + \varepsilon \Delta x \quad \text{(iii)}$$
for some infinitesimal ε.

Similarly, for the first summand we need to show that
$$f(a + \Delta x, b + \Delta y) - f(a + \Delta x, b) = f_y(a,b)\Delta y + \delta \Delta y \quad \text{(iv)}$$
for some infinitesimal δ. Then combining (ii)–(iv) will give
$$\Delta f = f_x(a,b)\Delta x + f_y(a,b)\Delta y + \varepsilon \Delta x + \delta \Delta y,$$
which is the desired result.

Now the left side of equation (iv) could be described as the increment in the function $y \mapsto f(a + \Delta x, y)$ at b corresponding to the infinitesimal Δy. This is not a *real* function, because of the hyperreal parameter $a + \Delta x$, so the incremental equation 8.2.2 does not apply directly to it. To overcome this we will examine the family of functions $y \mapsto f(a + x_0, y)$ for *real* x_0, and consider their increments corresponding to *real* increments y_0 in y. This will give a statement about x_0 and y_0 to which we can apply transfer and then replace x_0 and y_0 by Δx and Δy.

The technical details of this are as follows. Since f_x and f_y are continuous at (a, b), f must be defined on an open disk D around (a, b) of some real radius r. Then if x_0, y_0 are real numbers such that $(a + x_0, b + y_0) \in D$, the function $y \mapsto f(a + x_0, y)$ is defined on the interval $[b, b + y_0]$ and is subject to the one-variable mean value theorem. Hence there is some real c_0 between b and $b + y_0$ such that the derivative of this one-variable function at c_0 is given as

$$f_y(a + x_0, c_0) = \frac{f(a + x_0, b + y_0) - f(a + x_0, b)}{b + y_0 - b},$$

and so

$$f(a + x_0, b + y_0) - f(a + x_0, b) = f_y(a + x_0, c)y_0. \quad\text{(v)}$$

This obtains for all real x_0, y_0 such that $(a + x_0, b + y_0)$ is within r of (a, b). That is, for all such x_0, y_0 there exists $c_0 \in [b, b + y_0]$ such that (v) holds. Symbolically,

$(\forall x_0, y_0 \in \mathbb{R})$
$$\left(\sqrt{x_0^2 + y_0^2} < r \to (\exists c_0 \in \mathbb{R})\, [b \le c_0 \le b + y_0 \text{ and (v) holds}] \right).$$

But $(a + \Delta x, b + \Delta y)$ is within r of (a, b), since $\Delta x, \Delta y$ are infinitesimal, so by transfer there exists some hyperreal c between b and $b + \Delta y$ such that

$$f(a + \Delta x, b + \Delta y) - f(a + \Delta x, b) = f_y(a + \Delta x, c)\Delta y. \quad\text{(vi)}$$

Then $c \simeq b$, so $(a + \Delta x, c) \simeq (a, b)$, and hence by continuity of f_y at (a, b),

$$f_y(a + \Delta x, c) \simeq f_y(a, b).$$

Therefore the difference

$$\delta = f_y(a + \Delta x, c) - f_y(a, b)$$

is infinitesimal, with $f_y(a + \Delta x, c) = f_y(a, b) + \delta$. Applying this to (vi) yields (iv) and completes the proof. □

8.8 Exercises on Partial Derivatives

(1) Show that if f is smooth at (a, b), then it is continuous at (a, b).

(2) Let f be smooth at (a, b). Given infinitesimals $\Delta x, \Delta y$ show that the difference between Δf and df is itself infinitely smaller than the infinitesimal distance $\Delta l = \sqrt{\Delta x^2 + \Delta y^2}$ between (a, b) and $(a + \Delta x, b + \Delta y)$, in the sense that

$$\frac{\Delta f - df}{\Delta l} \simeq 0.$$

8.9 Taylor Series

Let f be a real function and a a real number. The *Taylor series of f at $x \in \mathbb{R}$, centred on a*, is the series

$$f(a) + f'(a)(x-a) + \frac{f''(a)}{2!}(x-a)^2 + \cdots + \frac{f^{(k)}(a)}{k!}(x-a)^k + \cdots,$$

or more briefly, $\sum_0^\infty \frac{f^{(k)}(a)}{k!}(x-a)^k$, where $f^{(k)}$ is the kth derivative of f. For this to be defined, f must be differentiable infinitely often at a, but even if $f^{(k)}(a)$ exists for all $k \in \mathbb{N}$, the series need not converge. Even if it does converge, the sum need not be equal to $f(x)$. A well-known example is the function $f(x) = e^{-1/x^2}$ with $f(0) = 0$. This is so "flat" at the centre $a = 0$ that all its derivatives $f^{(k)}(0)$ there are equal to 0. Hence the associated Taylor series converges at all real x, but converges to $f(x)$ only when $x = 0$.

The partial sums of a Taylor series are the *Taylor polynomials*. The nth polynomial is

$$\begin{aligned} p_n(x) &= \sum_0^n \frac{f^{(k)}(a)}{k!}(x-a)^k \\ &= f(a) + f'(a)(x-a) + \frac{f''(a)}{2!}(x-a)^2 + \cdots + \frac{f^{(n)}(a)}{n!}(x-a)^n. \end{aligned}$$

For any given x, the sequence $\langle p_n(x) : n \in \mathbb{N} \rangle$ extends to a hypersequence, so $p_n(x)$ is defined for all $n \in {}^*\mathbb{N}$. Then from our earlier work on sequences and series (Chapter 6) we see that

- the Taylor series for f at x converges to a real number L if and only if $p_n(x) \simeq L$ for all unlimited n.

The difference between $f(x)$ and $p_n(x)$ is the *nth remainder at x*:

$$R_n(x) = f(x) - p_n(x). \tag{vii}$$

If f is infinitely differentiable at a, then (vii) defines $R_n(x)$ for all $n \in \mathbb{N}$. The sequence $\langle R_n(x) : n \in \mathbb{N} \rangle$ then extends to a hypersequence, and by transfer (vii) holds for all hypernatural n. Then:

- the Taylor series for f at x converges to $f(x)$ if and only if $R_n(x)$ is infinitesimal for all unlimited n.

If the derivatives $f^{(n)}$ exist for all $n \in \mathbb{N}$ on some open interval J containing a, then the sequence of functions $\langle f^{(n)} : n \in \mathbb{N} \rangle$ extends to a hypersequence $\langle f^{(n)} : n \in {}^*\mathbb{N} \rangle$ of functions defined on $^*J$ in the manner described in Section 7.12. Formally, we put $F(n, x) = f^{(n)}(x)$ for $n \in \mathbb{N}$ and $x \in J$ and then by extension get $f^{(n)}(x) = {}^*F(n, x)$ for $n \in {}^*\mathbb{N}$ and $x \in {}^*J$. Then results like

$$p_n(x) - p_{n-1}(x) = \frac{f^{(n)}(a)}{n!}(x-a)^n$$

continue to hold for unlimited n, by transfer.

Now, the *Lagrange form of the remainder* stipulates that if f can be differentiated $n+1$ times on some open interval containing a, then for each x in that interval there is some real number c between a and x such that

$$R_n(x) = \frac{f^{(n+1)}(c)}{(n+1)!}(x-a)^{n+1}.$$

Thus if f is infinitely differentiable on some open interval J containing a, then for every $n \in \mathbb{N}$ and every $x \in J$ we have

$$f(x) - p_n(x) = \frac{f^{(n+1)}(c)}{(n+1)!}(x-a)^{n+1} \qquad \text{(viii)}$$

for some c between a and x. Hence by transfer, for every $n \in {}^*\mathbb{N}$ and $x \in {}^*J$, the *Taylor formula* (viii) holds for some hyperreal c between a and x (c may no longer be real). If we can show for a real x that the right side of (viii) is infinitesimal whenever n is unlimited, it will follow that the Taylor series of f at x converges to $f(x)$.

Let us illustrate this with the case of the function $f(x) = \cos x$, analysing its *Maclaurin* series, which is the Taylor series at the centre $a = 0$. For any $x \in {}^*\mathbb{R}$ and $n \in {}^*\mathbb{N}$ we have

$$R_n(x) = \frac{\cos^{(n+1)} c}{(n+1)!} x^{n+1}$$

for some c with $|c| \leq x$. Now, if $n \in \mathbb{N}$ and $c \in \mathbb{R}$, then $\cos^{(n+1)} c$ is $\pm \sin c$ or $\pm \cos c$, and so in all cases lies between -1 and 1. This fact then holds by transfer for any $n \in {}^*\mathbb{N}$ and $c \in {}^*\mathbb{R}$, so $\cos^{(n+1)} c$ is always limited. But if $x \in \mathbb{R}$ and n is unlimited,

$$\frac{x^{n+1}}{(n+1)!}$$

is infinitesimal (Exercise 6.11(10)). It follows in this case that $R_n(x)$ is infinitesimal, and therefore (cf. (vii)) $f(x) \simeq p_n(x)$. This shows that the Maclaurin series for the cosine function converges to $\cos x$ at all real x.

Exercise 8.9.1
Verify that the Maclaurin series for e^x converges to e^x at any $x \in \mathbb{R}$ by proving that the remainder $R_n(x)$ is infinitesimal when n is unlimited.

8.10 Incremental Approximation by Taylor's Formula

The incremental equation of Theorem 8.2.2 approximates the value $f(x + \Delta x)$ by a linear function $f(x) + f'(x)\Delta x$ of the increment Δx, with an error $\varepsilon \Delta x$ that is infinitely smaller than Δx. We will now see that there are similar approximations by higher-order polynomials in Δx (quadratics, cubics, quartics, etc.).

Fix a real number x and a positive integer $n \in \mathbb{N}$. Consider polynomials centred at x itself. If the nth derivative $f^{(n)}$ exists on an open interval J containing x, then the Taylor formula with Lagrange remainder (viii) stipulates that for real numbers of the form $x + \Delta x$ in J,

$$\begin{aligned} f(x + \Delta x) &= p_{n-1}(x + \Delta x) + R_{n-1}(x + \Delta x) \\ &= \sum_0^{n-1} \frac{f^{(k)}(x)}{k!} \Delta x^k + \frac{f^{(n)}(c)}{n!} \Delta x^n \end{aligned} \quad \text{(ix)}$$

for some c between x and $x + \Delta x$. By transfer this holds whenever $x + \Delta x$ belongs to $^*J$, and so it holds for any *infinitesimal* Δx, in which case $c \simeq x$. Now (ix) can be modified to give

$$f(x + \Delta x) = \sum_0^n \frac{f^{(k)}(x)}{k!} \Delta x^k + \frac{f^{(n)}(c) - f^{(n)}(x)}{n!} \Delta x^n,$$

and if $f^{(n)}$ is continuous at x, then from $c \simeq x$ we infer $f^{(n)}(c) \simeq f^{(n)}(x)$, implying that the number

$$\varepsilon = \frac{f^{(n)}(c) - f^{(n)}(x)}{n!}$$

is infinitesimal. Altogether:

Theorem 8.10.1 *If the nth derivative $f^{(n)}$ exists on an open interval containing the real number x, and $f^{(n)}$ is continuous at x, then for any infinitesimal Δx,*

$$f(x + \Delta x) = f(x) + f'(x)\Delta x + \frac{f''(x)}{2!} \Delta x^2 + \cdots + \frac{f^{(n)}(x)}{n!} \Delta x^n + \varepsilon \Delta x^n$$

for some infinitesimal ε. □

In other words, the difference between $f(x + \Delta x)$ and the nth-order polynomial
$$f(x) + f'(x)\Delta x + \frac{f''(x)}{2!}\Delta x^2 + \cdots + \frac{f^{(n)}(x)}{n!}\Delta x^n$$
in Δx is the infinitesimal $\varepsilon \Delta x^n$, which is, as Leibniz would put it (Section 1.2), *infinitely small in comparison with* Δx^n.

Exercise 8.10.2
There are forms for the Taylor remainder other than Lagrange's. One of these is
$$R_n(x) = \frac{f^{(n)}(c) - f^{(n)}(a)}{(c-a)(n+1)!}(x-a)^{n+1}$$
for some c between a and x when $f^{(n+1)}$ exists between a and x.

Apply this form for $R_{n-1}(x)$ to show that Theorem 8.10.1 holds without the hypothesis of continuity of $f^{(n)}$.

8.11 Extending the Incremental Equation

The equation
$$f(x + \Delta x) = f(x) + f'(x)\Delta x + \varepsilon \Delta x$$
holds for any *real* number x at which f is differentiable. It is natural to ask whether a similar formula holds for nonreal x, and it turns out that this is intimately connected with the question of the continuity of the derivative function f'.

Let us say that a hyperreal x is *well inside* an interval (y, z) if $y < x < z$ but x is not infinitely close to either of the end points y and z. Equivalently, this means that the halo of x is included in the interval, so that $y < x + \Delta x < z$ for all infinitesimals Δx.

Theorem 8.11.1 *Let f be differentiable on an interval (a, b) in $\mathbb{R}$. Then the derivative f' is continuous on (a, b) if and only if for each hyperreal x that is well inside *(a, b) and each infinitesimal Δx,*
$$f(x + \Delta x) = f(x) + f'(x)\Delta x + \varepsilon \Delta x$$
for some infinitesimal ε.

Proof. Assume that the incremental equation holds at points well inside *(a, b). To prove continuity of f', let c be a real point in (a, b) and suppose $x \simeq c$. We want $f'(x) \simeq f'(c)$.

Now, if $\Delta = (x - c) \simeq 0$, then using Theorem 8.2.2 we get
$$f(x) = f(c + \Delta) = f(c) + f'(c)\Delta + \varepsilon \Delta$$

for some $\varepsilon \simeq 0$. But x is well inside *(a,b), since $a < c < b$ and $x \simeq c$, so by the assumed incremental equation at x, applied to the infinitesimal $-\Delta$, we have

$$f(c) = f(x + (-\Delta)) = f(x) + f'(x)(-\Delta) + \varepsilon'(-\Delta)$$

for some $\varepsilon' \simeq 0$. Combining these equations leads to

$$f'(x) - f'(c) = \varepsilon - \varepsilon' \simeq 0,$$

giving our desired conclusion $f'(x) \simeq f'(c)$.

The proof that continuity of f' implies the incremental equation at points well inside *(a, b) is indicated in the following exercises, which also give an example to show what can happen when continuity fails. □

8.12 Exercises on Increments and Derivatives

(1) Let f be differentiable and have f' continuous on $(a, b) \subseteq \mathbb{R}$. Let x be well inside *(a, b) and $\Delta x \simeq 0$.

   (a) By an argument involving transfer of the standard mean value theorem (Section 8.5), the shadow of x, and the continuity of f', prove that
   $$f'(x) \simeq \frac{f(x + \Delta x) - f(x)}{\Delta x}.$$

   (b) Hence show that for some $\varepsilon \simeq 0$,
   $$f(x + \Delta x) = f(x) + f'(x)\Delta x + \varepsilon \Delta x. \tag{x}$$

(2) Let
$$f(x) = \begin{cases} x^2 \sin \frac{1}{x} & \text{if } x \neq 0, \\ 0 & \text{if } x = 0. \end{cases}$$

   (a) Prove that f' exists at 0 but is not continuous there.

   (b) Let $x = 1/(2\pi N)$ with N unlimited. Show that there is an infinitesimal Δx such that equation (x) of Exercise (1b) fails for any $\varepsilon \simeq 0$.

In terms of Theorem 8.11.1, note that in Exercise (2), f' is continuous on the interval $(0, 1)$ while the infinitesimal $1/(2\pi N)$ at which the incremental equation fails is *not* well inside *$(0, 1)$.

9

The Riemann Integral

The definite integral $\int_a^b f(x)dx$ represents the area under the graph of the function $y = f(x)$ between $x = a$ and $x = b$. The standard way to define this is to partition the interval $[a, b]$ into a finite number of subintervals, approximate the desired area by sums of areas of rectangles based on these subintervals, and then take the limit as the number of subintervals is increased.

The hyperreal perspective suggests the alternative procedure of partitioning $[a, b]$ into subintervals of infinitesimal width, in line with Leibniz's conception of the expression $\int y\, dx$—with $\int$ as an elongated "S" for "sum"—as meaning the sum of all the infinitely thin rectangles of size $y \times dx$. In order to develop this approach we will first review the standard definition of the integral that is associated with Riemann.

9.1 Riemann Sums

Let f be a function that is bounded on $[a, b]$ in $\mathbb{R}$. A *partition* of $[a, b]$ is a finite set $P = \{x_0, \ldots, x_n\}$ with $a = x_0 < \cdots < x_n = b$. Let M_i and m_i be the least upper bound and greatest lower bound of f on $[x_{i-1}, x_i]$, respectively, and $\Delta x_i = x_i - x_{i-1}$. Define the

- upper Riemann sum: $U_a^b(f, P) = \sum_{i=1}^n M_i \Delta x_i$;
- lower Riemann sum: $L_a^b(f, P) = \sum_{i=1}^n m_i \Delta x_i$;
- ordinary Riemann sum: $S_a^b(f, P) = \sum_{i=1}^n f(x_{i-1}) \Delta x_i$.

If M and m are the least upper bound and greatest lower bound of f on $[a,b]$, then

$$m(b-a) \leq L_a^b(f,P) \leq S_a^b(f,P) \leq U_a^b(f,P) \leq M(b-a).$$

Also, by using refinements of partitions it is shown that any lower sum is less than or equal to any upper sum:

$$L_a^b(f,P_1) \leq U_a^b(f,P_2).$$

We say that f is *Riemann integrable* on $[a,b]$ with integral $\int_a^b f(x)dx$ if the latter is a real number equal to the least upper bound of the lower sums $L_a^b(f,P)$ and also to the greatest lower bound of the upper sums $U_a^b(f,P)$ taken over all partitions P of $[a,b]$. This holds iff

(1) $L_a^b(f,P) \leq \int_a^b f(x)dx \leq U_a^b(f,P)$ for all partitions P; and

(2) for any real $\varepsilon > 0$ there is a partition P with $U_a^b(f,P) - L_a^b(f,P) < \varepsilon$.

For a given positive real Δx, let $P_{\Delta x} = \{x_0, \ldots, x_n\}$ be the partition

$$[a, x_1], \ldots, [x_{n-2}, x_{n-1}]$$

of $[a,b]$ into subintervals of equal width Δx, together with a (possibly smaller) last subinterval $[x_{n-1}, b]$. This is given by taking n to be the least integer such that $a + n\Delta x \geq b$, and $x_k = a + k\Delta x$ for $k < n$. The partition $P_{\Delta x}$ is uniquely determined by the number Δx (observe that if $\Delta x \geq b - a$, then $n = 1$, and we just get $P_{\Delta x} = \{a, b\}$).

Now let $U_a^b(f, \Delta x)$, $L_a^b(f, \Delta x)$, $S_a^b(f, \Delta x)$ be the upper, lower, and ordinary Riemann sums for this partition. These quantities can be regarded as functions of the real variable Δx, defined on $\mathbb{R}^+$. Hence these functions extend automatically to $^*\mathbb{R}^+$. In particular, they are defined for all positive infinitesimals, giving a hyperreal meaning to the notion of Riemann sums for infinitesimal width partitions. For instance, we may informally think of $S_a^b(f, \Delta x)$ as being the "sum"

$$f(x_0)\Delta x + f(x_1)\Delta x + \cdots + f(x_{n-2})\Delta x + f(x_{n-1})(b - x_{n-1}),$$

where n is the least hyperinteger such that $a + n\Delta x \geq b$. When Δx is infinitesimal, this n will be unlimited. (In Section 12.7 we will analyse this informal view further, and express $S_a^b(f, \Delta x)$ as a "hyperfinite sum" over a "hyperfinite partition".)

The relationships

$$m(b-a) \leq L_a^b(f, \Delta x) \leq S_a^b(f, \Delta x) \leq U_a^b(f, \Delta x) \leq M(b-a)$$

hold for all $\Delta x \in \mathbb{R}^+$, and hence by universal transfer hold for all positive hyperreal Δx, including positive infinitesimals. Similarly we have

$$L_a^b(f, \Delta x) \leq U_a^b(f, \Delta y)$$

for all $\Delta x, \Delta y \in {^*\mathbb{R}^+}$.

Theorem 9.1.1 *If f is continuous on the real interval $[a,b]$, then for any positive infinitesimal Δx, $L_a^b(f, \Delta x) \simeq U_a^b(f, \Delta x)$.*

Proof. The key to this is the fact that
$$U_a^b(f, \Delta x) - L_a^b(f, \Delta x) = \sum_{i=1}^n (M_i - m_i)\Delta x_i.$$

We will find an upper bound of the right side of this equation that has the form
$$[f(c) - f(d)](b - a),$$
where c, d are two numbers that are smaller than Δx. When Δx is infinitesimal, the continuity of f then ensures that $f(c) \simeq f(d)$, making our upper bound infinitesimal.

By a method that is now becoming familiar, we first formalise the real version of this construction, and then apply transfer. For positive *real* Δx, let $\mu(\Delta x)$ be the maximum of the numbers $M_i - m_i$ for $1 \le i \le n$ in the partition determined by Δx. $M_i - m_i$ is the *oscillation* of f on the ith interval, so $\mu(\Delta x)$ is the largest oscillation on any subinterval of the partition.

If $\mu(\Delta x) = M_j - m_j$, let $c(\Delta x)$ and $d(\Delta x)$ be the points in $[x_{j-1}, x_j]$ where M_j and m_j are attained. Existence of these points is guaranteed by the extreme value theorem 7.6, because f is continuous on the closed interval $[x_{j-1}, x_j]$. Then
$$\mu(\Delta x) = f(c(\Delta x)) - f(d(\Delta x)),$$
and
$$|c(\Delta x) - d(\Delta x)| \le \Delta x. \qquad (i)$$

Hence
$$\begin{aligned} U_a^b(f, \Delta x) - L_a^b(f, \Delta x) &= \sum_{i=1}^n (M_i - m_i)\Delta x_i \\ &\le \sum_{i=1}^n \mu(\Delta x)\Delta x_i \\ &= \mu(\Delta x)\sum_{i=1}^n \Delta x_i \\ &= \mu(\Delta x)(b - a), \end{aligned}$$

and so
$$U_a^b(f, \Delta x) - L_a^b(f, \Delta x) \le [f(c(\Delta x)) - f(d(\Delta x))](b - a). \qquad (ii)$$

Thus we have shown that for all real $\Delta x > 0$ there exist $c(\Delta x), d(\Delta x) \in [a, b]$ such that (i) and (ii) hold. But then this transfers to *$\mathbb{R}$. Choosing Δx to be a positive *infinitesimal*, the transfer of (i) gives $c(\Delta x) \simeq d(\Delta x)$ in *$[a, b]$, so by taking their shadow we get a real $r \in [a, b]$ with
$$c(\Delta x) \simeq r \simeq d(\Delta x).$$

Then by continuity of f,

$$f(c(\Delta x)) \simeq f(r) \simeq f(d(\Delta x)),$$

so $f(c(\Delta x)) - f(d(\Delta x))$ is infinitesimal (this is just a repetition of the proof from Theorem 7.7.2 that f is *uniformly* continuous on $[a,b]$). Since $(b-a)$ is limited, transfer of (ii) then implies that $U_a^b(f, \Delta x) - L_a^b(f, \Delta x)$ is infinitesimal, as desired. □

Exercise 9.1.2 (Monotonic Functions)
Suppose that f is *nondecreasing* on $[a,b]$ in the sense that $f(x) \leq f(y)$ whenever $x \leq y$. Show that for $\Delta x \in \mathbb{R}^+$,

$$U_a^b(f, \Delta x) - L_a^b(f, \Delta x) \leq \Delta x (f(b) - f(a)).$$

Similarly, if f is *nonincreasing* in the sense that $f(x) \geq f(y)$ whenever $x \leq y$, show that

$$U_a^b(f, \Delta x) - L_a^b(f, \Delta x) \leq \Delta x (f(a) - f(b)).$$

A function is *monotonic* on $[a,b]$ if it is either nondecreasing or else nonincreasing. Prove that Theorem 9.1.1 holds for monotonic functions as well as for continuous ones.

9.2 The Integral as the Shadow of Riemann Sums

It will now be shown that any continuous or monotonic function on a closed interval in $\mathbb{R}$ is Riemann integrable (note that any such function is bounded).

Let Δx_1 and Δx_2 be positive infinitesimals, with associated lower sums L_1, L_2 and upper sums U_1, U_2 for a *continuous* or *monotonic* function f on $[a,b]$. Then $L_1 \leq U_1$ and $L_2 \leq U_2$. But since real upper sums dominate all real lower sums, and this continues to be true in the hyperreal case by transfer, we also have $L_2 \leq U_1$ and $L_1 \leq U_2$. Thus the possible relationships are

$$L_1 \leq L_2 \leq U_2 \leq U_1,$$

or

$$L_1 \leq L_2 \leq U_1 \leq U_2,$$

or the corresponding statements with the subscripts interchanged. But $L_1 \simeq U_1$ and $L_2 \simeq U_2$ by Theorem 9.1.1 or Exercise 9.1.2, so it follows that in any case

$$L_1 \simeq L_2 \simeq U_1 \simeq U_2.$$

9.2 The Integral as the Shadow of Riemann Sums

Also, since the associated ordinary Riemann sums lie between their corresponding upper and lower sums, these are also infinitely close:

$$S_a^b(f, \Delta x_1) \simeq S_a^b(f, \Delta x_2).$$

Altogether then, the Riemann sums determined by arbitrary positive infinitesimals are all infinitely close to each other, and moreover are bounded above and below by the real numbers $m(b-a)$ and $M(b-a)$, so are all limited, and hence have the same shadow. Thus we may conveniently *define*

$$\int_a^b f(x)dx = \text{sh}\big(S_a^b(f, \Delta x)\big) \qquad \text{(iii)}$$

for any positive infinitesimal Δx.

Now we show that this definition fulfills the characterising conditions (1) and (2) of Section 9.1 for Riemann integrability. First observe that if P is any standard partition of $[a, b]$, then taking an infinitesimal Δx yields

$$L_a^b(f, P) \leq U_a^b(f, \Delta x) \simeq \int_a^b f(x)dx \simeq L_a^b(f, \Delta x) \leq U_a^b(f, P),$$

and so as $L_a^b(f, P)$ and $U_a^b(f, P)$ are real,

$$L_a^b(f, P) \leq \int_a^b f(x)dx \leq U_a^b(f, P).$$

Secondly, given a positive $\varepsilon \in \mathbb{R}$ then by Theorem 9.1.1 there exists a hyperreal Δx (namely any positive infinitesimal) such that

$$U_a^b(f, \Delta x) - L_a^b(f, \Delta x) < \varepsilon,$$

and so by existential transfer this holds for some *real* Δx.

This completes our proof using infinitesimals that a continuous or monotonic function f is Riemann integrable on $[a, b]$, with integral defined as in (iii).

Notice that (iii) implies the standard characterisation of the integral as a limit:

$$\int_a^b f(x)dx = \lim_{\Delta x \to 0^+} S_a^b(f, \Delta x).$$

Here $S_a^b(f, \Delta x)$ was obtained *formally* as the extension of a standard function. In Section 12.7 we will see how to obtain it by a more explicit summation of terms $f(x)\Delta x$, with Δx infinitesimal, over a "hyperfinite" partition of $[a, b]$.

If a function f is Riemann integrable on $[a, b]$, then in the standard theory it is shown that the upper and lower sums approximate each other arbitrarily closely, by showing that for any given $\varepsilon \in \mathbb{R}^+$ there exists a $\delta \in \mathbb{R}^+$ such that

$$(\forall \Delta x \in \mathbb{R}^+)\big(\Delta x < \delta \text{ implies } U_a^b(f, \Delta x) - L_a^b(f, \Delta x) < \varepsilon\big).$$

Transferring this and taking Δx to be infinitesimal, we get $U_a^b(f, \Delta x) - L_a^b(f, \Delta x) < \varepsilon$. Since ε is an arbitrary member of $\mathbb{R}^+$ here, it follows that $U_a^b(f, \Delta x)$ and $L_a^b(f, \Delta x)$ are infinitely close. Hence

110   9. The Riemann Integral

$U_a^b(f, \Delta x) \simeq L_a^b(f, \Delta x)$ for all positive infinitesimals Δx.

But it was just this property that enabled us to obtain $\int_a^b f(x)dx$ by the definition (iii). The property is therefore *equivalent* to Riemann integrability of a bounded function f on $[a, b]$.

Exercise 9.2.1
For each (standard) $n \in \mathbb{N}$, let $U_a^b(f, n)$, $L_a^b(f, n)$, $S_a^b(f, n)$ be the upper, lower, and ordinary Riemann sums for the partition determined by the number $\Delta x = \frac{b-a}{n}$. Prove that if $n \in {}^*\mathbb{N}$ is unlimited, then

$$L_a^b(f, n) \simeq S_a^b(f, n) \simeq U_a^b(f, n).$$

Show how the definition and proof of existence for the Riemann integral could be developed just using these functions of (hyper)natural numbers.

9.3 Standard Properties of the Integral

If f and g are integrable on $[a, b]$ in $\mathbb{R}$, then

- $\int_a^b cf(x)dx = c \int_a^b f(x)dx$.

- $\int_a^b f(x) + g(x)dx = \int_a^b f(x)dx + \int_a^b g(x)dx$.

- $\int_a^b f(x)dx = \int_a^c f(x)dx + \int_c^b f(x)dx$ if $a \leq c \leq b$.

- $\int_a^b f(x)dx \leq \int_a^b g(x)dx$ if $f(x) \leq g(x)$ on $[a, b]$.

- $m(b-a) \leq \int_a^b f(x)dx \leq M(b-a)$ if $m \leq f(x) \leq M$ on $[a, b]$.

Here is a concise proof via infinitesimals of the third (*juxtaposition*) property. If $\Delta x = (c-a)/n$ with $n \in \mathbb{N}$, then it is readily seen that

$$S_a^b(f, \Delta x) = S_a^c(f, \Delta x) + S_c^b(f, \Delta x). \qquad (iv)$$

Hence by transfer, this equation holds when $\Delta x = (c-a)/N$ with N unlimited in ${}^*\mathbb{N}$. But then Δx is infinitesimal, so applying the shadow map to (iv) and invoking (iii) gives the result.

Exercise 9.3.1
Derive proofs in this vein for the other properties of the integral listed above.

9.4 Differentiating the Area Function

Integration and differentiation are processes springing from quite different intuitive sources, but they are intimately related and, as every calculus student knows, are in a sense inverses of each other: differentiating the integral gives back the original integrand. This fundamental result is explained by examining the area function F, defined by

$$F(x) = \int_a^x f(t)dt$$

for $x \in [a, b]$, where f is continuous on $[a, b]$. The key to the relationship between differentiation and integration is the following fact.

Theorem 9.4.1 *The function $F(x) = \int_a^x f(t)dt$ is differentiable on $[a, b]$, and its derivative is f.*

(This includes the right- and left-hand derivatives at the end points of the interval.)

There is a very intuitive explanation of why this relationship should hold. The increment

$$\Delta F = F(x + \Delta x) - F(x)$$

of F at x corresponding to a positive infinitesimal Δx is closely approximated by the area of the rectangle of height $f(x)$ and width Δx, i.e, by $f(x)\Delta x$. Thus the quotient $\frac{\Delta F}{\Delta x}$ should be closely approximated by $f(x)$ itself.

Does "closely approximate" here mean $\frac{\Delta F}{\Delta x} \simeq f(x)$? Well, observe that ΔF is bounded above and below by $f(x_1)\Delta x$ and $f(x_2)\Delta x$, where x_1 and x_2 are points where f has its greatest and least values between x and $x + \Delta x$, so $\frac{\Delta F}{\Delta x}$ lies between $f(x_1)$ and $f(x_2)$. But x_1 and x_2 are infinitely close to x, hence by continuity $f(x_1)$ and $f(x_2)$ are infinitely close to $f(x)$, and therefore so is $\frac{\Delta F}{\Delta x}$.

We will now use transfer to legalise this intuitive approximation argument. First, if Δx is a positive *real* number less than $b - x$, then by juxtaposition of the integrals,

$$F(x + \Delta x) - F(x) = \int_x^{x+\Delta x} f(t)dt.$$

But on the interval $[x, x + \Delta x]$, f attains maximum and minimum values at some points x_1 and x_2, and so

$$f(x_2)\Delta x \leq \int_x^{x+\Delta x} f(t)dt \leq f(x_1)\Delta x.$$

Hence

$$f(x_2) \leq \frac{F(x + \Delta x) - F(x)}{\Delta x} \leq f(x_1). \qquad \text{(v)}$$

Thus for all real $\Delta x \in (0, b-x)$ there exist x_1, x_2 such that $x \le x_1, x_2 \le x+\Delta x$, and (v) is true. Hence by transfer, if Δx is any positive infinitesimal, then there are hyperreal $x_1, x_2 \in [x, x+\Delta x]$ for which (v) holds. But now $x \simeq x + \Delta x$, so $x_1 \simeq x \simeq x_2$, and hence by the continuity of f at x, $f(x_1) \simeq f(x) \simeq f(x_2)$, from which (v) yields

$$\frac{F(x+\Delta x) - F(x)}{\Delta x} \simeq f(x).$$

Similarly, this conclusion can be derived for any negative infinitesimal Δx. It follows (Theorem 8.1.1) that $f(x)$ is the derivative of F at x, proving Theorem 9.4.1.

Theorem 9.4.2 Fundamental Theorem of Calculus. *If a function G has a continuous derivative f on $[a,b]$, then $\int_a^b f(x)dx = G(b) - G(a)$.*

Proof. This follows from Theorem 9.4.1 by standard arguments that require no ideas of limits or infinitesimals. For if $F(x) = \int_a^x f(t)dt$, then on $[a,b]$ we have $(G(x) - F(x))' = f(x) - f(x) = 0$, so there is a constant c with $G(x) - F(x) = c$. This implies $G(b) - G(a) = F(b) - F(a)$. But $F(b) - F(a) = \int_a^b f(t)dt$. □

9.5 Exercise on Average Function Values

Let f be continuous on $[a,b] \subseteq \mathbb{R}$. Define the "sample average" function Av by putting, for each $n \in \mathbb{N}$,

$$Av(n) = \frac{f(x_0) + \cdots + f(x_{n-1})}{n},$$

where $x_i = a + i(b-a)/n$.

Prove that if $N \in {}^*\mathbb{N}$ is unlimited, then

$$Av(N) \simeq \frac{1}{b-a} \int_a^b f(x)dx$$

(i.e., the average value of f on $[a,b]$ is given by the shadow of $Av(N)$).

10
Topology of the Reals

Abstract topology studies the the notions of *nearness* and *proximity* of points by axiomatising the concept of an *open neighbourhood* of a point. Intuitively, an open set is one with the property that if it contains a point x, then it contains all points near x. In the hyperreal context we can make this idea quite explicit by taking "near" to mean "infinitely close". As we shall see, this leads to a very natural formulation and treatment of many topological ideas.

10.1 Interior, Closure, and Limit Points

If $A \subseteq \mathbb{R}$ and $r \in \mathbb{R}$, then the following are standard definitions:

- r is an *interior point* of A if $(r - \varepsilon, r + \varepsilon) \subseteq A$ for some real $\varepsilon > 0$. The *interior of A* is the set A° of interior points of A.

- r is a *closure point* of A if $(r - \varepsilon, r + \varepsilon)$ intersects A for every real $\varepsilon > 0$. The *(topological) closure of A* is the set $\overline{A}$ of closure points of A.

- r is a *limit point* of A if for every real $\varepsilon > 0$, $(r - \varepsilon, r + \varepsilon)$ intersects A in a point other than r. The set A' of limit points of A is the *derived set* of A.

It follows readily from these definitions that

$$A^\circ \subseteq A \subseteq \overline{A} = A \cup A'.$$

Now, r is an interior point of A if all points within some positive real distance of r belong to A. Our discussion of permanence in Section 7.10 suggests that this property should be equivalent to requiring that all points infinitely close to r should belong to *A. This is confirmed by the first part of the next result.

Theorem 10.1.1 *If $A \subseteq \mathbb{R}$ and $r \in \mathbb{R}$,*

(1) *r is interior to A if and only if $r \simeq x$ implies $x \in {}^*A$, i.e., iff* $\mathrm{hal}(r) \subseteq {}^*A$.

(2) *r is a limit point of A if and only if there is an $x \neq r$ such that $r \simeq x \in {}^*A$, i.e., iff* $\mathrm{hal}(r) \cap {}^*A$ *contains a point other than r.*

(3) *r is a closure point of A if and only if r is infinitely close to some $x \in {}^*A$, i.e., iff* $\mathrm{hal}(r) \cap {}^*A$ *is nonempty.*

Proof.

(1) Let $r \in A^\circ$. Then $(r - \varepsilon, r + \varepsilon) \subseteq A$ for some real $\varepsilon > 0$. Then the sentence

$$(\forall x \in \mathbb{R})\,(|r - x| < \varepsilon \to x \in A) \tag{i}$$

is true. But now if $r \simeq x$ in *$\mathbb{R}$, then $|r - x| < \varepsilon$, so by universal transfer of (i), $x \in {}^*A$. This shows that $\mathrm{hal}(r) \subseteq {}^*A$. (An alternative way of putting this is to observe that $\mathrm{hal}(r) \subseteq {}^*(r - \varepsilon, r + \varepsilon) \subseteq {}^*A$.)

Conversely, if $\mathrm{hal}(r) \subseteq {}^*A$, then the sentence

$$(\exists \varepsilon \in {}^*\mathbb{R}^+)\,(\forall x \in {}^*\mathbb{R})\,(|r - x| < \varepsilon \to x \in {}^*A)$$

is seen to be true by interpreting ε as any positive infinitesimal. But then by existential transfer there is some *real* $\varepsilon > 0$ for which (i) holds, so $(r - \varepsilon, r + \varepsilon) \subseteq A$ and hence $r \in A^\circ$.

(2) If $r \in A'$, then the sentence

$$(\forall \varepsilon \in \mathbb{R}^+)\,(\exists x \in \mathbb{R})(x \neq r \land |r - x| < \varepsilon \land x \in A) \tag{ii}$$

is true. Now take ε to be a positive infinitesimal. Then by transfer of (ii), there is a hyperreal $x \neq r$ with $|r - x| < \varepsilon$, whence $r \simeq x$, and $x \in {}^*A$.

Conversely, suppose there exists $x \in \mathrm{hal}(r) \cap {}^*A$ with $x \neq r$. Then if $\varepsilon > 0$ is real, $|r - x| < \varepsilon$, since $r \simeq x$, and thus the sentence

$$(\exists x \in {}^*\mathbb{R})\,(x \neq r \land |r - x| < \varepsilon \land x \in {}^*A)$$

is true. By transfer then, there exists a real number distinct from r that belongs to $(r - \varepsilon, r + \varepsilon) \cap A$. This shows that r is a limit point of A.

(3) If $r \in \overline{A}$, then either $r \in A$, in which case $r \in \text{hal}(r) \cap {}^*A$, or else $r \in A'$, in which case $\text{hal}(r) \cap {}^*A \neq \emptyset$, again by part (2).

Conversely, if there exists $x \in \text{hal}(r) \cap {}^*A$, then either $x = r$, so $r \in {}^*A \cap \mathbb{R} = A$, or else $x \neq r$, and so $r \in A'$ by (2). Thus $r \in A \cup A' = \overline{A}$. □

10.2 Open and Closed Sets

If $A \subseteq \mathbb{R}$, then

- A is *open* if all its points are interior to it, i.e., $A^\circ = A$.

- A is *(topologically) closed* if it contains all its closure points, i.e., $A = \overline{A}$. Since $\overline{A} = A \cup A'$, this is equivalent to requiring that A contain all its limit points, i.e., $A' \subseteq A$.

In view of Theorem 10.1.1, it follows that

- A is open if and only if for all $r \in A$, if x is infinitely close to r, then $x \in {}^*A$;

- A is closed if and only if for all real r, if r is infinitely close to some $x \in {}^*A$, then $r \in A$.

Theorem 10.2.1

(1) *A is open in $\mathbb{R}$ if and only if its complement $A^c = \mathbb{R} - A$ is closed in $\mathbb{R}$.*

(2) *The collection of open sets is closed under finite intersections and arbitrary unions.*

(3) *The collection of topologically closed sets is closed under finite unions and arbitrary intersections.*

Proof.

(1) Observe that $x \in {}^*(A^c)$ iff $x \notin {}^*A$, by transfer of

$$(\forall x \in \mathbb{R})\,(x \in A^c \leftrightarrow x \notin A).$$

Suppose A^c is closed. To show A is open, let $x \simeq r \in A$. Then we must show $x \in {}^*A$.

Now, if $x \in {}^*(A^c)$, then we would have $r \simeq x \in {}^*(A^c)$, making r a closure point of A^c, so as A^c is closed, $r \in A^c$, contradicting $r \in A$. Thus we must have $x \notin {}^*(A^c)$, implying $x \in {}^*A$, as desired, by the above observation.

The converse is similar, and given as an exercise.

(2) Let $A_1, \ldots, A_n$ be open. If $x \simeq r \in A_1 \cap \cdots \cap A_n$, then for each i, $x \simeq r \in A_i$, and so $x \in {}^*A_i$. Hence

$$x \in {}^*A_1 \cap \cdots \cap {}^*A_n = {}^*(A_1 \cap \cdots \cap A_n).$$

This shows that $A_1 \cap \cdots \cap A_n$ is open.

Now let $\{A_i : i \in I\}$ be a collection of open sets. If $x \simeq r \in \bigcup_{i \in I} A_i$, then for some j, $x \simeq r \in A_j$, so $x \in {}^*A_j \subseteq {}^*(\bigcup_{i \in I} A_i)$. Hence $\bigcup_{i \in I} A_i$ is open.

(3) Exercise.

□

Theorem 10.2.2 *For any real number r,*

$$\mathrm{hal}(r) = \bigcap \{{}^*A : r \in A \text{ and } A \text{ is open}\}.$$

Proof. We have already observed that if $r \in A \subseteq \mathbb{R}$ and A is open, then $\mathrm{hal}(r) \subseteq {}^*A$. On the other hand, if $x \notin \mathrm{hal}(r)$, then $x \not\simeq r$, so there must exist some real $\varepsilon > 0$ such that $|r - x| > \varepsilon$. Put $A = (r - \varepsilon, r + \varepsilon) \subseteq \mathbb{R}$. Then $r \in A$ and A is open, but $x \notin {}^*A = \{y \in {}^*\mathbb{R} : |r - y| < \varepsilon\}$.

□

A *topology* on a set X is defined axiomatically to be a collection of subsets of X that includes $\emptyset$ and X and is closed under finite intersections and arbitrary unions. The members of this collection are declared to be the *open* sets, and their complements are called *closed*. In such a setting it is possible to study any topological idea that can be characterised by properties of open and closed sets, even in the absence of a notion of numerical distance between points. For instance, the halo of a point $r \in X$ could be defined by the equation of Theorem 10.2.2, leading to a nonnumerical account of "infinite closeness" of points.

Exercise 10.2.3
Show that the proof of 10.2.1(2) does not work for infinite intersections by showing that

$$ {}^*\left(\bigcap_{n \in \mathbb{N}} \left(-\tfrac{1}{n}, \tfrac{1}{n}\right)\right) \neq \bigcap_{n \in \mathbb{N}} {}^*\left(-\tfrac{1}{n}, \tfrac{1}{n}\right).$$

10.3 Compactness

A set $B \subseteq \mathbb{R}$ is *compact* if every open cover of B has a finite subcover, i.e., if whenever $B \subseteq \bigcup_{i \in I} A_i$ and each A_i is open in $\mathbb{R}$, then there is a finite $J \subseteq I$ such that $B \subseteq \bigcup_{i \in J} A_i$.

This concept does not appear out of thin air. It emerged from studies in the nineteenth century of bounded and closed intervals in the real line, leading to a proof that such intervals are compact in the sense just defined (Heine–Borel theorem). Since the definition refers only to open sets, it becomes the appropriate one to use for an abstract topological space where there is no notion of numerical distance to specify the notion of boundedness.

Robinson's Compactness Criterion. *B is a compact subset of $\mathbb{R}$ if and only if every member of *B is infinitely close to some member of B, i.e., iff*
$$^*B \subseteq \bigcup_{r \in B} \mathrm{hal}(r).$$

Since the members of B are all real, the only such member that a given $x \in {}^*B$ could be infinitely close to is its shadow. Thus another way to state Robinson's criterion is

if $x \in {}^*B$, then x is limited and $\mathrm{sh}(x) \in B$.

This criterion gives an intuitively appealing and useful characterisation of the notion of compactness. Constructions involving open covers are replaced by elementary reasoning about hyperreal points. For instance:

- The open interval $(0,1) \subseteq \mathbb{R}$ is not compact, because if ε is a positive infinitesimal, then $\varepsilon \in {}^*(0,1)$ as $0 < \varepsilon < 1$, but ε is not infinitely close to any member of $(0,1)$ because its shadow is $0 \notin (0,1)$.

- Any closed interval $[a,b] \subseteq \mathbb{R}$ is compact, because if $x \in {}^*[a,b]$, then $a \leq x \leq b$, so x is limited and its shadow r must also satisfy $a \leq r \leq b$. Thus $x \simeq r \in [a,b]$.

- Any finite set is compact, because if B is finite, then $^*B = B$, so each member of $^*B$ is infinitely close to itself in B.

- If $B \subseteq \mathbb{R}$ is unbounded above, in the sense that
$$(\forall x \in \mathbb{R})\,(\exists y \in B)\,(x < y),$$
then B cannot be compact: taking any unlimited $x \in {}^*\mathbb{R}$, by transfer there exists $y > x$ with $y \in {}^*B$. Then y is unlimited, so cannot be infinitely close to any member of B. Similarly, B cannot be compact if it is unbounded below.

  Altogether then, a compact set must be bounded above and below.

- If B is not closed, then B cannot be compact: it must have a closure point r that does not belong to B. As a closure point, r is infinitely close to some $x \in {}^*B$. But then x is not infinitely close to any member of B, since $\mathrm{sh}(x) = r \notin B$.

Hence a compact set must be closed.

Proof of Robinson's Criterion.

We will show that Robinson's criterion fails if and only if the standard definition of compactness fails.

If Robinson's criterion fails, there is a hyperreal $b \in {}^*B$ that is not infinitely close to any member of B. Then for each $r \in B$, $b \not\simeq r$, so there must be a real $\varepsilon_r > 0$ such that $|b-r| \geq \varepsilon_r$. Then $\{(r - \varepsilon_r, r + \varepsilon_r) : r \in B\}$ is an open cover of B. But this cover can have no finite subcover: if, say,

$$B \subseteq (r_1 - \varepsilon_{r_1}, r_1 + \varepsilon_{r_1}) \cup \cdots \cup (r_n - \varepsilon_{r_n}, r_n + \varepsilon_{r_n}),$$

then by properties of enlargements of sets (3.10),

$${}^*B \subseteq {}^*(r_1 - \varepsilon_{r_1}, r_1 + \varepsilon_{r_1}) \cup \cdots \cup {}^*(r_n - \varepsilon_{r_n}, r_n + \varepsilon_{r_n}).$$

Since $b \in {}^*B$, it then follows that $b \in {}^*(r_i - \varepsilon_{r_i}, r_i + \varepsilon_{r_i})$, and hence $|b-r_i| < \varepsilon_{r_i}$ for some i, contradicting the definition of ε_{r_i}. Thus compactness fails for B.

For the converse, suppose B is not compact, so that there is an open cover $\mathcal{C} = \{A_i : i \in I\}$ of B that has no finite subcover. Each r in B belongs to A_j for some $j \in I$, and hence $r \in (r - \varepsilon, r + \varepsilon) \subseteq A_j$ for some $\varepsilon \in \mathbb{R}^+$ because A_j is open. But then using the density of the rationals we can find some rational numbers p, q with $r \in (p, q) \subseteq A_j$. This shows that there is an open cover $\mathcal{C}'$ of B by intervals with rational end points, each of which is included in a member of $\mathcal{C}$. Because the rationals are countable, there are only countably many intervals with rational end points, so we can enumerate $\mathcal{C}'$ and write

$$\mathcal{C}' = \{(p_n, q_n) : n \in \mathbb{N}\},$$

where $\langle p_n : n \in \mathbb{N} \rangle$ and $\langle q_n : n \in \mathbb{N} \rangle$ are sequences of rational numbers.

Now, $\mathcal{C}'$ includes no finite subcovering of B, or else this would lead to a finite subcover from $\mathcal{C}$, since each member of $\mathcal{C}'$ is included in a member of $\mathcal{C}$. Thus for each $k \in \mathbb{N}$,

$$B \not\subseteq (p_1, q_1) \cup \cdots \cup (p_k, q_k).$$

We can express this fact by the following true sentence:

$$(\forall k \in \mathbb{N})\,(\exists x \in B)\,(\forall n \in \mathbb{N})\,[n \leq k \rightarrow \neg(p_n < x < q_n)].$$

Now take an unlimited $K \in {}^*\mathbb{N}$. Then by transfer there exists some hyperreal $x \in {}^*B$ such that the statement $p_n < x < q_n$ is false for all $n \in {}^*\mathbb{N}$ with $n \leq K$. In particular, $p_n < x < q_n$ is false for all standard n. But now x cannot be infinitely close to any $r \in B$, because such a point r belongs to (p_n, q_n) for some $n \in \mathbb{N}$, and so if $x \simeq r$, then $p_n < x < q_n$. Hence Robinson's criterion fails, and the proof is complete. □

Robinson's criterion can be established for abstract compact topological spaces (and leads to a beautifully simple proof of Tychonoff's theorem that the product of compact spaces is compact). In that context the reduction to a countable cover via the density of $\mathbb{Q}$ in $\mathbb{R}$ is not generally applicable, and instead a special principle of hyperreal analysis, known as *enlargement* (cf. Chapter 14), is needed to establish the criterion.

Theorem 10.3.1 (Heine–Borel) *A set $B \subseteq \mathbb{R}$ is compact if and only if it is closed and bounded.*

Proof. We have already seen that if B satisfies Robinson's criterion, then it is closed and bounded (above and below).

Conversely, if B is closed and bounded, then there is some real b such that
$$(\forall x \in B)\,(|x| \leq b).$$
Now, to prove Robinson's criterion, suppose $x \in {}^*B$. Then by transfer, $|x| \leq b \in \mathbb{R}$. Hence x is limited, and so has a shadow $r \in \mathbb{R}$. Then $r \simeq x \in {}^*B$, and so $r \in B$ because B is closed. Thus we have shown that x is infinitely close to the member r of B, proving that B is compact. □

Exercise 10.3.2
Use Robinson's criterion to prove that in $\mathbb{R}$ a closed subset of a compact set is compact.

10.4 Compactness and (Uniform) Continuity

Compactness is an inherently topological notion, being preserved by continuous transformations. Here is a simple hyperreal proof of that fact.

Theorem 10.4.1 *The continuous image of a compact set is compact.*

Proof. Let f be a continuous real function, and B a compact subset of $\mathbb{R}$ included in the domain of f. Now, it is true, by definition of $f(B)$, that
$$(\forall y \in f(B))\,(\exists x \in B)\,(y = f(x)).$$
Thus by transfer, if $y \in {}^*(f(B))$, then $y = f(x)$ for some $x \in {}^*B$. Since B is compact, $x \simeq r$ for some $r \in B$. Then by continuity of f, $f(x) \simeq f(r)$, i.e., y is infinitely close to $f(r) \in f(B)$. This shows by Robinson's criterion that $f(B)$ is compact. □

In Theorem 7.7.2 it was shown that a continuous function on a closed interval $[a, b]$ is uniformly continuous. We can now see that it is the compactness of $[a, b]$ that accounts for this phenomenon:

Theorem 10.4.2 *If f is continuous on a compact set $B \subseteq \mathbb{R}$, then f is uniformly continuous on B.*

Proof. By Theorem 7.7.1 we have to show that for all $x, y \in {}^*B$,
$$x \simeq y \quad \text{implies} \quad f(x) \simeq f(y).$$
But if $x, y \in {}^*B$, then by compactness $x \simeq r \in B$ and $y \simeq s \in B$ for some r, s. Thus if $x \simeq y$, then $r \simeq s$, and so $r = s$, as both are real. Hence by continuity of f at $r \in B$, $f(x) \simeq f(r)$ and $f(y) \simeq f(r)$, whence $f(x) \simeq f(y)$ as desired. □

10.5 Topologies on the Hyperreals

There is no canonical way to extend the definitions of interior, closure, and limit point—and hence the definitions of open and closed sets—to general subsets of ${}^*\mathbb{R}$. These definitions depend on the concept of an open neighbourhood $(r - \varepsilon, r + \varepsilon)$ of a point r, and one option would be to allow r to be any member of ${}^*\mathbb{R}$ but to continue to require that the radius ε be a positive *real* number. Here $(r - \varepsilon, r + \varepsilon)$ is the hyperreal interval $\{x \in {}^*\mathbb{R} : r - \varepsilon < x < r + \varepsilon\}$, and we call it a *real-radius neighbourhood* of r when ε is a real number.

Thus a subset of ${}^*\mathbb{R}$ will be called *real-open* if it is a union of real-radius neighbourhoods. Examples of real-open sets include the sets of limited numbers, unlimited numbers, positive (respectively negative) unlimited numbers, appreciable numbers, positive (respectively negative) appreciable numbers, and any galaxy (Section 5.4).

The class of real-open sets is closed under arbitrary unions, but is not a topology on ${}^*\mathbb{R}$ because it is not closed under finite intersections. For instance, if $(r - \varepsilon, r + \varepsilon)$ and $(s - \delta, s + \delta)$ are real-radius neighbourhoods that overlap in such a way that $s - \delta$ is infinitely close to $r + \varepsilon$, then the intersection $(s - \delta, r + \varepsilon)$ has infinitesimal width and so does not contain any real-radius neighbourhoods, hence is not real-open.

This suggests that the overlaps between real-radius neighbourhoods are in some sense "too small". One way to remedy this is to modify the neighbourhoods by removing those members that are infinitely close to the end points, retaining those that are *well inside* the interval (as defined in Section 8.11), thereby forcing any overlaps to be appreciable. To formalise this, put

$$\begin{aligned}((r - \varepsilon, r + \varepsilon)) &= \{x \in {}^*\mathbb{R} : x \text{ is well inside } (r - \varepsilon, r + \varepsilon)\} \\ &= \{x \in {}^*\mathbb{R} : \mathrm{hal}(x) \subseteq (r - \varepsilon, r + \varepsilon)\}.\end{aligned}$$

A set of the form $((r - \varepsilon, r + \varepsilon))$ with real ε will be called an *S-neighbourhood*, and an *S-open set* is one that is a union of S-neighbourhoods. The S-open sets form the *S-topology* on ${}^*\mathbb{R}$, first introduced by Abraham Robinson.

(The "S-" prefix here is for "standard" and is typically used when a standard concept, or the nonstandard characterisation of some standard concept, is applied more widely to nonstandard entities. In Theorem 11.14.4, S-openness of a set B will be related to the notion of B being open when it includes the halo of each of its points.)

Note that when $\varepsilon \in \mathbb{R}^+$, each point of $((r - \varepsilon, r + \varepsilon))$ is of appreciable distance from $r - \varepsilon$ and from $r + \varepsilon$. Alternatively, we can define $((r - \varepsilon, r + \varepsilon))$ as consisting of those points x whose distance from r is *appreciably less than* ε, in the sense that

$$\text{sh}|r - x| < \varepsilon,$$

i.e., $\varepsilon - |r - x|$ is appreciable. The intersection of two S-neighbourhoods $((r - \varepsilon, r + \varepsilon))$ and $((s - \delta, s + \delta))$ is S-open, because if t belongs to this intersection, then

$$((t - \gamma, t + \gamma)) \subseteq ((r - \varepsilon, r + \varepsilon)) \cap ((s - \delta, s + \delta)),$$

where γ is any positive real number such that

$$\gamma \leq \min\{\varepsilon - \text{sh}|r - t|, \delta - \text{sh}|s - t|\}.$$

From this it follows that the intersection of S-open sets is S-open.

Exercise 10.5.1
Show that

(i) any S-open set is real-open;

(ii) each S-open set is a union of halos, but a union of halos need not be S-open;

(ii) no real-radius neighbourhood can be S-open. □

Another topology on *$\mathbb{R}$ is obtained if neighbourhoods are allowed to have any positive hyperreal (possibly infinitesimal) as radius. Thus any hyperreal open interval (a, b) can be a neighbourhood, and we define a set $A \subseteq {}^*\mathbb{R}$ to be *interval-open* if it is a union of intervals (a, b) with $a, b \in {}^*\mathbb{R}$. The interval-open sets form the *interval topology* on *$\mathbb{R}$. It is immediate that all real-open sets are interval-open, but the converse is not true. There are many hyperreal intervals (a, b) that are interval-open but not real-open, for instance any having $a \simeq b$. Also, any halo becomes interval-open but is not real-open.

Consider furthermore the construction

$$\bigcap\{A : r \in A \text{ and } A \text{ is open}\}.$$

If "open" here means real-open (or S-open), then this gives the halo of r. If it means interval-open, then the result is just $\{r\}$.

Any hyperreal interval (a, b) can be readily constructed as the intersection of two real-radius neighbourhoods (in many ways). Therefore any topology that includes the real-radius neighbourhoods must include all hyperreal intervals, and hence all interval-open sets.

Exercise 10.5.2

Let A be an open subset of $\mathbb{R}$.

(i) Show that $^*A$ is interval-open in $^*\mathbb{R}$.

(ii) Suppose A is the union of a sequence $\langle A_n : n \in \mathbb{N} \rangle$ of pairwise disjoint open intervals in $\mathbb{R}$, with the length of A_n being less than $\frac{1}{n}$. Use transfer to show that some element of $^*A$ is infinitely close to something not in $^*A$. Deduce that $^*A$ is not S-open.

(iii) Show further that $^*A$ contains a point that does not belong to any real-radius neighbourhood that is included in $^*A$. Hence deduce the stronger result that $^*A$ is not real-open.

The relationships between various topologies on $^*\mathbb{R}$ will be explored further in Section 11.14.

Part III

Internal and External Entities

11
Internal and External Sets

In the construction of *ℝ as an ultrapower in Chapter 3, each *sequence of points* $r = \langle r_n : n \in \mathbb{N} \rangle$ in ℝ gives rise to the *single point* $[r]$ of *ℝ, which we also denote by the more informative symbol $[r_n]$. Equality of *ℝ-points is given by

$$[r_n] = [s_n] \quad \text{iff} \quad \{n \in \mathbb{N} : r_n = s_n\} \in \mathcal{F}.$$

This description works for other kinds of entities than points. We will now see that a sequence of *subsets* of ℝ determines a single subset of *ℝ. In the next chapter we will see that a sequence of functions on ℝ determines a single function on *ℝ.

11.1 Internal Sets

Given a sequence $\langle A_n : n \in \mathbb{N} \rangle$ of subsets $A_n \subseteq \mathbb{R}$, define a subset $[A_n] \subseteq {}^*\mathbb{R}$ by specifying, for each $[r_n] \in {}^*\mathbb{R}$,

$$[r_n] \in [A_n] \quad \text{iff} \quad \{n \in \mathbb{N} : r_n \in A_n\} \in \mathcal{F}.$$

Of course it must be checked that this is a well-defined notion that does not depend on how points are named, which means that if $[r_n] = [s_n]$, then

$$\{n \in \mathbb{N} : r_n \in A_n\} \in \mathcal{F} \quad \text{iff} \quad \{n \in \mathbb{N} : s_n \in A_n\} \in \mathcal{F}.$$

This is a slight extension of the argument given in Section 3.9.

The subsets of *ℝ that are produced by this construction are called *internal*. Here are some examples:

- If $\langle A_n \rangle$ is a constant sequence with $A_n = A \subseteq \mathbb{R}$ for all $n \in \mathbb{N}$, then the internal set $[A_n]$ is just the enlargement *A of A defined in Section 3.9. Hence we may also denote *A as $[A]$.

    Thus the enlargement of any subset of $\mathbb{R}$ is an internal subset of *$\mathbb{R}$. In particular, we see that *$\mathbb{N}$, *$\mathbb{Z}$, and *$\mathbb{Q}$ and *$\mathbb{R}$ itself are all internal, as is any *finite* subset $A \subseteq \mathbb{R}$, since in that case $A =$ *A.

- More generally, any finite set $X = \{[r_n^1], \ldots, [r_n^k]\}$ of hyperreals is internal, for then $X = [A_n]$, where $A_n = \{r_n^1, \ldots, r_n^k\}$.

- If $a < b$ in *$\mathbb{R}$, then the hyperreal open interval
$$(a, b) = \{x \in {}^*\mathbb{R} : a < x < b\}$$
is internal. Indeed, if $a = [a_n]$ and $b = [b_n]$, then (a, b) is the internal set defined by the sequence $\langle (a_n, b_n) : n \in \mathbb{N} \rangle$ of real intervals $(a_n, b_n) \subseteq \mathbb{R}$. This follows because
$$[a_n] < [r_n] < [b_n] \quad \text{iff} \quad \{n \in \mathbb{N} : a_n < r_n < b_n\} \in \mathcal{F}.$$
Similarly, the hyperreal intervals $(a, b], [a, b), [a, b], \{x \in {}^*\mathbb{R} : a < x\}$ are internal. Notice that if a is unlimited, then each of these intervals is disjoint from $\mathbb{R}$, so none of them can be the enlargement *A of a set $A \subseteq \mathbb{R}$, since *A always includes the (real) members of A.

- If $N \in {}^*\mathbb{N}$, then the set
$$\{k \in {}^*\mathbb{N} : k \leq N\} = \{1, 2, \ldots, N\}$$
is internal. If $N = [N_n]$, then this is the internal set $[A_n]$, where
$$A_n = \{k \in \mathbb{N} : k \leq N_n\} = \{1, 2, \ldots, N_n\}$$
(since $N \in {}^*\mathbb{N}$, we have $\{n : N_n \in \mathbb{N}\} \in \mathcal{F}$, so we may as well assume $N_n \in \mathbb{N}$ for all $n \in \mathbb{N}$).

- If $N = [N_n] \in {}^*\mathbb{N}$, then the set
$$\left\{\frac{k}{N} : k \in {}^*\mathbb{N} \cup \{0\} \text{ and } k \leq N\right\} = \left\{0, \frac{1}{N}, \frac{2}{N}, \ldots, \frac{N-1}{N}, 1\right\}$$
is the internal set $[A_n]$, where
$$A_n = \left\{0, \frac{1}{N_n}, \frac{2}{N_n}, \ldots, \frac{N_n - 1}{N_n}, 1\right\}.$$

These last two examples illustrate the notion of *hyperfinite* set, which will be studied in the next chapter.

11.2 Algebra of Internal Sets

(1) *The class of internal sets is closed under the standard finite set operations $\cap$, $\cup$, and $-$, with*

$$[A_n] \cap [B_n] = [A_n \cap B_n],$$
$$[A_n] \cup [B_n] = [A_n \cup B_n],$$
$$[A_n] - [B_n] = [A_n - B_n].$$

(2) $[A_n] \subseteq [B_n]$ *iff* $\{n \in \mathbb{N} : A_n \subseteq B_n\} \in \mathcal{F}$.

(3) $[A_n] = [B_n]$ *iff* $\{n \in \mathbb{N} : A_n = B_n\} \in \mathcal{F}$.

(4) $[A_n] \neq [B_n]$ *iff* $\{n \in \mathbb{N} : A_n \neq B_n\} \in \mathcal{F}$.

Proof.

(1) Exercise.

(2) If $[A_n] \not\subseteq [B_n]$, then there is some hyperreal $[r_n] \in [A_n] - [B_n]$, so by (1) we have

$$I = \{n \in \mathbb{N} : r_n \in A_n - B_n\} \in \mathcal{F}.$$

But if

$$J = \{n \in \mathbb{N} : A_n \subseteq B_n\},$$

then $I \subseteq J^c$, so $J^c \in \mathcal{F}$ and hence $J \notin \mathcal{F}$.

Conversely, if $J \notin \mathcal{F}$, then $J^c \in \mathcal{F}$, so choosing $r_n \in A_n - B_n$ for each $n \in J^c$ and r_n arbitrary for $n \in J$, the argument reverses to give a point $[r_n] \in [A_n] - [B_n]$.

(3) This follows from (2) and closure properties of $\mathcal{F}$ (note that the result is not a matter of the definition of $[A_n]$ via $\mathcal{F}$, since equality of $[A_n]$ and $[B_n]$ is defined independently of $\mathcal{F}$ to mean "having the same members").

(4) Exercise.

□

Part (3) above is important for what it says about the sequence $\langle A_n : n \in \mathbb{N}\rangle$ that determines a certain internal set. We can replace this sequence by another $\langle B_n : n \in \mathbb{N}\rangle$ without changing the resulting internal set, provided that $A_n = B_n$ for $\mathcal{F}$-almost all n. Thus we are free to alter A_n arbitrarily when n is outside a set that belongs to $\mathcal{F}$. For instance, if $[A_n]$ is nonempty, then as $\emptyset = [\emptyset]$, we can assume that $A_n \neq \emptyset$ for *every* $n \in \mathbb{N}$ (11.2(4)), while if $[A_n]$ is a subset of *$\mathbb{N}$, then as *$\mathbb{N} = [\mathbb{N}]$, we can assume that $A_n \subseteq \mathbb{N}$ for every n (11.2(2)). Moreover, we can combine finitely many such conditions, using the closure of $\mathcal{F}$ under finite intersections. So if $[A_n]$ is a nonempty subset of *$\mathbb{N}$, we can assume that $\emptyset \neq A_n \subseteq \mathbb{N}$ for every $n \in \mathbb{N}$.

Subsets of Internal Sets

The fact that the intersection of two internal sets is internal allows us to prove now that

- *if a set A of real numbers is internal, then so is every subset of A.*

Proof. Let $X \subseteq A$. Then *X is internal, so if A is internal, then so is $A \cap {}^*X$. But since $A \subseteq \mathbb{R}$,

$$A \cap {}^*X = A \cap {}^*X \cap \mathbb{R} = A \cap X$$

(cf. Ex. 3.10.5), so $X = A \cap X$ is internal. □

This result will be used in Section 11.7 to show that actually the only internal subsets of $\mathbb{R}$ are the finite ones.

11.3 Internal Least Number Principle and Induction

A characteristic feature of $\mathbb{N}$ is that each of its nonempty subsets has a least member (indeed this holds for any subset of $\mathbb{Z}$ that has a lower bound). The same is not true, however, for *$\mathbb{N}$: the set *$\mathbb{N} - \mathbb{N}$ of unlimited hypernaturals has no least member, for if N is unlimited, then so is $N - 1$ (why?). But we do have

Theorem 11.3.1 *Any nonempty* **internal** *subset of* *$\mathbb{N}$ *has a least member.*

Proof. Let $[A_n]$ be a nonempty internal subset of *$\mathbb{N}$. Then by the observations above we can assume that for each $n \in \mathbb{N}$,

$$\emptyset \neq A_n \subseteq \mathbb{N},$$

and so A_n has a least member r_n. This defines a point $[r_n] \in {}^*\mathbb{R}$ with

$$\{n \in \mathbb{N} : r_n \in A_n\} = \mathbb{N} \in \mathcal{F},$$

so $[r_n] \in [A_n]$. Moreover, if $[s_n] \in [A_n]$, then

$$\{n \in \mathbb{N} : s_n \in A_n\} \in \mathcal{F} \quad \text{and} \quad \{n \in \mathbb{N} : s_n \in A_n\} \subseteq \{n \in \mathbb{N} : r_n \leq s_n\},$$

leading to the conclusion $[r_n] \leq [s_n]$ in *$\mathbb{R}$. Hence $[A_n]$ indeed has a least member, namely the hyperreal number $[r_n]$ determined by the sequence of least members of the sets A_n.

Writing "min X" for the least element of a set X, this construction can be expressed concisely by the equation

$$\min [A_n] = [\min A_n].$$

□

Now, the least number principle for ℕ is equivalent to the principle of *induction*:

> A subset of ℕ that contains 1 and is closed under the successor function $n \mapsto n+1$ must be equal to ℕ.

The corresponding assertion about subsets of *ℕ is not in general true, and can only be derived for internal sets:

Theorem 11.3.2 (Internal Induction) *If X is an **internal** subset of *ℕ that contains 1 and is closed under the successor function $n \mapsto n+1$, then $X = {}^*\mathbb{N}$.*

Proof. Let $Y = {}^*\mathbb{N} - X$. Then Y is internal (11.2(1)), so if it is nonempty, it has a least element n. Then $n \neq 1$, as $1 \in X$, so $n - 1 \in {}^*\mathbb{N}$. But now $n - 1 \notin Y$, as n is least in Y, so $n - 1 \in X$, and therefore $n = (n-1)+1$ is in X by closure under successor. This contradiction forces us to conclude that $Y = \emptyset$, and so $X = {}^*\mathbb{N}$. □

11.4 The Overflow Principle

The set ℕ cannot be internal, or else by internal induction it would be equal to *ℕ. Thus if an internal set X contains all members of ℕ, then since X cannot be equal to ℕ, it must "overflow" into *ℕ − ℕ. Indeed, we will see that X must contain the initial segment of *ℕ up to some unlimited hypernatural. In fact, a slightly stronger statement than this can be demonstrated by assuming only that X contains "almost all" members of ℕ:

Theorem 11.4.1 *Let X be an internal subset of *ℕ and $k \in \mathbb{N}$. If $n \in X$ for all $n \in \mathbb{N}$ with $k \leq n$, then there is an unlimited $K \in {}^*\mathbb{N}$ with $n \in X$ for all $n \in {}^*\mathbb{N}$ with $k \leq n \leq K$.*

Proof. If all unlimited hypernaturals are in X, then any unlimited $K \in {}^*\mathbb{N}$ will do. Otherwise there are unlimited hypernaturals not in X. If we can show that there is a least such unlimited number H, then all unlimited numbers smaller than H will be in X, giving the desired result.

To spell this out: if *ℕ − X has unlimited members, then these must be greater than k, and so the set

$$Y = \{n \in {}^*\mathbb{N} : k < n \in {}^*\mathbb{N} - X\}$$

is nonempty. But Y is internal, by the algebra of internal sets, since it is equal to

$$({}^*\mathbb{N} - \{1, \ldots, k\}) \cap ({}^*\mathbb{N} - X).$$

Hence Y has a least element H by the internal least number principle. Then H is a hypernatural that is greater than k but not in X, so it must be the case that $H \notin \mathbb{N}$, because of our hypothesis that all limited $n \geq k$ are in X. Thus H is unlimited. Then $K = H - 1$ is unlimited and meets the requirements of the theorem: H is the least hypernatural greater than k that is not in X, so every $n \in {}^*\mathbb{N}$ with $k \leq n \leq H - 1$ does belong to X. □

Exercise 11.4.2
Show that overflow is equivalent to the following statement:

> *If an internal subset X of $^*\mathbb{N}$ contains arbitrarily small unlimited members, then it is unbounded in $\mathbb{N}$, i.e., contains arbitrarily large limited members.*

(Hint: consider $^*\mathbb{N} - X$.) □

The overflow principle implies that any cofinite subset of $\mathbb{N}$ is external: if an internal $A \subseteq \mathbb{N}$ were cofinite, then it would contain $\{n \in \mathbb{N} : k \leq n\}$ for some $k \in \mathbb{N}$, so by overflow A would contain some unlimited number, contradicting $A \subseteq \mathbb{N}$.

11.5 Internal Order-Completeness

The principle of order-completeness, attributed to Dedekind, asserts that every nonempty subset of $\mathbb{R}$ with an upper bound in $\mathbb{R}$ must have a least upper bound in $\mathbb{R}$. The corresponding statement about $^*\mathbb{R}$ is false. In fact, $\mathbb{R}$ itself is a nonempty subset of $^*\mathbb{R}$ that is bounded but has no least upper bound. This is because the upper bounds of $\mathbb{R}$ in $^*\mathbb{R}$ are precisely the positive unlimited numbers, and there is no least positive unlimited number.

Just as for the least number principle, order-completeness is preserved in passing from $\mathbb{R}$ to $^*\mathbb{R}$ for internal sets:

Theorem 11.5.1 *If a nonempty internal subset of $^*\mathbb{R}$ is bounded above/below, then it has a least upper/greatest lower bound in $^*\mathbb{R}$.*

Proof. We treat the case of upper bounds. In effect, the point of the proof is to show that the least upper bound of a bounded internal set $[A_n]$ is the hyperreal number determined by the sequence of least upper bounds of the A_n's:

$$\text{lub}[A_n] = [\text{lub} A_n].$$

More precisely, it is enough to require that $\mathcal{F}$-almost all A_n's have least upper bounds to make this work.

Suppose then that a nonempty internal set $[A_n]$ has an upper bound $[r_n]$. Write $A_n \leq x$ to mean that x is an upper bound of A_n in $\mathbb{R}$, and put

$$J = \{n \in \mathbb{N} : A_n \leq r_n\}.$$

We want $J \in \mathcal{F}$. If not, then $J^c \in \mathcal{F}$. But if $n \in J^c$, there exists some a_n with $r_n < a_n \in A_n$. This leads to the conclusion $[r_n] < [a_n] \in [A_n]$, contradicting the fact that $[r_n]$ is an upper bound of $[A_n]$.

It follows that $J \in \mathcal{F}$. Since $[A_n] \neq \emptyset$, this then implies

$$J' = \{n \in \mathbb{N} : \emptyset \neq A_n \leq r_n\} \in \mathcal{F}.$$

Now, if $n \in J'$, then A_n is a nonempty subset of $\mathbb{R}$ bounded above (by r_n), and so by the order-completeness of $\mathbb{R}$, A_n has a least upper bound $s_n \in \mathbb{R}$. Then if $[b_n] \in [A_n]$,

$$\{n \in \mathbb{N} : b_n \in A_n\} \cap J' \subseteq \{n \in \mathbb{N} : b_n \leq s_n\},$$

leading to $[b_n] \leq [s_n]$, and showing that $[s_n]$ is an upper bound of $[A_n]$. Finally, if $[t_n]$ is any other upper bound of $[A_n]$, then $\{n : A_n \leq t_n\} \in \mathcal{F}$ by the same argument as for $[r_n]$, and

$$\{n \in \mathbb{N} : A_n \leq t_n\} \cap J' \subseteq \{n \in \mathbb{N} : s_n \leq t_n\},$$

so we get $[s_n] \leq [t_n]$. This shows that $[s_n]$ is indeed the least upper bound of $[A_n]$ in *$\mathbb{R}$. □

Exercise 11.5.2
Let X be an internal subset of *$\mathbb{R}$. Prove the following.

(i) If X has arbitrarily large limited members, then it has a positive unlimited member.

(ii) If X has only limited members, then there is some *real* r such that X is included in the interval $[-r, r]$ in *$\mathbb{R}$.

(iii) If X has arbitrarily small positive unlimited members, then it has a positive limited member.

(iv) If X has no limited members, then there is some *unlimited* b such that $X \subseteq \{x \in {}^*\mathbb{R} : x < -b \text{ or } b < x\}$.

11.6 External Sets

A subset of *$\mathbb{R}$ is *external* if it is not internal. Many of the properties that are special to the structure of *$\mathbb{R}$ define external sets:

- *Unlimited Hypernaturals.* Since *$\mathbb{N} - \mathbb{N}$ has no least member, the internal least number principle (Theorem 11.3.1) implies that it cannot be internal.

- *Limited Hypernaturals.* If $\mathbb{N}$ were internal, then by 11.2(1) so too would be *$\mathbb{N} - \mathbb{N}$, which we have just seen to be false.

  Alternatively, by the internal induction principle, Theorem 11.3.2, if $\mathbb{N}$ were internal, it would be equal to *$\mathbb{N}$.

- *Real Numbers.* $\mathbb{R}$ is external, for if it were internal, then so too would be $\mathbb{R} \cap {}^*\mathbb{N} = \mathbb{N}$.

  Alternatively, as noted at the beginning of Section 11.5, $\mathbb{R}$ is bounded but has no least upper bound in *$\mathbb{R}$, so must fail to be internal by the internal order-completeness property, Theorem 11.5.1.

  The fact that $\mathbb{N}$ is external will be used in the next section to show that all infinite subsets of $\mathbb{R}$ are external.

- *Limited Hyperreals.* The set $\mathbb{L}$ of limited numbers is external for the same reason $\mathbb{R}$ is: it is bounded above by all members of *$\mathbb{R}_\infty^+$, but has no least upper bound. Since

$$\mathbb{L} = \bigcap \{(-b, b) : b \text{ is unlimited}\},$$

it follows that the intersection of an *infinite* family of internal sets can fail to be internal.

Observe that if X is an internal set that includes $\mathbb{L}$, then $X \neq \mathbb{L}$, and so X must contain unlimited members (cf. (i) and (ii) of Exercise 11.5.2). In fact, by considering lower and upper bounds of *$\mathbb{R}_\infty^+ - X$ and *$\mathbb{R}_\infty^- - X$, respectively, we can show that if X is an internal set with $\mathbb{L} \subseteq X$, then $[-b, b] \subseteq X$ for some unlimited b.

- *Infinitesimals.* The set $\mathbb{I} = \text{hal}(0)$ of infinitesimals is bounded above (by any positive real), so if it were internal, it would have a least upper bound $b \in {}^*\mathbb{R}$. Such a b would have to be positive but less than every positive real, forcing $b \simeq 0$. But then $b < 2b \in \mathbb{I}$, so b cannot be an upper bound of $\mathbb{I}$ after all.

  By similar reasoning, any halo $\text{hal}(r)$ is seen to be an external set, as are its "left and right halves" $\{x > r : x \simeq r\}$ and $\{x < r : x \simeq r\}$.

  More strongly, this type of reasoning shows that if X is any internal subset of $\mathbb{I}$, then the least upper bound and greatest lower bound of X must be infinitesimal, and so $X \subseteq [-\varepsilon, \varepsilon]$ for some $\varepsilon \simeq 0$.

Exercise 11.6.1
Show that the following form external sets:

the positive real numbers $\mathbb{R}^+$,

the integers $\mathbb{Z}$,

the rational numbers $\mathbb{Q}$,

the (positive/negative) unlimited hyperreals,

the appreciable numbers.

11.7 Defining Internal Sets

In proving the internal least number and order-completeness properties, we reverted once more to ultrafilter calculations, so it is natural to ask whether such results could be obtained instead by a logical transfer. The assertion that a nonempty set A of natural numbers has a least element can be expressed by the $\mathcal{L}_\Re$-sentence

$$[(\exists x \in \mathbb{R})(x \in A) \wedge (\forall x \in A)(x \in \mathbb{N})] \rightarrow (\exists x \in A)(\forall y \in A)(x \leq y).$$

This sentence transforms to

$$[(\exists x \in {}^*\mathbb{R})(x \in {}^*A) \wedge (\forall x \in {}^*A)(x \in {}^*\mathbb{N})] \rightarrow (\exists x \in {}^*A)(\forall y \in {}^*A)(x \leq y),$$

which asserts the least number principle for the enlarged set $^*A$ in $^*\mathbb{R}$. But what we saw in Section 11.3 was that this transformed sentence is true when $^*A$ is replaced by any internal set $X \subseteq {}^*\mathbb{N}$. The same observation applies to the assertion

$$(\exists x \in \mathbb{R})(\forall y \in A)(y \leq x) \rightarrow$$
$$(\exists x \in \mathbb{R})\Big[(\forall y \in A)(y \leq x) \wedge (\forall z \in \mathbb{R})[(\forall y \in A)(y \leq z) \rightarrow x \leq z]\Big],$$

which expresses the order-completeness property that if A has an upper bound, then it has a least upper bound. The transform of this last sentence is also true when $^*A$ is replaced by any nonempty internal set $X \subseteq {}^*\mathbb{R}$ (Theorem 11.5.1).

Let us write $\varphi(A)$ to indicate that φ is an $\mathcal{L}_\Re$-sentence containing the set symbol A. Then $^*\varphi(X)$ denotes the sentence obtained by putting X in place of $^*A$ in $^*\varphi$. The examples just discussed suggest the following transfer principle:

(†) *If $\varphi(A)$ is true whenever A is taken as an arbitrary subset of $\mathbb{R}$, then $^*\varphi(X)$ is true whenever X is taken as an arbitrary* **internal** *subset of $^*\mathbb{R}$.*

To understand what is happening here we need to look more widely at formulae that may have free variables. Let $\varphi(x, y, A)$ be a formula with free variables x and y as well as the set symbol A; for example the formula

$$(\forall z \in \mathbb{R})\,(x < z < y \rightarrow z \in A).$$

We can replace x and y in ${}^*\varphi$ by elements of ${}^*\mathbb{R}$. Thus ${}^*\varphi([r_n], [s_n], [A_n])$ would be the sentence

$$(\forall z \in {}^*\mathbb{R})\,([r_n] < z < [s_n] \rightarrow z \in [A_n]).$$

It can be shown that this sentence is true if and only if

$$\{n \in \mathbb{N} : \forall z \in \mathbb{R}\,(r_n < z < s_n \rightarrow z \in A_n)\} \in \mathcal{F}.$$

The general situation here is that

$$\quad {}^*\varphi([r_n], [s_n], [A_n]) \text{ is true} \qquad\qquad (i)$$

if and only if
$$\{n \in \mathbb{N} : \varphi(r_n, s_n, A_n) \text{ is true}\} \in \mathcal{F}. \qquad (ii)$$

This fact can then be used to derive the transfer principle (†). But it also leads to a new way of defining internal sets: holding the hyperreal $[s_n]$ and the internal set $[A_n]$ fixed, and allowing the value of x to range over ${}^*\mathbb{R}$, define

$$X = \{b \in {}^*\mathbb{R} : {}^*\varphi(b, [s_n], [A_n]) \text{ is true}\}.$$

Correspondingly, for each $n \in \mathbb{N}$ put

$$B_n = \{r \in \mathbb{R} : \varphi(r, s_n, A_n) \text{ is true}\}.$$

Then the equivalence of (i) and (ii) amounts to saying that for any hyperreal $[r_n]$,

$$[r_n] \in X \quad \text{iff} \quad \{n \in \mathbb{N} : r_n \in B_n\} \in \mathcal{F}.$$

But this shows that X is the *internal* set $[B_n]$ determined by the sequence of real subsets $\langle B_n : n \in \mathbb{N} \rangle$.

Expressing this phenomenon in the most general form available at this stage, we have the following statement.

11.7.1 Internal Set Definition Principle. *Let*

$$\varphi(x_0, x_1, \ldots, x_n, A_1, \ldots, A_k)$$

be an $\mathcal{L}_{\mathfrak{R}}$-formula with free variables $x_0, \ldots, x_n$ and set symbols $A_1, \ldots, A_k$. Then for any hyperreals $c_1, \ldots, c_n$ and any internal sets $X_1, \ldots, X_k$,

$$\{b \in {}^*\mathbb{R} : {}^*\varphi(b, c_1, \ldots, c_n, X_1, \ldots, X_k)\}$$

11.7 Defining Internal Sets

is an internal subset of *$\mathbb{R}$. □

Note that this statement applies in particular to formulae that have only number variables and no set symbols. It provides a ready means of demonstrating that various sets are internal, including the examples from Section 11.1:

- Taking $\varphi(x_0, x_1, \ldots, x_n)$ as the formula $(x_0 = x_1 \vee \cdots \vee x_0 = x_n)$ shows that any finite set

$$\{c_1, \ldots, c_n\} = \{b \in {}^*\mathbb{R} : {}^*\varphi(b, c_1, \ldots, c_n)\}$$

of hyperreals is internal.

- Taking $\varphi(x_0, x_1, x_2)$ as $(x_1 < x_0 < x_2)$ yields that any open hyperreal interval

$$(c_1, c_2) = \{b \in {}^*\mathbb{R} : c_1 < b < c_2\} = \{b \in {}^*\mathbb{R} : {}^*\varphi(b, c_1, c_2)\}$$

is internal.

We observed at the end of Section 11.4 that the overflow principle implies that any cofinite subset of $\mathbb{N}$ is external. But much more strongly than this, we can now use internal set definition to show (as promised earlier, in Section 11.2) that

- *every infinite set of real numbers is external.*

In other words, if $A \subseteq \mathbb{R}$ is internal, then A must be finite.

Proof: if such an A were infinite, then it would contain an infinite sequence, i.e., there would be an injective function $f : \mathbb{N} \to A$. Put $X = \{f(n) : n \in \mathbb{N}\}$. Then X is internal, since it is a subset of the internal set A, and we saw at the end of Section 11.2 that any subset of an internal set of *real* numbers is internal.

Now, X is a bijective copy of $\mathbb{N}$, so we should be able to show that $\mathbb{N}$ is internal if X is, thereby getting a contradiction because we already know that $\mathbb{N}$ is external. To make this work requires the internal set definition principle, applied with $\varphi(x, A)$ as the formula $(x \in \mathbb{N} \wedge f(x) \in A)$. This implies that the set

$$B = \{n \in {}^*\mathbb{R} : {}^*\varphi(n, X)\}$$

is internal. Observe that

$$B = \{n \in {}^*\mathbb{N} : {}^*f(n) \in X\} = {}^*f^{-1}(X).$$

However, as f is injective, ${}^*f : {}^*\mathbb{N} \to {}^*A$ is an *injective* extension of f (by transfer), from which it follows that B is just $\mathbb{N}$ itself, so we have the contradiction. □

We will reconsider the internal set definition principle in a stronger formulation in Section 13.15.

Exercise 11.7.1 Use the internal set definition principle to show that all hyperreal intervals of the form

$$(c,d], \quad [c,d), \quad [c,d], \quad \{b \in {}^*\mathbb{R} : c \le b\}, \quad \{b \in {}^*\mathbb{R} : b < c\}$$

etc. are internal, as are the sets

$$\{1, 2, \ldots, N\}, \quad \{0, \tfrac{1}{N}, \tfrac{2}{N}, \ldots, \tfrac{N-1}{N}, 1\},$$

for all $N \in {}^*\mathbb{N}$.

11.8 The Underflow Principle

This is the order-theoretic dual of the overflow principle of Theorem 11.4.1, but its proof requires the additional reasoning power provided by the internal set definition principle.

Theorem 11.8.1 *Let X be an internal subset of ${}^*\mathbb{N}$, and let $K \in {}^*\mathbb{N}$ be unlimited. If every unlimited hypernatural $H \le K$ belongs to X, then there is some $k \in \mathbb{N}$ such that every limited n with $k \le n$ belongs to X.*

Proof. For $M, N \in {}^*\mathbb{N}$ with $M \le N$, let

$$\lfloor M, N \rfloor = \{z \in {}^*\mathbb{N} : M \le z \le N\}$$

be the interval in ${}^*\mathbb{N}$ between M and N. Our hypothesis is that $\lfloor H, K \rfloor \subseteq X$ for all unlimited hypernatural $H \le K$. What we want to show is that $\lfloor k, K \rfloor \subseteq X$ for some $k \in \mathbb{N}$. To put this more symbolically, we want to show that the set

$$Y = \{k \in {}^*\mathbb{N} : \lfloor k, K \rfloor \subseteq X\}$$

has a limited member.

Now, if Y is internal, then by the internal least number principle it has a least element k, and such a k must belong to $\mathbb{N}$, because if it were unlimited, then $k - 1$ would be unlimited, so by our hypothesis $k - 1$ would also be in Y but less than k.

It thus suffices to show that Y is internal. But if $\varphi(x, y, A)$ is the formula

$$x \in \mathbb{N} \wedge x \le y \wedge \forall z \in \mathbb{N}\, (x \le z \le y \to z \in A),$$

expressing "$x \in \mathbb{N}$ and $\lfloor x, y \rfloor \subseteq A$", then by the internal set definition principle the set

$$\{k \in {}^*\mathbb{R} : {}^*\varphi(k, K, X)\}$$
$$= \{k \in {}^*\mathbb{N} : k \le K \text{ and } \forall z \in {}^*\mathbb{N}\,(k \le z \le K \to z \in X)\}$$

is internal. This set is just Y. □

Exercise 11.8.2
Show that underflow is equivalent to the following statement:

> If an internal subset X of *$\mathbb{N}$ contains arbitrarily large limited members, then it contains arbitrarily small unlimited members.

Deduce that if A is any infinite subset of $\mathbb{N}$, then *A contains arbitrarily small unlimited members.

11.9 Internal Sets and Permanence

In Section 7.10 it was shown that any property that is expressible by an $\mathcal{L}_\Re$-formula $\varphi(x)$ and holds for all points infinitely close to a real number c must in fact hold for all points within some real distance of c. In other words, *φ cannot be true exactly of the members of the halo of c, and so hal(c) cannot be defined by the transform of any $\mathcal{L}_\Re$-formula.

Now, the set $B = \{b \in {}^*\mathbb{R} : {}^*\varphi(b)\}$ is internal, by the internal set definition principle, so cannot be equal to hal(c) because the latter is external. Viewed in this way, the result of Section 7.10 is seen to be a manifestation of the fact that external sets are not internal.

Similar observations hold for the overflow and underflow principles, which are related to the fact that an internal set cannot be equal to the external set $\mathbb{N}$, and the "spillover" results of Exercise 11.5.2 and Section 11.6, which relate to the fact that an internal set cannot be equal to $\mathbb{L}$. In general then, we see that a *permanence principle* is typically a statement to the effect that a property that defines an internal set cannot hold just of the members of an external set like hal(r), $\mathbb{N}$, $\mathbb{N} - \{1, \ldots, k\}$, *$\mathbb{N} - \mathbb{N}$, $\mathbb{L}$, etc. If such an internal property holds for all members of an external set E, then it must continue to hold throughout a larger internal set strictly containing E.

Notice that universal transfer (Section 4.5) is a permanence assertion. It states that if a certain kind of property holds for all members of the external set $\mathbb{R}$, then it must continue to hold throughout the internal extension *$\mathbb{R}$.

Here now is a stronger version of the result of Section 7.10, which gives the promised extension to arbitrary hyperreal numbers. We will consider an even stronger form of the result in Section 15.1.

Theorem 11.9.1 *If X is an internal subset of *$\mathbb{R}$ that contains all points that are infinitely close to $b \in {}^*\mathbb{R}$, then there is a positive real ε such that X contains all points that are within ε of b.*

Proof. Our hypothesis is that hal$(b) \subseteq X$. For $k \in {}^*\mathbb{N}$, let $(b - \frac{1}{k}, b + \frac{1}{k})$ be the hyperreal interval

$$\{z \in {}^*\mathbb{R} : |z - b| < \tfrac{1}{k}\}.$$

Now,
$$(b - \tfrac{1}{k}, b + \tfrac{1}{k}) \subseteq X$$
whenever k is unlimited, because in this case $\tfrac{1}{k}$ is infinitesimal, and so
$$(b - \tfrac{1}{k}, b + \tfrac{1}{k}) \subseteq \mathrm{hal}(b) \subseteq X$$
by our hypothesis. Thus the set
$$Y = \{k \in {}^*\mathbb{N} : (b - \tfrac{1}{k}, b + \tfrac{1}{k}) \subseteq X\}$$
contains all unlimited members of *$\mathbb{N}$. Hence by underflow we could conclude that $(b - \tfrac{1}{k}, b + \tfrac{1}{k}) \subseteq X$ for some $k \in \mathbb{N}$, and thereby complete the proof by putting $\varepsilon = \tfrac{1}{k}$, *provided that Y is internal*. But applying internal set definition with $\varphi(x, y, A)$ as the formula
$$x \in \mathbb{N} \wedge (\forall z \in \mathbb{R})(|z - y| < \tfrac{1}{x} \to z \in A)$$
(expressing "$x \in \mathbb{N}$ and $(y - \tfrac{1}{x}, y + \tfrac{1}{x}) \subseteq A$") shows that the set
$$\{k \in {}^*\mathbb{R} : {}^*\varphi(k, b, X)\} = \{k \in {}^*\mathbb{N} : (\forall z \in {}^*\mathbb{R})\,(|z - b| < \tfrac{1}{k} \to z \in X)\}$$
is internal, and this set is just Y. □

11.10 Saturation of Internal Sets

The internal sets form a very special collection whose members are related to each other in remarkable ways. For instance, it is impossible to construct a nested sequence of internal sets whose intersection is empty. This fact, which we now prove, is known as *countable saturation*. (The use of "saturation" is explained at the beginning of the next section.)

Theorem 11.10.1 *The intersection of a decreasing sequence*
$$X^1 \supseteq X^2 \supseteq \cdots \supseteq X^k \supseteq \cdots$$
of nonempty internal sets is always nonempty :
$$\bigcap_{k \in \mathbb{N}} X^k \neq \emptyset.$$

Proof. This is a delicate analysis of the ultrapower construction, involving a kind of diagonalisation argument, that is not easy to motivate intuitively.

For each $k \in \mathbb{N}$, let $X^k = [A_n^k]$, so that X^k is the internal set defined by the sequence $\langle A_n^k : n \in \mathbb{N} \rangle$ of subsets of $\mathbb{R}$. Then by Section 11.2 the sets $\{n \in \mathbb{N} : A_n^k \neq \emptyset\}$ and $\{n \in \mathbb{N} : A_n^k \supseteq A_n^{k+1}\}$ belong to $\mathcal{F}$. Hence if
$$J^k = \{n \in \mathbb{N} : A_n^1 \supseteq \cdots \supseteq A_n^k \neq \emptyset\},$$

then by closure of $\mathcal{F}$ under finite intersections it follows for each $k \in \mathbb{N}$ that $J^k \in \mathcal{F}$. Note that $J^1 \supseteq J^2 \supseteq \cdots$.

We want to define a hyperreal $[s_n]$ that belongs to every X^k. This will require that for each k we have $s_n \in A_n^k$ for $\mathcal{F}$-almost all n. We will arrange this to work for almost all $n \geq k$, in the sense that

$$\{n \in \mathbb{N} : k \leq n\} \cap J^k \subseteq \{n \in \mathbb{N} : s_n \in A_n^k\}. \qquad \text{(iii)}$$

But the set $\{n \in \mathbb{N} : k \leq n\}$ is cofinite in $\mathbb{N}$, and so belongs to the nonprincipal ultrafilter $\mathcal{F}$. Since also $J^k \in \mathcal{F}$, (iii) then yields $\{n \in \mathbb{N} : s_n \in A_n^k\} \in \mathcal{F}$, and therefore $[s_n] \in X^k$ as desired.

It thus remains to define s_n fulfilling (iii). For $n \in J^1$ let

$$k_n = \max\{k : k \leq n \text{ and } n \in J^k\}. \qquad \text{(iv)}$$

Then $n \in J^{k_n}$, so by the definition of J^{k_n} we can choose some $s_n \in A_n^{k_n}$, and hence

$$s_n \in A_n^1 \cap \cdots \cap A_n^{k_n}. \qquad \text{(v)}$$

For $n \notin J^1$ let s_n be arbitrary. Now, to prove (iii), observe that if $k \leq n$ and $n \in J^k$, then by (iv), $k \leq k_n$, and so by (v), $s_n \in A_n^k$.
□

Countable saturation has some important consequences for the nature of countable unions and intersections of internal sets:

Corollary 11.10.2 *If $\{X_n : n \in \mathbb{N}\}$ is a collection of internal sets and X is internal, then:*

(1) $\bigcap_{n \in \mathbb{N}} X_n \neq \emptyset$ *if $\{X_n : n \in \mathbb{N}\}$ has the finite intersection property.*

(2) *If $X \subseteq \bigcup_{n \in \mathbb{N}} X_n$, then $X \subseteq \bigcup_{n \leq k} X_n$ for some $k \in \mathbb{N}$.*

(3) *If $\bigcap_{n \in \mathbb{N}} X_n \subseteq X$, then $\bigcap_{n \leq k} X_n \subseteq X$ for some $k \in \mathbb{N}$.*

(4) *If $\bigcup_{n \in \mathbb{N}} X_n$ is internal, then it is equal to $\bigcup_{n \leq k} X_n$ for some $k \in \mathbb{N}$.*

(5) *If $\bigcap_{n \in \mathbb{N}} X_n$ is internal, then it is equal to $\bigcap_{n \leq k} X_n$ for some $k \in \mathbb{N}$.*

Proof.

(1) Let $Y^k = X_1 \cap \cdots \cap X_k$. Then $Y^1 \supseteq Y^2 \supseteq \cdots$, and each Y^k is internal by 11.2(1). The finite intersection property implies that $Y^k \neq \emptyset$, so by the above theorem there is some hyperreal that belongs to every Y^k, and hence to every X_k.

(2) Suppose that for all $k \in \mathbb{N}$, $X \not\subseteq \bigcup_{n \leq k} X_n$ and hence

$$\bigcap_{n \leq k}(X - X_n) = X - \left(\bigcup_{n \leq k} X_n\right) \neq \emptyset.$$

Then $\{X - X_n : n \in \mathbb{N}\}$ is a collection of internal sets with the finite intersection property, so by (1) there is some x with

$$x \in \bigcap_{n \in \mathbb{N}} (X - X_n) = X - (\bigcup_{n \in \mathbb{N}} X_n);$$

hence $X \not\subseteq \bigcup_{n \in \mathbb{N}} X_n$.

(3) Exercise.

(4) Put $X = \bigcup_{n \in \mathbb{N}} X_n$ in (2).

(5) Similarly, from (3). □

Result (4) of Corollary 11.10.2 plays a crucial role in the nonstandard approach to measure theory discussed in Chapter 16 (cf. example 6 of Section 16.1 and example 3 of Section 16.2).

Exercise 11.10.3
Show that the union of a strictly decreasing, or strictly increasing, sequence of internal sets is external.

What is the corresponding result about intersections of internal sets?

11.11 Saturation Creates Nonstandard Entities

The use of the term "saturation" is intended to convey that *$\mathbb{R}$ is "full of elements". Countable saturation legislates into existence those elements that can be characterised as belonging to the intersection of a decreasing sequence of internal sets. For example, if we take X_n to be the hyperreal interval $(0, \frac{1}{n})$, then $\langle X_n : n \in \mathbb{N} \rangle$ is a decreasing sequence of nonempty internal sets. Its (nonempty) intersection $\bigcap_{n \in \mathbb{N}} X_n$ is precisely the set of *positive infinitesimals*.

Exercise 11.11.1
Use countable saturation to infer the existence of positive unlimited and negative unlimited members of *$\mathbb{R}$. □

Another interesting consequence of saturation is the property that

- *every sequence of infinitesimals has an infinitesimal upper bound.*

To see this, take $\langle \varepsilon_n : n \in \mathbb{N} \rangle$ with $\varepsilon_n \simeq 0$ for all $n \in \mathbb{N}$. If X_n is the internal hyperreal interval $[\varepsilon_n, \frac{1}{n})$, then the collection $\{X_n : n \in \mathbb{N}\}$ has the finite intersection property. For in general, if ε is the maximum element of $\{\varepsilon_{n_1}, \ldots, \varepsilon_{n_k}\}$, then

$$\varepsilon \in [\varepsilon_{n_1}, \tfrac{1}{n_1}) \cap \cdots \cap [\varepsilon_{n_k}, \tfrac{1}{n_k}).$$

But any member of $\bigcap_{n \in \mathbb{N}} X_n$ is an upper bound of the ε_n's that is smaller than $\frac{1}{n}$ for all $n \in \mathbb{N}$, and hence is infinitesimal.

Dually, we can use saturation to show that

- *every sequence $\langle s_n : n \in \mathbb{N}\rangle$ of unlimited hypernatural numbers has an unlimited hypernatural lower bound,*

by considering the internal intervals $X_n = (n, s_n]$. In this case if x belongs to $\bigcap_{n\in\mathbb{N}} X_n$, then x is a positive unlimited lower bound of the s_n's. But then (by transfer) we can take a member of *$\mathbb{N}$ between $x - 1$ and x to get an unlimited hypernatural number that is less that s_n for all $n \in \mathbb{N}$. (Alternatively, put $X_n = (n, s_n] \cap$ *$\mathbb{N}$ in this argument.)

Exercise 11.11.2
Show that the two properties

- every sequence of infinitesimals has an infinitesimal upper bound,
- every sequence of unlimited hypernatural numbers has an unlimited hypernatural lower bound,

imply each other *without using saturation*. □

We will consider this property from another perspective at the end of Section 15.4.

11.12 The Size of an Internal Set

Countable saturation implies that *$\mathbb{R}$ has so many elements that any big internal set is *very* big. Such a set cannot be countably infinite:

- *Every internal set is either finite or uncountable.*

We already have an argument for this in the case of subsets of $\mathbb{R}$: any internal set of reals must be finite. In proving this in Section 11.7 we showed in effect that an internal subset of $\mathbb{R}$ cannot be put in one-to-one correspondence with $\mathbb{N}$. But now we can demonstrate this for any internal set whatsoever.

To see why this is so, let $X = \{x_n : n \in \mathbb{N}\}$ be a countable internal set. We remove all the points from X in turn, by defining for each n the set $X_n = X - \{x_1, \ldots, x_n\}$, which is internal. But the X_n's form a decreasing sequence, so if they were all nonempty, countable saturation would imply that their intersection would be nonempty, which is false. We must therefore conclude that there is an n for which $X_n = \emptyset$ and so $X = \{x_1, \ldots, x_n\}$.

This shows that any countable internal set must be finite. Hence an infinite internal set must be uncountable.

This observation has an interesting bearing on the structure of the set *$\mathbb{N}$ of hypernatural numbers (Section 5.10). If N is an unlimited hypernatural, then the initial segment $\{1, 2, \ldots, N\}$ of *$\mathbb{N}$ is internal, and is certainly infinite, since it includes all of $\mathbb{N}$, so is uncountable. It follows that there must be uncountably many *unlimited* members of *ℕ that are less than N. The set of all unlimited hypernaturals is partitioned into *ℕ-galaxies, each of which looks like a copy of ℤ. If N is unlimited, then there are uncountably many of these *ℕ-galaxies between ℕ and N.

11.13 Closure of the Shadow of an Internal Set

For any $X \subseteq {}^*\mathbb{R}$, let

$$\mathrm{sh}(X) = \{\mathrm{sh}(x) : x \in X \text{ and } x \text{ is limited}\}.$$

For example, if X is an interval (a, b) in *ℝ, then if a, b are limited, $\mathrm{sh}(X)$ is the closed interval $[\mathrm{sh}(a), \mathrm{sh}(b)]$ in ℝ, while if a is limited but b unlimited, then $\mathrm{sh}(X) = [\mathrm{sh}(a), +\infty) \subseteq \mathbb{R}$, again a topologically closed subset of ℝ.

Theorem 11.13.1 *If X is internal, then $\mathrm{sh}(X)$ is closed.*

Proof. Let $r \in \mathbb{R}$ be a closure point of $\mathrm{sh}(X)$. We need to show that $r \in \mathrm{sh}(X)$, i.e., r is the shadow of some $y \in X$.

Now, for each $n \in \mathbb{N}$, the hyperreal open interval $(r - \frac{1}{n}, r + \frac{1}{n})$ meets $\mathrm{sh}(X)$ in some *real* point s_n that must be the shadow of some $x_n \in X$. Hence $x_n \simeq s_n \in (r - \frac{1}{n}, r + \frac{1}{n})$, so the *internal* set

$$X_n = X \cap \left(r - \tfrac{1}{n}, r + \tfrac{1}{n}\right)$$

contains x_n and is thereby nonempty. The X_n's form a decreasing sequence, so by countable saturation there is a point y in their intersection. Then $y \in X$ and $|y - r| < \frac{1}{n}$ for all $n \in \mathbb{N}$, so $y \simeq r$. Hence $r = \mathrm{sh}(y) \in \mathrm{sh}(X)$.

This shows that $\mathrm{sh}(X)$ contains all its closure points and so is closed. □

Topological closure of the shadow of an internal set plays an important role in the hyperreal "reconstruction" of Lebesgue measure. This will be explained in Chapter 16 (cf. the proof of Theorem 16.8.2).

Exercise 11.13.2
Apply the internal set definition principle to show that if X is internal, then for any hyperreal number r the set

$$\{n \in {}^*\mathbb{N} : X \cap (r - \tfrac{1}{n}, r + \tfrac{1}{n}) \neq \emptyset\}$$

is internal. Use this fact together with overflow to give an alternative proof of Theorem 11.13.1 that does not appeal to countable saturation. □

The connection between countable saturation and overflow indicated by this last exercise will emerge again in Chapter 15 (cf. the proof of Theorem 15.4.2).

11.14 Interval Topology and Hyper-Open Sets

In Section 10.5 we introduced the following notions:

- A set A of hyperreals is *interval-open* if each of its points belongs to some hyperreal open interval (a,b) that is included in A. Thus the interval-open sets are precisely those that are unions of hyperreal open intervals. The class of interval-open sets is the *interval topology* on *$\mathbb{R}$.

- A *real-open* set, on the other hand, is one that is a union of hyperreal open neighbourhoods $(r-\varepsilon, r+\varepsilon)$ having *real* radius ε. Equivalently, a real-open set is a union of hyperreal open intervals of *appreciable* length. Each real-open set is interval-open, but not conversely: the real-open sets are not a topology on *$\mathbb{R}$, since they are not closed under intersection.

- An S-*open* set is a union of S-*neighbourhoods* $((r-\varepsilon, r+\varepsilon))$ having *real* radius ε, where

$$((r-\varepsilon, r+\varepsilon)) = \{x \in {}^*\mathbb{R} : \mathrm{hal}(x) \subseteq (r-\varepsilon, r+\varepsilon)\}.$$

The S-open sets form the S-topology on *$\mathbb{R}$. Every S-open set is real-open, but not conversely. Every S-open set is a union of halos, but not conversely.

The example of the set

$$\mathbb{L} = \bigcup_{n \in \mathbb{N}} (-n, n)$$

of limited numbers shows that while a real-open set is always a union of internal sets (namely, open intervals), it may itself be external.

We now introduce a further class of subsets of *$\mathbb{R}$: an internal set $[A_n]$ will be called *hyper-open* if

$$\{n \in \mathbb{N} : A_n \text{ is open in } \mathbb{R}\} \in \mathcal{F}.$$

Each hyperreal interval (a,b) is hyper-open, as we saw in Section 11.1: if $a = [a_n]$ and $b = [b_n]$, then (a,b) is the internal set defined by the sequence $\langle A_n : n \in \mathbb{N} \rangle$, where A_n is the real interval (a_n, b_n), which is indeed open in $\mathbb{R}$.

Lemma 11.14.1 *Every hyper-open set is a union of hyperreal open intervals.*

Proof. Let $A = [A_n]$ be hyper-open. Take a point $r = [r_n]$ in A. Then we find that the set

$$J = \{n \in \mathbb{N} : r_n \in A_n \text{ and } A_n \text{ is open in } \mathbb{R}\}$$

belongs to the ultrafilter $\mathcal{F}$. Our task is to show that r belongs to some hyperreal interval (a, b) that is included in A.

Now, if $n \in J$, then there is some real interval $(a_n, b_n) \subseteq \mathbb{R}$ with

$$r_n \in (a_n, b_n) \subseteq A_n.$$

Since $J \in \mathcal{F}$, this is enough to specify a as the hyperreal number $[a_n]$ and b as $[b_n]$. Working with the properties of $\mathcal{F}$, in a now familiar way, we can then show that $[a_n] < [r_n] < [b_n]$, and also that $[s_n] \in [A_n]$ whenever $[a_n] < [s_n] < [b_n]$, so that

$$r \in (a, b) \subseteq A$$

as desired. □

This lemma implies that every hyper-open set is interval-open. But there are interval-open sets, like the set $\mathbb{L}$ of limited numbers, that are not hyper-open, simply because they are external, whereas hyper-open sets are always internal by definition. The example of $\mathbb{L}$ shows that the class of hyper-open sets is not a topology, because it is not closed under infinite unions. Instead, it is what is known as a *base* for the interval topology, because every interval-open set is a union of hyper-open sets (open intervals).

Exercise 11.14.2
Show that the class of hyper-open sets is closed under *finite* unions. □

However, for *internal* sets, hyper-openness does prove to be equivalent to interval-openness:

Exercise 11.14.3
If $A = [A_n]$ is an internal set, let $B = [B_n]$, where B_n is the interior of A_n in $\mathbb{R}$ (cf. Section 10.1). Show that B is interval-open, and is in fact the interior of A in the interval topology, i.e., B is the union of all interval-open subsets of A.

Deduce from this that *an internal set is interval-open iff it is hyper-open.*
□

The classes of real-open sets and hyper-open sets are incomparable: $\mathbb{L}$ is real-open (indeed S-open) but not hyper-open, while any infinitesimal-length open interval is hyper-open but not real-open. This latter example shows that even for internal sets the two classes remain distinguishable.

There is a characterisation of S-openness of internal sets that corresponds to the nonstandard characterisation of openness of subsets of $\mathbb{R}$, and involves an interesting application of underflow:

Theorem 11.14.4 *If B is an internal set, then B is S-open if and only if it contains the halo of each of its points.*

Proof. We have already observed that an S-open set is a union of halos.

11.14 Interval Topology and Hyper-Open Sets

Conversely, assume that hal(r) $\subseteq B$ whenever $r \in B$. For such an r, consider the set

$$X = \{n \in {}^*\mathbb{N} : (\forall x \in {}^*\mathbb{R})\,(\,|r - x| < \tfrac{1}{n} \to x \in B)\}.$$

Since B is internal, it follows by the internal set definition principle that X is internal. Moreover, since hal(r) $\subseteq B$, it follows that X contains every unlimited hypernatural n, because for such an n, $|r - x| < \tfrac{1}{n}$ implies $x \in$ hal(r). Hence by underflow, X must contain some standard $n \in \mathbb{N}$, so B includes the real-radius interval $(r - \tfrac{1}{n}, r + \tfrac{1}{n})$. But then since $\tfrac{1}{n}$ is real,

$$r \in ((r - \tfrac{1}{n}, r + \tfrac{1}{n})) \subseteq (r - \tfrac{1}{n}, r + \tfrac{1}{n}) \subseteq B.$$

This shows that B is the union of S-neighbourhoods, and is thereby S-open. □

12

Internal Functions and Hyperfinite Sets

The method used to construct an internal subset of *ℝ out of a sequence of subsets of ℝ will now be adapted to build hyperreal-valued functions out of sequences of real-valued functions.

12.1 Internal Functions

Let $\langle f_n : n \in \mathbb{N} \rangle$ be a sequence of functions $f_n : A_n \to \mathbb{R}$, with domains A_n included in ℝ. Then a *ℝ-valued function $[f_n]$ is defined on the internal set $[A_n]$ by putting
$$[f_n]([r_n]) = [f_n(r_n)].$$
Observe that if $[r_n] \in [A_n]$, then the set $J = \{n \in \mathbb{N} : r_n \in A_n\}$ belongs to $\mathcal{F}$, and for each $n \in J$, $f_n(r_n)$ is defined. This is enough to make $[f_n]([r_n])$ well-defined. We have
$$\operatorname{dom}[f_n] = [\operatorname{dom} f_n].$$
Functions $f : X \to {}^*\mathbb{R}$ that are obtained by this construction are called *internal*. In the case that $\langle f_n \rangle$ is a constant sequence, with $f_n = f : A \to \mathbb{R}$ for all n, then $[f_n]$ is just the function $^*f : {}^*A \to {}^*\mathbb{R}$ extending f, as defined in Section 3.13.

The following result shows that we only need to specify *almost all* of the real functions f_n in order to determine the internal function $[f_n]$.

Theorem 12.1.1 *Let $\langle f_n : n \in \mathbb{N} \rangle$ and $\langle g_n : n \in \mathbb{N} \rangle$ be sequences of partial functions from ℝ to ℝ. Then the internal functions $[f_n]$ and $[g_n]$ are equal*

if and only if
$$\{n \in \mathbb{N} : f_n = g_n\} \in \mathcal{F}.$$

Proof. Let $J_{fg} = \{n \in \mathbb{N} : f_n = g_n\}$, and suppose $J_{fg} \in \mathcal{F}$. Now in general, two functions are equal precisely when they have the same domain and assign the same values to all members of that domain. Thus

$$J_{fg} \subseteq \{n \in \mathbb{N} : \operatorname{dom} f_n = \operatorname{dom} g_n\},$$

leading by 11.2(3) to the conclusion that the internal sets $[\operatorname{dom} f_n]$ and $[\operatorname{dom} g_n]$ are equal, i.e., $\operatorname{dom}[f_n] = \operatorname{dom}[g_n]$. But for $[r_n] \in \operatorname{dom}[f_n]$,

$$J_{fg} \cap \{n \in \mathbb{N} : r_n \in \operatorname{dom} f_n\} \subseteq \{n \in \mathbb{N} : f_n(r_n) = g_n(r_n)\},$$

which leads to $[f_n]([r_n]) = [g_n]([r_n])$. Hence $[f_n] = [g_n]$.

For the converse, suppose that $J_{fg} \notin \mathcal{F}$. Now, J_{fg}^c is a subset of the union

$$\{n \in \mathbb{N} : \operatorname{dom} f_n \neq \operatorname{dom} g_n\} \cup \{n \in \mathbb{N} : \operatorname{dom} f_n = \operatorname{dom} g_n \text{ but } f_n \neq g_n\},$$

so either $\{n : \operatorname{dom} f_n \neq \operatorname{dom} g_n\} \in \mathcal{F}$, whence $\operatorname{dom}[f_n] \neq \operatorname{dom}[g_n]$ and so $[f_n] \neq [g_n]$, or else

$$J = \{n \in \mathbb{N} : \operatorname{dom} f_n = \operatorname{dom} g_n \text{ but } f_n \neq g_n\} \in \mathcal{F}.$$

But for $n \in J$ there exists some r_n with $f_n(r_n) \neq g_n(r_n)$. This leads to $[f_n]([r_n]) \neq [g_n]([r_n])$, and so $[f_n] \neq [g_n]$. □

12.2 Exercises on Properties of Internal Functions

(1) *The image of an internal set under an internal function is internal*: if $f = [f_n]$ is an internal function, and $A = [A_n]$ is any internal subset of $\operatorname{dom} f$, then the image set

$$f(A) = \{f(a) : a \in A\}$$

is in fact the internal set $[f_n(A_n)]$.

(2) *The inverse image of an internal set under an internal function is internal*: if $f = [f_n]$ is an internal function, and $B = [B_n]$ is an internal set, then the inverse-image set

$$f^{-1}(B) = \{a \in \operatorname{dom} f : f(a) \in B\}$$

is the internal set $[f_n^{-1}(B_n)]$.

(3) *The composition of internal functions is internal*: if f and g are internal functions, with the range of f included in the domain of g, then $g \circ f$ is an internal function.

(4) An internal function $[f_n]$ is injective iff $\{n \in \mathbb{N} : f_n \text{ is injective}\} \in \mathcal{F}$.

(5) *The inverse of an internal function is internal*: if $[f_n]$ is injective, then $[f_n]^{-1}$ is the internal function $[f_n^{-1}]$ (which is well-defined by the previous exercise and Theorem 12.1.1).

(6) If f and g are internal functions (with the same domain), then so are the functions $f + g$, $f \cdot g$, and cf for any hyperreal c.

(7) Let f be an internal function that takes only infinitesimal values: $f(x) \simeq 0$ whenever $f(x)$ is defined. Show that the range $\{f(x) : x \in \text{dom } f\}$ of f has an infinitesimal least upper bound.

(8) Give an alternative proof that *every infinite subset of $\mathbb{R}$ is external* (Section 11.7), by using Exercise 1 in place of the internal set definition principle.

12.3 Hyperfinite Sets

If A_n is finite for (almost) all $n \in \mathbb{N}$, then $[A_n]$ may nevertheless be infinite (and then in fact uncountable!) but will have many properties that are similar to those of finite sets. Thus an internal set $A = [A_n]$ is called *hyperfinite* if almost all A_n's are finite, i.e., if

$$\{n \in \mathbb{N} : A_n \text{ is finite}\} \in \mathcal{F}.$$

In that case, by 11.2(3) we may as well assume that all A_n's are finite and have finite integer size $|A_n|$. The *internal cardinality* (or *size*) of A is then defined to be the hyperinteger

$$|A| = [\langle |A_n| : n \in \mathbb{N} \rangle].$$

(More succinctly, $|[A_n]| = [|A_n|]$.) For example:

- Let $A_n = \{1, \ldots, n\} \subseteq \mathbb{N}$. The resulting hyperfinite set A includes $\mathbb{N}$. Being internal, it must therefore be an uncountable subset of *$\mathbb{N}$. To see that $\mathbb{N} \subseteq A$, observe that if $m \in \mathbb{N}$, then the set $\{n \in \mathbb{N} : m \in A_n\}$ is cofinite, being equal to $\{m, m+1, \ldots\}$, so belongs to $\mathcal{F}$. Hence *$m \in A$.

- Refining the previous example, we see that if B is any countable subset of $\mathbb{R}$, then there exists a hyperfinite set A with $B \subseteq A \subseteq {}^*B$.

For if $B = \{x_n : n \in \mathbb{N}\}$, let $A = [A_n]$ where $A_n = \{x_1, \ldots, x_n\}$. In this case the internal size of A is $\omega = [\langle 1, 2, 3, \ldots \rangle]$.

Later we will see that the restriction to countability here can be removed: *any* subset B of $\mathbb{R}$ has a "hyperfinite approximation" A satisfying $B \subseteq A \subseteq {}^*B$ (cf. Sections 14.1 and 14.2).

- Any finite set of hyperreals is hyperfinite: as observed in Section 11.1, if $X = \{[r_n^1], \ldots, [r_n^k]\} \subseteq {}^*\mathbb{R}$, then X is the hyperfinite set $[A_n]$, where $A_n = \{r_n^1, \ldots, r_n^k\}$.

- If $N = [N_n] \in {}^*\mathbb{N}$, then the set
$$\{k \in {}^*\mathbb{N} : k \leq N\} = \{1, 2, \ldots, N\}$$
discussed in Section 11.1 is hyperfinite and has internal cardinality N, since it is equal to $[A_n]$, where $A_n = \{1, 2, \ldots, N_n\}$ and $|A_n| = N_n$.

- If $N = [N_n] \in {}^*\mathbb{N}$, then the set
$$\left\{\frac{k}{N} : k \in {}^*\mathbb{Z} \text{ and } 0 \leq k \leq N\right\} = \left\{0, \frac{1}{N}, \frac{2}{N}, \ldots, \frac{N-1}{N}, 1\right\}$$
is hyperfinite of internal cardinality $N + 1$, since it is equal to $[A_n]$, where
$$A_n = \left\{0, \frac{1}{N_n}, \frac{2}{N_n}, \ldots, \frac{N_n - 1}{N_n}, 1\right\}.$$

- The last example is a special case of the fact that for any hyperreals a, b, and any $N \in {}^*\mathbb{N}$, the uniform partition
$$\left\{a + k\frac{(b-a)}{N} : k \in {}^*\mathbb{Z} \text{ and } 0 \leq k \leq N\right\}$$
is hyperfinite of internal cardinality $N + 1$.

12.4 Exercises on Hyperfiniteness

(1) Any hyperfinite set has a greatest and a least element.

(2) The union and intersection of any two hyperfinite sets X and Y are hyperfinite, with
$$|X \cup Y| = |X| + |Y| - |X \cap Y|.$$

(3) Any internal subset of a hyperfinite set is hyperfinite.

12.5 Counting a Hyperfinite Set

The above results are indicative of ways in which hyperfinite sets behave like finite sets. More fundamentally, a finite set can be defined as one that has n elements for some $n \in \mathbb{N}$, and so is in bijective correspondence with the set $\{1,\ldots,n\}$. Correspondingly, for hyperfinite sets we have

Theorem 12.5.1 *An internal set A is hyperfinite with internal cardinality N if and only if there is an internal bijection $f : \{1,\ldots,N\} \to A$.*

Proof. Let $A = [A_n]$. If A is hyperfinite with internal cardinality $N = [N_n]$, then we may suppose that for each $n \in \mathbb{N}$, A_n is a finite set of cardinality N_n. Thus there is a bijection $f_n : \{1,\ldots,N_n\} \to A_n$. Let $f = [f_n]$. Then f is an internal function with domain $\{1,\ldots,N\}$ that is injective (12.2(4)) and has range A (12.2(1)).

Conversely, suppose that $f = [f_n]$ is an internal bijection from $\{1,\ldots,N\}$ onto A. Then

$$[\text{dom } f_n] = \text{dom } [f_n] = \{1,\ldots,N\} = [\{1,\ldots,N_n\}],$$

so for $\mathcal{F}$-almost all n,

$$\text{dom } f_n = \{1,\ldots,N_n\}. \tag{i}$$

Also, as A is the image of $\{1,\ldots,N\}$ under $[f_n]$, Exercise 12.2(1) implies that $A = [f_n(\{1,\ldots,N_n\})]$, so

$$f_n(\{1,\ldots,N_n\}) = A_n \tag{ii}$$

for $\mathcal{F}$-almost all n. Finally, by 12.2(4),

$$f_n \text{ is injective} \tag{iii}$$

for $\mathcal{F}$-almost all n. Then the set J of those $n \in \mathbb{N}$ satisfying (i)–(iii) must belong to $\mathcal{F}$. But for $n \in J$, A_n is finite of cardinality N_n. Hence A is hyperfinite of internal cardinality N. □

An important feature of this result is that it gives a characterisation of hyperfinite sets that makes no reference to the ultrafilter $\mathcal{F}$, but requires only the hypernatural numbers *$\mathbb{N}$ and the notion of an internal function. This approach will be revisited in Section 13.17.

12.6 Hyperfinite Pigeonhole Principle

One classical way to distinguish the finite from the infinite is to characterise the infinite sets as those that are equinumerous with a proper subset of themselves. Thus A is infinite iff there is an injection $f : A \to A$ whose range $f(A)$ is a proper subset of A. Equivalently, A is finite iff every injection $f : A \to A$ mapping A into itself is surjective, i.e., has $f(A) = A$ (this latter statement is known as the *pigeonhole principle*). Correspondingly, we have the following characterisation of hyperfiniteness.

Theorem 12.6.1 *An internal set $A = [A_n]$ is hyperfinite if and only if every injective **internal** function f whose domain includes A and has $f(A) \subseteq A$ must in fact have $f(A) = A$.*

Proof. Suppose A is hyperfinite. Let $f = [f_n]$ be an internal injective function with $A \subseteq \text{dom } f$ and $f(A) \subseteq A$. Then each of the following is true for $\mathcal{F}$-almost all $n \in \mathbb{N}$:

A_n is finite,
$A_n \subseteq \text{dom } f_n$,
$f_n(A_n) \subseteq A_n$,   cf. 12.2(1), 11.2(2)
f_n is injective.

Thus the set J of those $n \in \mathbb{N}$ satisfying all of these conditions must belong to $\mathcal{F}$. But
$$J \subseteq \{n \in \mathbb{N} : f_n(A_n) = A_n\}$$
by the standard pigeonhole principle, and so $f(A) = [f_n(A_n)] = [A_n] = A$.

For the converse, suppose A is not hyperfinite. It follows that
$$J' = \{n \in \mathbb{N} : A_n \text{ is infinite}\} \in \mathcal{F}.$$
But for each $n \in J'$ there is an injective function $f_n : A_n \to A_n$ and some $r_n \in A_n - f_n(A_n)$. Let $f = [f_n]$. This makes f an internal function with domain A that is injective (12.2(4)) and has $f(A) = [f_n(A_n)] \subseteq A$, while $[r_n] \in A - f(A)$ and so $f(A) \neq A$. □

Here now is an example of a noninternal function:
$$f(n) = \begin{cases} 2n & \text{if } n \in \mathbb{N}, \\ n & \text{if } n \notin \mathbb{N}. \end{cases}$$
This function maps the internal set $^*\mathbb{N}$ injectively into, but not onto, itself. Hence by the hyperfinite pigeonhole principle (Theorem 12.6.1), f cannot be internal.

12.7 Integrals as Hyperfinite Sums

The operation of forming the sum of finitely many numbers can be extended to hyperfinitely many. More generally, we can define the sum over a hyperfinite set of the values of an internal function.

To see this, if X is a finite set, let the symbol
$$\sum_{x \in X} g(x)$$
denote the sum of the members of $\{g(x) : x \in X\}$. Then if $A = [A_n]$ is a hyperfinite set included in the domain of an internal function $f = [f_n]$, define $\sum_{x \in A} f(x)$ to be the hyperreal number $[r_n]$ given by
$$r_n = \sum_{x \in A_n} f_n(x).$$
This makes sense because for $\mathcal{F}$-almost all n we have A_n a finite subset of dom f_n. Thus
$$\sum_{x \in [A_n]} [f_n](x) = \left[\sum_{x \in A_n} f_n(x)\right].$$
This operation has many of the properties familiar from finite summations: if f, g are internal functions, A, B are hyperfinite sets, and $c \in {}^*\mathbb{R}$, then:

- $\sum_{x \in A} cf(x) = c\left(\sum_{x \in A} f(x)\right).$
- $\sum_{x \in A} f(x) + g(x) = \sum_{x \in A} f(x) + \sum_{x \in A} g(x).$
- $\sum_{x \in A \cup B} f(x) = \sum_{x \in A} f(x) + \sum_{x \in B} f(x)$ if A and B are disjoint.
- $\sum_{x \in A} f(x) \leq \sum_{x \in A} g(x)$ if $f(x) \leq g(x)$ on A.

These are analogues of familiar properties of *integrals* (cf. Section 9.3). We will now see that standard integrals can be realised as hyperfinite Riemann sums over hyperfinite partitions.

Let $f : [a, b] \to \mathbb{R}$ be an integrable function on the closed interval $[a, b] \subseteq \mathbb{R}$. Take a positive infinitesimal $\Delta x = [\varepsilon_n]$. Then for each $n \in \mathbb{N}$ we may assume that ε_n is a positive real number less than $b - a$. Let $P_n \cup \{b\}$ be the finite partition of $[a, b]$ into subintervals of width ε_n. Thus if P_n is of size N_n, we have the description
$$P_n = \{a + k\varepsilon_n : k \in \mathbb{Z} \text{ and } 0 \leq k < N_n\}.$$
Let P be the hyperfinite set $[P_n]$, of internal size $N = [N_n] \in {}^*\mathbb{N}$. Then in fact,
$$P = \{a + k\Delta x : k \in {}^*\mathbb{Z} \text{ and } 0 \leq k < N\},$$
so $P \cup \{b\}$ is a hyperfinite partition of $[a, b]$ into subintervals of infinitesimal width Δx.

Now, the original function f lifts to the internal function $[f_n] : {}^*[a, b] \to {}^*\mathbb{R}$ determined by the constant sequence of functions $f_n = f$. We continue to use the symbol "f" for this extended function. Its domain includes P, so the hyperfinite sum $\sum_{x \in P} f(x)$ is specified as the hyperreal number
$$\left[\left\langle \sum_{x \in P_1} f(x), \sum_{x \in P_2} f(x), \ldots \right\rangle\right].$$

Thus
$$\sum_{x\in P} f(x) = \left[\sum_{x\in P_n} f(x)\right].$$
The ordinary Riemann sum for the real partition $P_n \cup \{b\}$ was defined in Section 9.1 as the number
$$S_a^b(f, \varepsilon_n) = \sum_{x\in P_n} f(x)\varepsilon_n = \left(\sum_{x\in P_n} f(x)\right)\varepsilon_n.$$
But the sequence of numbers $\langle S_a^b(f, \varepsilon_n) : n \in \mathbb{N}\rangle$ determines a hyperreal, which by definition is the extension of the function $S_a^b(f, -)$ to the hyperreal $[\varepsilon_n] = \Delta x$:
$$S_a^b(f, \Delta x) = [S_a^b(f, \varepsilon_n)].$$
Thus we calculate
$$\begin{aligned}S_a^b(f, \Delta x) &= \left[\left(\sum_{x\in P_n} f(x)\right)\varepsilon_n\right] \\ &= \left[\sum_{x\in P_n} f(x)\right][\varepsilon_n] \\ &= \left(\sum_{x\in P} f(x)\right)\Delta x \\ &= \sum_{x\in P} f(x)\Delta x,\end{aligned}$$
showing that the hyperreal number $S_a^b(f, \Delta x)$, defined formally by the extension process of Section 3.13, can be viewed as the (extended) ordinary Riemann sum of the hyperfinite partition P.

Finally, from the analysis in Section 9.2 of the Riemann integral as a shadow of Riemann sums we get that for any positive infinitesimal Δx,
$$\int_a^b f(x)\,dx = \operatorname{sh}\left(S_a^b(f, \Delta x)\right) = \operatorname{sh}\left(\sum_{x\in P} f(x)\Delta x\right).$$

Exercise 12.7.1
Verify that each member of the hyperfinite set P has the form $a + k\Delta x$ for some nonnegative hyperinteger $k < N$.

Part IV

Nonstandard Frameworks

13
Universes and Frameworks

The discussion of internal sets and functions in the previous two chapters raises some fundamental conceptual issues:

- In proving internal versions of induction, the least number principle, order-completeness, etc., we reverted once more to ultrafilter calculations. Could we instead obtain these results by a logical transfer principle, involving an extended version of the formal language of Chapter 4? A limited extension of this kind is provided by the statement (†) of Section 11.7, but perhaps this can be taken further by using a more powerfully expressive language that would allow the quantifiers $\forall, \exists$ to range over collections of sets or functions rather than just collections of numbers (cf. Section 4.7.)

- Now that we see how to identify certain subsets and functions in *$\mathbb{R}$ as being *internal*, can we do the same for other more complex entities? Are there internal topologies on *$\mathbb{R}$? Or internal measures? If $A \subseteq$ *$\mathbb{R}$ is hyperfinite of internal cardinality N, does it follow that the power set of A is hyperfinite of cardinality 2^N? Or is it the collection of *internal* subsets of A that should be hyperfinite? This would seem to require the notion of an internal function of the type $\{1, \ldots, 2^N\} \to \mathcal{P}(A)$.

It is time in fact to consider just how widely the methodology we have been developing can be applied. To address this we will work with the entire set-theoretic universe that can be erected on a set like $\mathbb{R}$ by forming sets of sets, sets of sets of sets, etc., and then consider how this universe may be "enlarged" to admit nonstandard entities, by analogy with the enlargement of $\mathbb{R}$ to $^*\mathbb{R}$. Ultimately this will provide a framework that allows the methodology of nonstandard analysis to be applied to any kind of mathematical structure (function spaces, measure spaces, infinite-dimensional Hilbert spaces, ...). It will also cause us to review what we have been doing so far from a more abstract set-theoretic standpoint.

13.1 What Do We Need in the Mathematical World?

In developing a mathematical theory, or analysing a particular structure, access may be needed to a wide range of entities: sets, members of sets, sequences, relations, functions, etc. We will posit the existence of a "universe" $\mathbb{U}$ that contains all such entities that might be required. This will have an associated formal language $\mathcal{L}_\mathbb{U}$ whose sentences express properties of the members of $\mathbb{U}$. Then $\mathbb{U}$ will be enlarged to another universe $^*\mathbb{U}$ that contains certain new (nonstandard) entities whose behaviour can be used to establish results about $\mathbb{U}$ by the use of transfer and other principles.

Here now is some more detailed discussion about the entities and closure properties that $\mathbb{U}$ should have.

- *Individuals.* Although a real number might be viewed as a set of Cauchy sequences, or a pair of sets of rationals (Section 1.3), when studying real analysis we generally regard real numbers as *individuals*, i.e., as "points" or entities that have no internal structure. The same applies to the basic elements of any other structure that might concern us, be they elements of an algebraic number field, complex numbers, vectors in some Hilbert space, and so on.

  The universe $\mathbb{U}$ will contain a set $\mathbb{X}$ of entities that are viewed as individuals in this way. An element of $\mathbb{X}$ will be taken to have no members within $\mathbb{U}$. It will be assumed that $\mathbb{R} \subseteq \mathbb{X}$.

- *Functions.* If two sets A and B belong to $\mathbb{U}$, then we may wish to have all functions $f : A \to B$ available in $\mathbb{U}$, along with the range of f, the f-image $f(C) \subseteq B$ of any $C \subseteq A$, and the inverse image of any subset of B under f. Moreover, the set B^A of all functions from A to B should itself be in $\mathbb{U}$.

  Also, we should be able to compose functions in $\mathbb{U}$.

- *Relations.* An m-ary relation is a set of m-tuples $\langle a_1, \ldots, a_m \rangle$, and is usually presented as a subset of some Cartesian product $A_1 \times \cdots \times A_m$, the latter being the set of *all* such m-tuples that have

$a_1 \in A_1, \ldots, a_m \in A_m$. Thus $\mathbb{U}$ should be closed under the formation of tuples, and of Cartesian products and their subsets.

For binary relations ($m = 2$) the domain and range should be available, and the operations of composing and inverting relations should be possible within our universe.

- *Set Operations.* All the usual set operations of intersection $A \cap B$, union $A \cup B$, difference $A - B$, and power set $\mathcal{P}(A)$, when performed on sets in $\mathbb{U}$, should produce entities that belong to $\mathbb{U}$. In fact, some important constructions will require the union $\bigcup Y$ and intersection $\bigcap Y$ of any (possibly infinite) collection $Y \in \mathbb{U}$ to be available. Also, if a set A belongs to $\mathbb{U}$, then all subsets of A should too.

- *Transitivity.* If a set A is in $\mathbb{U}$, we will want all members of A to be present in $\mathbb{U}$ as well, i.e., $A \subseteq \mathbb{U}$. This condition is usually called *transitivity* of $\mathbb{U}$, because it takes the form

$$a \in A \in \mathbb{U} \quad \text{implies} \quad a \in \mathbb{U}.$$

This has an important bearing on the interpretation of a bounded quantifier ($\forall x \in A$). We naturally read this as "for all x in A", but when used to express a property of an entity of $\mathbb{U}$, there is a potential issue as to whether this means "for all x in A that belong to $\mathbb{U}$", or whether the variable x is ranging over all members of A absolutely. When $\mathbb{U}$ is transitive, this is not an issue: the members of A that belong to $\mathbb{U}$ are simply all the members of A that there are. Transitivity thus ensures that quantified variables always range over members of $\mathbb{U}$ when given their natural interpretation.

Subset and Relation Closure

Transitivity of $\mathbb{U}$ together with closure under the power set operation will guarantee that $\mathbb{U}$ has the property mentioned above of closure under subsets of its members. For then if $A \subseteq B \in \mathbb{U}$, we get $A \in \mathcal{P}(B) \in \mathbb{U}$, and hence $A \in \mathbb{U}$ by transitivity.

Then closure of $\mathbb{U}$ under Cartesian products will lead to closure under relations between given sets in general. Thus if A, B are sets in $\mathbb{U}$ and $R \subseteq A \times B$, then if $A \times B \in \mathbb{U}$, it follows that $R \in \mathbb{U}$ by the argument just given for subset closure.

13.2 Pairs Are Enough

The more we assume about the entities that exist and constructions that can be performed within $\mathbb{U}$, the more powerful will be this universe as a tool for applications. On the other hand, for demonstrating properties of U itself or showing that it exists (and *U does too), it is desirable to have very few primitive concepts, so that we can minimize the number of cases and the amount and complexity of work required in carrying out proofs.

Studies of the foundations of mathematics have shown that these opposing tendencies can be effectively balanced by basing our conceptual framework on set theory. To see this we will first show that apart from purely set-theoretic operations, the other notions just described in Section 13.1 can be reduced to the construction of sets of ordered pairs:

- *Functions.* A function $f : A \to B$ can be identified with the set of pairs
$$\{\langle a, b \rangle : b = f(a)\},$$
which is a subset of the Cartesian product set $A \times B$. Set-theoretically, we *define* a function from A to B to be a set f of pairs satisfying
  (i) if $\langle a, b \rangle \in f$ then $a \in A$ and $b \in B$;
  (ii) if $\langle a, b \rangle, \langle a, c \rangle \in f$, then $b = c$ (functionality);
  (iii) for each $a \in A$ there exists $b \in B$ with $\langle a, b \rangle \in f$ (the domain of f is A).

- *m-Tuples.* Given a construction for ordered pairs (2-tuples), the case $m > 2$ can be handled by defining
$$\langle a_1, \ldots, a_m \rangle = \{\langle 1, a_1 \rangle, \ldots, \langle m, a_m \rangle\}.$$
Thus an m-tuple becomes a set of ordered pairs (and actually is a function with domain $\{1, \ldots, m\}$).

  Note: an alternative approach would be to inductively put
$$\langle a_1, \ldots, a_{m+1} \rangle = \langle \langle a_1, \ldots, a_m \rangle, a_{m+1} \rangle,$$
so that an m-tuple beomes a pair of pairs of $\cdots$ of pairs. This works just as well, but would be more complex set-theoretically than the definition given.

- *Relations.* An m-ary relation is a set of m-tuples $\langle a_1, \ldots, a_m \rangle$, and hence becomes a set of sets of ordered pairs. The Cartesian product $A_1 \times \cdots \times A_m$ is a particular case of this, being the set of *all* such m-tuples that have $a_1 \in A_1, \ldots, a_m \in A_m$.

13.3 Actually, Sets Are Enough

But what is an ordered pair? Well, one of the most effective ways to explain a mathematical concept is to give an account of when two instances of the concept are equal, and for ordered pairs the condition is that

$$\langle a, b \rangle = \langle c, d \rangle \quad \text{iff} \quad a = c \text{ and } b = d.$$

In fact, this condition is all that is ever needed in handling pairs, and it can be fulfilled by putting

$$\langle a, b \rangle = \{\,\{a\}, \{a, b\}\,\}.$$

In this way pairs are represented as certain sets, and therefore so too are m-tuples, relations, and functions. When it comes to the study of a particular structure whose elements belong to some given set $\mathbb{X}$, all the entities we need can be obtained by applying set theory to $\mathbb{X}$. This demonstrates the power and elegance of set theory, and explains the sense in which it provides a foundation for mathematics.

Exercise 13.3.1

(i) Verify that $\{\,\{a\}, \{a, b\}\,\} = \{\,\{c\}, \{c, d\}\,\}$ iff $a = c$ and $b = d$.

(ii) Show that for $m \geq 2$,

$$\langle a_1, \ldots, a_m \rangle = \langle b_1, \ldots, b_m \rangle \quad \text{iff} \quad a_1 = b_1, \ldots, a_m = b_m.$$

Product Closure

Closure of $\mathbb{U}$ under Cartesian products can now be derived set-theoretically from transitivity and closure under unions and power sets. If $A, B \in \mathbb{U}$ and $\langle a, b \rangle \in A \times B$, then both $\{a\}$ and $\{a, b\}$ are subsets of $A \cup B$, i.e., members of $\mathcal{P}(A \cup B)$. Hence

$$\langle a, b \rangle = \{\,\{a\}, \{a, b\}\,\} \in \mathcal{PP}(A \cup B).$$

This shows that $A \times B \subseteq \mathcal{PP}(A \cup B)$, and so $A \times B \in \mathcal{PPP}(A \cup B)$. Closure under $\cup$ and $\mathcal{P}$ and transitivity of $\mathbb{U}$ then give $A \times B \in \mathbb{U}$.

13.4 Strong Transitivity

Before giving the axioms for a universe, there is a further important property to be explained, which we do with the following example.

If a binary relation R belongs to $\mathbb{U}$, then its domain $\operatorname{dom} R$ should be available in $\mathbb{U}$ as well. Now, if $a \in \operatorname{dom} R$, then there is some entity b with $\langle a, b \rangle \in R$. According to our new definition of pairs, we then have the "membership chain"

$$a \in \{a\} \in \langle a, b \rangle \in R \in \mathbb{U}.$$

Transitivity of $\mathbb{U}$ will ensure that it is closed downwards under such membership chains, giving $a \in \mathbb{U}$. But this leads only to the conclusion that $\operatorname{dom} R \subseteq \mathbb{U}$, whereas we want $\operatorname{dom} R \in \mathbb{U}$. Is $\operatorname{dom} R$ perhaps too "big" to be an element of $\mathbb{U}$?

Now, if R itself were transitive, we would get $a \in R$, showing $\operatorname{dom} R \subseteq R \in \mathbb{U}$, from which our desired conclusion would result by subset closure. But of course R need not be transitive. On the other hand, it is reasonable to suppose that R can be extended to a transitive set B that belongs to $\mathbb{U}$ (i.e., $R \subseteq B \in \mathbb{U}$). Then we can reason that $\operatorname{dom} R \subseteq B \in \mathbb{U}$, leading to $\operatorname{dom} R \in \mathbb{U}$, as desired, by subset closure.

The justification for this is that any set A has a *transitive closure* $Tr(A)$, whose members are precisely the members of members of $\cdots$ of members of A. $Tr(A)$ is the smallest transitive set that includes A: any transitive set including A will include $Tr(A)$. We are going to require that $\mathbb{U}$ be "big enough" to have room for the transitive closure of any set $A \in \mathbb{U}$. For this to hold it is enough that some transitive set including A belong to $\mathbb{U}$. Thus our requirement is

- **Strong Transitivity:** *for any set A in $\mathbb{U}$ there exists a transitive set $B \in \mathbb{U}$ with $A \subseteq B \subseteq \mathbb{U}$.*

Note that the stipulation that $B \subseteq \mathbb{U}$ is superfluous if $\mathbb{U}$ is transitive, since it then follows from $B \in \mathbb{U}$. But the definition of strong transitivity itself implies that $\mathbb{U}$ is transitive (since we get $A \subseteq \mathbb{U}$ when $A \in \mathbb{U}$ because $A \subseteq B \subseteq \mathbb{U}$), so this single statement captures all that is needed.

In a strongly transitive $\mathbb{U}$ we can assume that any set we are dealing with is located within a large transitive set. This will be the "key to the universe", as will become apparent.

13.5 Universes

In the light of the foregoing discussion, we now define a *universe* to be any *strongly transitive* set $\mathbb{U}$ such that

- if $a, b \in \mathbb{U}$, then $\{a, b\} \in \mathbb{U}$;

- if A and B are sets in $\mathbb{U}$, then $A \cup B \in \mathbb{U}$;

- if A is a set in $\mathbb{U}$, then $\mathcal{P}(A) \in \mathbb{U}$.

Such a $\mathbb{U}$ will be called a *universe over* $\mathbb{X}$ if $\mathbb{X}$ is a set that belongs to $\mathbb{U}$ ($\mathbb{X} \in \mathbb{U}$), and the members of $\mathbb{X}$ are regarded as *individuals* that are not sets and have no members:

$$(\forall x \in \mathbb{X})\,[x \neq \emptyset \wedge (\forall y \in \mathbb{U})\,(y \notin x)].$$

13.5 Universes

It will always be assumed further that a universe contains at least one set, and also contains the positive integers $1, 2, \ldots$ to ensure that m-tuple formation can be carried out. In practice we will be using universes that have $\mathbb{R} \in \mathbb{U}$, with each member of $\mathbb{R}$ being an individual, so these conditions will hold.

Here now is a list of the main closure properties of such universes, many of which have been indicated already. Uppercase letters A, B, A_i, etc. are reserved for members of $\mathbb{U}$ that are *sets*.

Set Theory

- If $a \in \mathbb{U}$, then $\{a\} \in \mathbb{U}$.
- $A_1, \ldots, A_m \in \mathbb{U}$ implies $A_1 \cup \cdots \cup A_m \in \mathbb{U}$.
- $\mathbb{U}$ contains all its finite subsets: if $A \subseteq \mathbb{U}$ and A is finite, then $A \in \mathbb{U}$.
- $A \subseteq B \in \mathbb{U}$ implies $A \in \mathbb{U}$.
- $\emptyset \in \mathbb{U}$.
- If $\{A_i : i \in I\} \subseteq A \in \mathbb{U}$, then $\bigcup_{i \in I} A_i \in \mathbb{U}$. (Note: this uses strong transitivity.)
- $\mathbb{U}$ is closed under unions of sets of sets: if $B = \{A_i : i \in I\} \in \mathbb{U}$ and each A_i is a set, then $\bigcup B = \bigcup_{i \in I} A_i \in \mathbb{U}$.
- $\mathbb{U}$ is closed under arbitrary intersections: if $\{A_i : i \in I\} \subseteq \mathbb{U}$, then $\bigcap_{i \in I} A_i \in \mathbb{U}$, whether or not the set $\{A_i : i \in I\}$ itself belongs to $\mathbb{U}$.

Relations and Functions

- If $a, b \in \mathbb{U}$, then $\langle a, b \rangle \in \mathbb{U}$.
- If $A, B \in \mathbb{U}$ and $R \subseteq A \times B$, then $R \in \mathbb{U}$.
- If $a_1, \ldots, a_m \in \mathbb{U}$ $(m > 2)$, then $\langle a_1, \ldots, a_m \rangle \in \mathbb{U}$.
- $\mathbb{U}$ is closed under finitary relations: if $A_1, \ldots, A_m \in \mathbb{U}$ and $R \subseteq A_1 \times \cdots \times A_m$, then $R \in \mathbb{U}$.
- If $R \in \mathbb{U}$ is a binary relation, then $\mathbb{U}$ contains the *domain* $\operatorname{dom} R$, the *range* $\operatorname{ran} R$, the *R-image* $R`(C)$ of any $C \subseteq \operatorname{dom} R$, and the *inverse* R^{-1}, where

$$\operatorname{dom} R = \{a : \exists b\, (\langle a, b \rangle \in R)\},$$
$$\operatorname{ran} R = \{b : \exists a\, (\langle a, b \rangle \in R)\},$$

$$R`(C) = \{b : \exists a \in C (\langle a,b\rangle \in R)\},$$
$$R^{-1} = \{\langle b,a\rangle : \langle a,b\rangle \in R\}.$$

- If $R, S \in \mathbb{U}$ are binary relations, then $\mathbb{U}$ contains their *composition*
$$R \circ S = \{\langle a,c\rangle : \exists b(\langle a,b\rangle \in R \text{ and } \langle b,c\rangle \in S)\}.$$

- If $f : A \to B$ is a function with $A, B \in \mathbb{U}$, then $f \in \mathbb{U}$. Moreover, for any $C \subseteq A$ and $D \subseteq B$, $\mathbb{U}$ contains the *image*
$$f`(C) = \{f(a) : a \in C\}$$
and the *inverse image*
$$f^{-1}(D) = \{a \in A : f(a) \in D\}.$$

- If $A, B \in \mathbb{U}$, then the set B^A of all functions from A to B belongs to $\mathbb{U}$.

- If $\{A_i : i \in I\} \in \mathbb{U}$ and $I \in \mathbb{U}$, then $\left(\prod_{i \in I} A_i\right) \in \mathbb{U}$.

13.6 Superstructures

It is time to demonstrate that there are such things as universes. Let $\mathbb{X}$ be a set with $\mathbb{R} \subseteq \mathbb{X}$. The *nth cumulative power set* $\mathbb{U}_n(\mathbb{X})$ of $\mathbb{X}$ is defined inductively by
$$\mathbb{U}_0(\mathbb{X}) = \mathbb{X},$$
$$\mathbb{U}_{n+1}(\mathbb{X}) = \mathbb{U}_n(\mathbb{X}) \cup \mathcal{P}(\mathbb{U}_n(\mathbb{X})),$$
so that
$$\mathbb{U}_0(\mathbb{X}) \subseteq \mathbb{U}_1(\mathbb{X}) \subseteq \cdots \subseteq \mathbb{U}_n(\mathbb{X}) \subseteq \cdots.$$
The *superstructure* over $\mathbb{X}$ is the union of all these cumulative power sets:
$$\mathbb{U}(\mathbb{X}) = \bigcup_{n=0}^{\infty} \mathbb{U}_n(\mathbb{X}).$$

The *rank* of an entity a is the least n such that $a \in \mathbb{U}_n(\mathbb{X})$. The rank 0 entities (members of $\mathbb{X}$) will be regarded as *individuals*:
$$(\forall x \in \mathbb{U}_0(\mathbb{X})) \, [x \neq \emptyset \wedge (\forall y \in \mathbb{U}(\mathbb{X})) \, (y \notin x)].$$

All other members of $\mathbb{U}(\mathbb{X})$ (those with positive rank) are sets, and so $\mathbb{U}(\mathbb{X})$ has just these two types of entity. We can show:

(1) $\mathbb{U}_{n+1}(\mathbb{X}) = \mathbb{X} \cup \mathcal{P}(\mathbb{U}_n(\mathbb{X}))$.

(2) $\mathbb{U}_n(\mathbb{X}) \in \mathbb{U}_{n+1}(\mathbb{X})$. Hence $\mathbb{U}_n(\mathbb{X}) \in \mathbb{U}(\mathbb{X})$, and in particular, $\mathbb{X} \in \mathbb{U}(\mathbb{X})$.

(3) $\mathbb{U}_{n+1}(\mathbb{X})$ is transitive. Indeed, $a \in B \in \mathbb{U}_{n+1}(\mathbb{X})$ implies $a \in \mathbb{U}_n(\mathbb{X})$.

(4) If $a, b \in \mathbb{U}_n(\mathbb{X})$, then $\{a, b\} \in \mathbb{U}_{n+1}(\mathbb{X})$.

(5) If $A, B \in \mathbb{U}_n(\mathbb{X})$, then $A \cup B \in \mathbb{U}_{n+1}(\mathbb{X})$.

(6) $A \in \mathbb{U}_n(\mathbb{X})$ implies $\mathcal{P}(A) \in \mathbb{U}_{n+2}(\mathbb{X})$.

From (3) it follows that $\mathbb{U}(\mathbb{X})$ is strongly transitive, since every element of $\mathbb{U}(\mathbb{X})$ belongs to some $\mathbb{U}_{n+1}(\mathbb{X})$. Properties (4)–(6) then ensure that $\mathbb{U}(\mathbb{X})$ is a universe, and by (2) it is a universe over $\mathbb{X}$.

In fact, $\mathbb{U}(\mathbb{X})$ is the smallest universe containing $\mathbb{X}$, in the sense that if any universe $\mathbb{U}$ has $\mathbb{X} \in \mathbb{U}$, then $\mathbb{U}(\mathbb{X}) \subseteq \mathbb{U}$. Another description of this superstructure over $\mathbb{X}$ is that it is the smallest transitive set that contains $\mathbb{X}$ and is closed under binary unions $A \cup B$ and power sets $\mathcal{P}(A)$.

Exercise 13.6.1
Verify results (1)–(6) above, and the observations that follow them. Show further that if $\mathbb{X} \subseteq \mathbb{Y}$, then $\mathbb{U}(\mathbb{R}) \subseteq \mathbb{U}(\mathbb{X}) \subseteq \mathbb{U}(\mathbb{Y})$. □

A universe is not closed under arbitrary subsets: if $A \subseteq \mathbb{U}$, it need not follow that $A \in \mathbb{U}$ (e.g., consider $A = \mathbb{U}$). In the case of a superstructure, A will belong to $\mathbb{U}(\mathbb{X})$ iff there is an upper bound $n \in \mathbb{N}$ on the ranks of the members of A, i.e., iff $A \subseteq \mathbb{U}_n(\mathbb{X})$ for some n. All the entities typically involved in studying the analysis of $\mathbb{X}$ can be obtained in $\mathbb{U}(\mathbb{X})$ using only rather low ranks. If $A, B \in \mathbb{U}_n(\mathbb{X})$, then any subset of $A \times B$, and in particular any function from A to B, is in $\mathbb{U}_{n+2}(\mathbb{X})$. So constructing a function between given sets increases the rank by at most 2. Using this, we see that:

- A topology on $\mathbb{X}$ is a subset of $\mathcal{P}(\mathbb{X})$, hence a subset of $\mathbb{U}_1(\mathbb{X})$, so belongs to $\mathbb{U}_2(\mathbb{X})$. Thus the set of all topologies on $\mathbb{X}$ is itself a member of $\mathbb{U}_3(\mathbb{X})$.

- An $\mathbb{R}$-valued measure on $\mathbb{X}$ is a function $\mu : \mathcal{A} \to \mathbb{R}$ with $\mathcal{A}$ a collection of subsets of $\mathbb{X}$, so $\mathcal{A}$ is of rank 2 and μ of rank 4. Thus the set of all measures on $\mathbb{X}$ is also an element of $\mathbb{U}(\mathbb{X})$, of rank 5.

- A metric on $\mathbb{X}$ is a function $d : \mathbb{X} \times \mathbb{X} \to \mathbb{R}$ of rank 5 (since $\mathbb{X} \times \mathbb{X}$ has rank 3). The set of all metrics on $\mathbb{X}$ has rank 6.

- The Riemann integral on a closed interval $[a, b]$ can be viewed as a function
$$\int_a^b : \mathcal{R}[a, b] \to \mathbb{R},$$

where $\mathcal{R}[a,b]$ is the set of integrable functions $f : [a,b] \to \mathbb{R}$. Such an f is of rank 3, since $[a,b]$ and $\mathbb{R}$ have rank 1, so $\mathcal{R}[a,b]$ has rank 4 and therefore the integral $\int_a^b$ is an entity of rank 6.

13.7 The Language of a Universe

Given a denumerable list of *variables*, a language $\mathcal{L}_\mathbb{U}$ associated with the universe $\mathbb{U}$ is generated as follows:

$\mathcal{L}_\mathbb{U}$-*Terms*

- Each variable is an $\mathcal{L}_\mathbb{U}$-term.

- Each member of $\mathbb{U}$ is a *constant* $\mathcal{L}_\mathbb{U}$-term.

- If $\tau_1, \ldots, \tau_m$ are $\mathcal{L}_\mathbb{U}$-terms ($m \geq 2$), then $\langle \tau_1, \ldots, \tau_m \rangle$ is an $\mathcal{L}_\mathbb{U}$-term, called a *tuple*.

- If τ and σ are $\mathcal{L}_\mathbb{U}$-terms, then $\tau(\sigma)$ is a *function-value* $\mathcal{L}_\mathbb{U}$-term.

Notice that our rules allow iterated formations of tuples of terms, such as

$$\langle \langle \tau, \sigma \rangle, \tau, \langle \tau_1, \ldots, \tau_m \rangle \rangle.$$

A term with no variables is *closed*, and will name a particular entity of $\mathbb{U}$ if it is defined (recall the discussion of undefined terms in Section 4.3.1).

The rules for determining when a closed tuple is defined, and what it names, are as follows:

- If $\tau_1, \ldots, \tau_m$ name elements $a_1, \ldots, a_m$, respectively, then $\langle \tau_1, \ldots, \tau_m \rangle$ names the the m-tuple $\langle a_1, \ldots, a_m \rangle$.

- $\langle \tau_1, \ldots, \tau_m \rangle$ is undefined if one of $\tau_1, \ldots, \tau_m$ is undefined.

For a closed function-value term, the rules are:

- If τ names a function f and σ names an entity a that belongs to the domain of f, then $\tau(\sigma)$ names the entity $f(a)$.

- $\tau(\sigma)$ is undefined if one of τ and σ is undefined, or if they are both defined but τ does not name a function, or if τ names a function but σ does not name a member of its domain.

The language $\mathcal{L}_\mathfrak{R}$ of Chapter 4 allowed formation of terms $f(\tau_1, \ldots, \tau_m)$ where f is an m-ary function. This is catered for here because of tuple formation. Any finitary function on $\mathbb{R}$ belongs to $\mathbb{U}$ because $\mathbb{R} \subseteq \mathbb{X}$, so f is a constant of $\mathcal{L}_\mathbb{U}$, and $f(\tau_1, \ldots, \tau_m)$ can be taken to be a simplified notation for the $\mathcal{L}_\mathbb{U}$-term $f(\langle \tau_1, \ldots, \tau_m \rangle)$. More generally, we can write $\sigma(\tau_1, \ldots, \tau_m)$ for $\sigma(\langle \tau_1, \ldots, \tau_m \rangle)$ when σ is an arbitrary $\mathcal{L}_\mathbb{U}$-term (this is in line with common practice: an m-ary function on a set A is just a one-placed function on A^m).

It follows that all $\mathcal{L}_\mathfrak{R}$-terms are $\mathcal{L}_\mathbb{U}$-terms.

Atomic $\mathcal{L}_\mathbb{U}$-Formulae

These have one of the forms

$$\tau = \sigma,$$
$$\tau \in \sigma,$$

where τ and σ are $\mathcal{L}_\mathbb{U}$-terms. For example, if $P \in \mathbb{U}$ is a k-ary relation, then there are atomic formulae

$$\langle \tau_1, \ldots, \tau_k \rangle \in P,$$

which may also be written $P(\tau_1, \ldots, \tau_k)$ as in the notation of Section 4.3, or, in the case $k = 2$, using infix notation, as in $\tau_1 < \tau_2$, $\tau_1 \neq \tau_2$, etc.

Since a function is a special kind of relation, symbols for functions may occur in atomic formulae in two ways. For example, the two formulae

$$f(\tau_1, \ldots, \tau_m) = \sigma \quad \text{and} \quad \langle \langle \tau_1, \ldots, \tau_m \rangle, \sigma \rangle \in f$$

have the same intended meaning.

Formulae

- Each atomic $\mathcal{L}_\mathbb{U}$-formula is an $\mathcal{L}_\mathbb{U}$-formula.

- If φ and ψ are $\mathcal{L}_\mathbb{U}$-formulae, then so are $\varphi \wedge \psi$, $\varphi \vee \psi$, $\neg \varphi$, $\varphi \to \psi$, $\varphi \leftrightarrow \psi$.

- If φ is an $\mathcal{L}_\mathbb{U}$-formula, then so are $(\forall x \in \tau)\varphi$ and $(\exists x \in \tau)\varphi$, where τ is any $\mathcal{L}_\mathbb{U}$-term and x is any variable symbol that does not occur in τ.

A **sentence** is, as usual, a formula in which every occurrence of a variable is within the scope of a quantifier for that variable.

If φ is a formula in which only the variable x occurs freely, and τ is a closed term denoting the set $A \in \mathbb{U}$, then the sentence $(\forall x \in \tau)\varphi$ asserts that $\varphi(a)$ is true for every $a \in A$, while $(\exists x \in \tau)\varphi$ asserts that $\varphi(a)$ is true for some such a. As explained in Section 13.1, transitivity of $\mathbb{U}$ ensures that quantified variables always range over members of $\mathbb{U}$ when given their natural interpretation.

As we will see later in applications of the language $\mathcal{L}_\mathbb{U}$ to mathematical reasoning, the term τ in a quantifier form $(\forall x \in \tau)$ or $(\exists x \in \tau)$ is usually a variable or a constant.

Having observed above that the $\mathcal{L}_\mathbb{U}$-terms include all $\mathcal{L}_\Re$-terms, and that $\mathcal{L}_\mathbb{U}$ allows the atomic formation $P(\tau_1, \ldots, \tau_k)$, we can now conclude that the $\mathcal{L}_\mathbb{U}$-formulae include all $\mathcal{L}_\Re$-formulae: any subset P of $\mathbb{R}$ is in $\mathbb{U}$, so the formation rules of $\mathcal{L}_\mathbb{U}$ admit the bounded quantifier forms $(\forall x \in P)$ and $(\exists x \in P)$.

13.8 Nonstandard Frameworks

Let $\mathbb{U} \xrightarrow{*} \mathbb{U}'$ be a mapping between two universes, taking each $a \in \mathbb{U}$ to an element $^*a$ of $\mathbb{U}'$. Then each $\mathcal{L}_\mathbb{U}$-term τ has an associated $*$-transform $^*\tau$, which is the $\mathcal{L}_{\mathbb{U}'}$-term obtained by replacing each constant symbol a by $^*a$.

A constant a occurring in an $\mathcal{L}_\mathbb{U}$-formula φ will do so as part of a term τ that appears either in an atomic formula or within one of the quantifier forms $(\forall x \in \tau)$ and $(\exists x \in \tau)$. Applying the replacement $a \mapsto {^*a}$ to all such constants transforms φ into an $\mathcal{L}_{\mathbb{U}'}$-formula $^*\varphi$. If φ is a sentence, then so too is $^*\varphi$.

A *nonstandard framework* for a set $\mathbb{X}$ comprises a universe $\mathbb{U}$ over $\mathbb{X}$ and a map $\mathbb{U} \xrightarrow{*} \mathbb{U}'$ satisfying:

- $^*a = a$ for all $a \in \mathbb{X}$.
- $^*\emptyset = \emptyset$.
- **Transfer:** an $\mathcal{L}_\mathbb{U}$-sentence φ is true if and only if $^*\varphi$ is true.

Such a map will be called a *universe embedding* or *transfer map*. It preserves many set-theoretic operations:

- $a = b$ iff $^*a = {^*b}$. Hence $a \mapsto {^*a}$ is injective.
- $a \in B$ iff $^*a \in {^*B}$.
- $A \subseteq B$ iff $^*A \subseteq {^*B}$.
- If $A \subseteq \mathbb{X}$, then $A \subseteq {^*A} \subseteq {^*\mathbb{X}}$. In particular, $\mathbb{X} \subseteq {^*\mathbb{X}}$.
- $^*(A \cap B) = {^*A} \cap {^*B}$.
- $^*(A \cup B) = {^*A} \cup {^*B}$.
- $^*(A - B) = {^*A} - {^*B}$.
- $^*\{a_1, \ldots, a_m\} = \{{^*a_1}, \ldots, {^*a_m}\}$. Thus $^*A = \{{^*a} : a \in A\}$ if A is finite.

13.8 Nonstandard Frameworks   169

- All members of *$\mathbb{X}$ are individuals in $\mathbb{U}'$.

- $\overset{*}{\to}$ preserves transitivity: if A is a transitive set in $\mathbb{U}$, then *A is transitive.

- *$\mathcal{P}(A) \subseteq \mathcal{P}($*$A)$.

- *$\langle a_1, \ldots, a_m \rangle = \langle $*$a_1, \ldots, $*$a_m \rangle$.

- $\langle a_1, \ldots, a_m \rangle \in R$ iff $\langle $*$a_1, \ldots, $*$a_m \rangle \in $*$R$.

- If $R \in \mathbb{U}$ is an m-ary relation, then so is *R.

- *$(A_1 \times \cdots \times A_m) = $*$A_1 \times \cdots \times $*$A_m$. Hence *$(A^m) = ($*$A)^m$ for $m \in \mathbb{N}$.

- If $R \in \mathbb{U}$ is a binary relation, then

$$\begin{aligned}
*(\operatorname{dom} R) &= \operatorname{dom} {}^*R, \\
*(\operatorname{ran} R) &= \operatorname{ran} {}^*R, \\
*(R^{-1}) &= (^*R)^{-1}, \\
*(R\text{'}(C)) &= (^*R)\text{'}(^*C) \quad \text{for } C \subseteq \operatorname{dom} R, \\
*(R^{-1}(C)) &= (^*R)^{-1}(^*C) \quad \text{for } C \subseteq \operatorname{ran} R.
\end{aligned}$$

- If R and S are binary relations, then *$(R \circ S) = $*$R \circ $*$S$.

- If a function $f : A \to B$ belongs to $\mathbb{U}$, then *f is a function from *A to *B, with *$(f(a)) = $*$f(^*a)$ for all $a \in A$. Also, f is injective/surjective iff *f is.

To show that $A \subseteq $*$A \subseteq $*$\mathbb{X}$ whenever $A \subseteq \mathbb{X}$, observe that if $A \subseteq \mathbb{X}$, then *$A \subseteq $*$\mathbb{X}$, and if $a \in A$, then $a \in \mathbb{X}$, and so $a = $*$a \in $*$A$. Also, by transfer (using *$\emptyset = \emptyset$) we have

$$(\forall x \in {}^*\mathbb{X}) [x \neq \emptyset \land \neg(\exists y \in x)(y \in x)]$$

true, so if $b \in $*$\mathbb{X}$, then b is not the empty set and has no members, and therefore is an individual.

For preservation of transitivity by $\overset{*}{\to}$, let $A \in \mathbb{U}$ be transitive, i.e.,

$$(\forall x \in A)(\forall y \in x) y \in A.$$

This transforms to

$$(\forall x \in {}^*A)(\forall y \in x) y \in {}^*A,$$

showing that any set belonging to *A is a subset of *A, i.e., *A is transitive as desired.

The fact that $^*\mathcal{P}(A) \subseteq \mathcal{P}(^*A)$ follows by transfer of

$$(\forall x \in \mathcal{P}(A))\,(\forall y \in x)\,(y \in A).$$

This shows that if $x \in {}^*\mathcal{P}(A)$, then $y \in x$ implies $y \in {}^*A$, and so $x \subseteq {}^*A$, whence $x \in \mathcal{P}({}^*A)$. The exact relationship between ${}^*\mathcal{P}(A)$ and $\mathcal{P}({}^*A)$ will be revealed in Section 13.12.

Exercise 13.8.1
Verify all the other properties of the transfer map listed above. □

If $\mathbb{R} \subseteq \mathbb{X}$, all of the standard operations and relations on $\mathbb{R}$ like $+, -, \times, |x|$, $\sin x, <, \geq, \neq$, etc. are entities in $\mathbb{U}$, and so have corresponding entities $^*+$, $^*\sin x$, $^*\neq$, etc. for $^*\mathbb{R}$ in $\mathbb{U}'$. We will continue the practice of dropping the $*$-prefix from such familiar notions when the intention is evident. However, while this is harmless for functions and relations between individuals, when entities of nonzero rank are involved, a transformed function $^*f$ need not agree with f where their domains overlap, so more caution is needed. In general, if $a \in \operatorname{dom} f$, then $^*f(^*a) = {}^*(f(a))$, but even when $^*a = a$ this will reduce to $^*f(a) = f(a)$ only when $^*(f(a))$ is equal to $f(a)$. For an example showing that this need not hold, let $f : \mathbb{R} \to \mathcal{P}(\mathbb{R})$ be defined by

$$f(r) = \{x \in \mathbb{R} : x > r\}.$$

For a given $r \in \mathbb{R}$, transfer of the sentence

$$(\forall x \in \mathbb{R})\,(x \in f(r) \leftrightarrow x > r)$$

shows (since $^*r = r$) that

$$^*f(r) = {}^*(f(r)) = \{x \in {}^*\mathbb{R} : x > r\}.$$

In particular, $f(0) = \mathbb{R}^+$, while $^*f(0) = {}^*\mathbb{R}^+$.

Exercise 13.8.2
If $\varphi(x_1, \ldots, x_m)$ is an $\mathcal{L}_{\mathbb{U}}$-formula and $A \in \mathbb{U}$, show that

$$^*\{\langle a_1, \ldots, a_m\rangle \in A^m : \varphi(a_1, \ldots, a_m)\} = \\ \{\langle b_1, \ldots, b_m\rangle \in {}^*A^m : {}^*\varphi(b_1, \ldots, b_m)\}.$$

Explain how various of the above results about preservation of properties by $\stackrel{*}{\to}$ can be derived from this general fact.

13.9 Standard Entities

The members of $\mathbb{U}'$ of the form $^*a$ with $a \in \mathbb{U}$, will be called *standard*. The other members of $\mathbb{U}'$ are *nonstandard*. Any element a of $\mathbb{X}$ is thus standard, since in that case $a = {}^*a$.

13.9 Standard Entities

The question of the existence of nonstandard entities will be taken up in earnest in the next chapter. For now we will just assume that $\mathbb{U}$ is a universe over a set $\mathbb{X}$ that includes $\mathbb{R}$, and that

- there exists an element $n \in {}^*\mathbb{N} - \mathbb{N}$.

Then $\frac{1}{n}$ will be infinitesimal, and using transfer instead of ultrafilter calculations we can derive in $\mathbb{U}'$ the arithmetical properties of limited, unlimited, appreciable, etc. numbers as in Chapter 5, and then develop the theory of convergence, continuity, differentiation, integration for $\mathbb{R}$-valued sequences and functions as in Chapters 6–9.

All standard members of ${}^*\mathbb{N}$ belong to $\mathbb{N}$, for if ${}^*a \in {}^*\mathbb{N}$, then by transfer $a \in \mathbb{N}$, and so ${}^*a = a \in \mathbb{N}$. Thus any member of ${}^*\mathbb{N} - \mathbb{N}$ must be nonstandard. More generally, this argument shows that if $A \subseteq \mathbb{X}$, then any member of ${}^*A - A$ will be nonstandard.

We see then that standard sets ${}^*A$ may have nonstandard members. In fact it turns out that ${}^*A$ has nonstandard members for *every* infinite set $A \in \mathbb{U}$ (cf. Section 13.14).

Examples of nonstandard *sets* are provided by initial segments of ${}^*\mathbb{N}$. Consider the sentence

$$(\forall n \in \mathbb{N})\,(\exists V \in \mathcal{P}(\mathbb{N}))\,(\forall x \in \mathbb{N})\,[x \in V \leftrightarrow x \leq n],$$

which expresses "for all $n \in \mathbb{N}$, $\{1,\ldots,n\} \in \mathcal{P}(\mathbb{N})$". By transfer it follows that for any $N \in {}^*\mathbb{N}$, the subset $\{1,\ldots,N\}$ of ${}^*\mathbb{N}$ belongs to the standard set ${}^*\mathcal{P}(\mathbb{N})$. If $N \in {}^*\mathbb{N} - \mathbb{N}$, then $\{1,\ldots,N\}$ cannot itself be a standard set. For if ${}^*A$ is any standard subset of ${}^*\mathbb{N}$, then $A \subseteq \mathbb{N}$, so either A is finite and hence ${}^*A = A$, or else A is unbounded in $\mathbb{N}$, and hence by transfer ${}^*A$ is unbounded in ${}^*\mathbb{N}$. In either case ${}^*A \neq \{1,\ldots,N\}$.

For another illustration, let $Int \in \mathbb{U}_2(\mathbb{X})$ be the set of open subintervals of the real line:

$$Int = \{(a,b) \subseteq \mathbb{R} : a,b \in \mathbb{R}\}.$$

It follows by a transfer argument that

$${}^*(a,b) = \{x \in {}^*\mathbb{R} : a < x < b\},$$

so the standard set ${}^*(a,b)$ will have nonstandard elements. By transfer of the statements

$$(\forall A \in Int)\,(\exists a,b \in \mathbb{R})\,(\forall x \in \mathbb{R})\,(x \in A \leftrightarrow a < x < b),$$
$$(\forall a,b \in \mathbb{R})\,(\exists A \in Int)\,(\forall x \in \mathbb{R})\,(x \in A \leftrightarrow a < x < b),$$

we get

$$(\forall A \in {}^*Int)\,(\exists a,b \in {}^*\mathbb{R})\, A = \{x \in {}^*\mathbb{R} : a < x < b\},$$
$$(\forall a,b \in {}^*\mathbb{R})\,(\exists A \in {}^*Int)\, A = \{x \in {}^*\mathbb{R} : a < x < b\},$$

172   13. Universes and Frameworks

so we see that the standard set $^*Int \in \mathbb{U}'$ consists precisely of the *hyperreal* intervals $(a,b) \subseteq {}^*\mathbb{R}$ for all $a,b \in {}^*\mathbb{R}$. When one of a,b is nonstandard, the corresponding interval is a nonstandard member of $^*Int$.

Exercise 13.9.1 Characterise exactly the standard elements of $^*Int$.

External Images of Infinite Sets

If $^*A$ has nonstandard members, then it is distinguishable from the set
$$^{im}A = \{{}^*a : a \in A\},$$
which will be called the *image* of A. We have already observed that when A is finite, then $\{{}^*a : a \in A\}$ is equal to $^*A$. If A is infinite, we will call ^{im}A the *external image of* A. The reason for this name will be given in Theorem 13.14.1. ^{im}A is a subset of $^*A$ (since $a \in A$ implies $^*a \in {}^*A$) that forms a copy of A, in the sense that the transfer map $a \mapsto {}^*a$ gives a bijection between A and ^{im}A. It turns out that ^{im}A is a *proper* subset of $^*A$ for any infinite $A \in \mathbb{U}$ (cf. Section 13.14).

Note that $^{im}\mathbb{N} = \mathbb{N}$, $^{im}\mathbb{R} = \mathbb{R}$, and in general $^{im}A = A$ whenever $A \subseteq \mathbb{X}$.

Theorem 13.9.2 ^{im}A *is the set of* **all** *standard members of* $^*A$:
$$^{im}A = \{b \in \mathbb{U}' : b \in {}^*A \text{ and } b \text{ is standard}\}.$$
Hence $^*A$ *and* ^{im}A *have the same standard elements.*

Proof. The members of ^{im}A are standard by definition, and were noted above to be members of $^*A$.

Conversely, if b is a standard member of $^*A$, then $b = {}^*a$ for some $a \in \mathbb{U}$, so then $^*a \in {}^*A$ and hence $a \in A$ by transfer, showing $b = {}^*a \in {}^{im}A$.   □

Thus the nonstandard members of $^*A$ are precisely the members of $^*A - {}^{im}A$.

Exercise 13.9.3
Prove that standard sets are uniquely determined by their standard members. In other words, if two standard sets have the same standard members, then they are equal.

13.10   Internal Entities

In a nonstandard framework, an entity of $\mathbb{U}'$ is called *internal* if it belongs to some standard set:

- a is internal if and only if $a \in {}^*A$ for some $A \in \mathbb{U}$.

The set of all internal entities will be denoted by $^*\mathbb{U}$. Observe that

- *every standard entity is internal,*

because in general $*a \in *\{a\}$, so the standard entity $*a$ belongs to the standard set $*\{a\}$ and hence is itself internal. In other words, $*a \in *\mathbb{U}$ for all $a \in \mathbb{U}$.

Members of $\mathbb{U}'$ that are not internal are *external*. The question of the existence of external entities will be clarified once we have explored the world of internal entities a little further: in Theorem 13.14.1 it will be shown that $^{\text{im}}A$ is an external subset of the standard set $*A$ whenever A is infinite.

Theorem 13.10.1 *Any internal set belongs to a standard set that is transitive. Hence $*\mathbb{U}$ is strongly transitive, and in particular, every member of an internal set is internal.*

Proof. Let A be internal, with $A \in *B$ for some $B \in \mathbb{U}$. By strong transitivity of $\mathbb{U}$ there is a transitive $T \in \mathbb{U}$ with $B \subseteq T$. But as we have seen, transitivity is preserved by the transfer map, so the standard set $*T \in *\mathbb{U}$ is transitive. Also, $*B \subseteq *T$, so $A \in *T$, establishing the first part of the theorem. But then $A \subseteq *T$, and every member of $*T$ is internal by definition, so belongs to $*\mathbb{U}$. Thus $A \subseteq *T \subseteq *\mathbb{U}$, completing the proof that $*\mathbb{U}$ is strongly transitive.

The assertion that every member of an internal set is internal is now just the statement that $*\mathbb{U}$ is transitive. □

We have already seen in Section 13.9 some interesting examples of nonstandard internal entities. Every initial segment $\{1, \ldots, N\}$ of the hypernaturals $*\mathbb{N}$ is internal, since it belongs to $*\mathcal{P}(\mathbb{N})$, while any open hyperreal interval $\{x \in *\mathbb{R} : a < x < b\}$ with $a, b \in *\mathbb{R}$ is internal, since it belongs to $*Int$. Also, any uniform partition

$$P_N^{ab} = \left\{a + k\tfrac{(b-a)}{N} : k \in *\mathbb{Z} \ \& \ 0 \le k \le N\right\}$$

with $a, b \in *\mathbb{R}$ and $N \in *\mathbb{N}$ is internal, since it belongs to $*\mathcal{P}(\mathbb{R}) \subseteq \mathcal{P}(*\mathbb{R})$ and hence is an internal subset of $*\mathbb{R}$. This follows by transfer of the statement

$$(\forall a, b \in \mathbb{R})\,(\forall n \in \mathbb{N})\,(\exists P \in \mathcal{P}(\mathbb{R}))\,(\forall x \in \mathbb{R})$$
$$\left[x \in P \leftrightarrow (\exists k \in \mathbb{Z})\left(0 \le k \le n \wedge x = a + k\tfrac{(b-a)}{n}\right)\right].$$

13.11 Closure Properties of Internal Sets

- If A and B are internal sets, then so are $A \cap B$, $A \cup B$, $A - B$, and $A \times B$.

- The union and intersection of any internal collection of sets are internal: if $\{A_i : i \in I\} \in *\mathbb{U}$ and each A_i is a set, then $\bigcup_{i \in I} A_i$ and $\bigcap_{i \in I} A_i$ are internal. (Note: $\{A_i : i \in I\} \subseteq *\mathbb{U}$ does not suffice here.)

- If $a_1, \ldots, a_m$ are internal, then so are the finite set $\{a_1, \ldots, a_m\}$ and the m-tuple $\langle a_1, \ldots, a_m \rangle$.

- If a binary relation R is internal, then so is its domain dom R, its range ran R, its inverse R^{-1}, and the image $R'(C)$ for any *internal* $C \subseteq \text{dom } R$.

- If binary relations R, S are internal, then so is their composition $R \circ S$.

- If a function f is internal and $a \in \text{dom } f$, then $f(a)$ is internal. Moreover, if C is an internal subset of dom f, then $f'(C)$ is internal, and if D is an internal subset of ran f then $f^{-1}(D)$ is internal.

Here is the proof that the union of two internal sets is internal, i.e., *U is closed under binary unions. The proof uses the closure of U under unions together with strong transitivity of U to include any such union in a transitive set.

So, let $A, B \in {}^*\mathbb{U}$, with $A \in {}^*C$ and $B \in {}^*D$. Now, in U there is a transitive set T including $C \cup D$. Then $A, B \in {}^*T$. Since $^*T$ is also transitive (cf. the proof of Theorem 13.10.1), A and B are then subsets of $^*T$.

Now consider the sentence

$$(\forall x, y \in T)\, (\exists z \in \mathcal{P}(T))\, (\forall u \in T)\, [u \in z \leftrightarrow u \in x \vee u \in y],$$

which asserts the existence in $\mathcal{P}(T)$ of the union $z = x \cup y$ when $x, y \in T$ (and hence $x, y \subseteq T$). By transferring this sentence and applying it to $A, B \in {}^*T$ we see that there is some $Z \in {}^*\mathcal{P}(T)$ such that Z and $A \cup B$ contain exactly the same elements of $^*T$:

$$Z \cap {}^*T = (A \cup B) \cap {}^*T.$$

Then $Z \in \mathcal{P}(^*T)$, so Z and $A \cup B$ are subsets of $^*T$, i.e., *all* members of Z and $A \cup B$ are in $^*T$. Therefore $A \cup B$ is equal to the internal set Z.

Exercise 13.11.1
Verify all the other closure properties of *U listed above.

13.12 Transformed Power Sets

*U falls short of being a universe in its own right because it is not closed under power sets. If it were, then $\mathcal{P}(^*\mathbb{N})$ would be internal, i.e., $\mathcal{P}(^*\mathbb{N}) \in {}^*\mathbb{U}$. By transitivity of *U this would imply that every subset of $^*\mathbb{N}$ was in *U too, and in particular $\mathbb{N}$ would be internal. However, this is incompatible with the internal induction principle when $^*\mathbb{N} - \mathbb{N} \neq \emptyset$, as will be seen in this section.

Now, for any set $A \in \mathbb{U}$ we have

$$^{\mathrm{im}}\mathcal{P}(A) = \{{^*B} : B \in \mathcal{P}(A)\} \subseteq {^*\mathcal{P}(A)} \subseteq \mathcal{P}({^*A}),$$

but these three sets derived from $\mathcal{P}(A)$ are not the same (for infinite A). $^{\mathrm{im}}\mathcal{P}(A)$ is the collection of all *standard* subsets of $^*A$, while $^*\mathcal{P}(A)$ consists of the *internal* subsets:

Theorem 13.12.1 $^*\mathcal{P}(A)$ *is the set of all internal subsets of $^*A$, i.e.,*

$$^*\mathcal{P}(A) = \mathcal{P}({^*A}) \cap {^*\mathbb{U}} = \{B \subseteq {^*A} : B \text{ is internal}\}.$$

Proof. We have $^*\mathcal{P}(A) \subseteq \mathcal{P}(^*A)$ in general. Moreover, if $B \in {^*\mathcal{P}(A)}$, then B belongs to a standard entity, so B is internal.

For the converse, let B be an internal subset of $^*A$. Then $B \in {^*C}$ for some $C \in \mathbb{U}$. Now, the sentence

$$(\forall x \in C)\,[(\forall y \in x)(y \in A) \to x \in \mathcal{P}(A)]$$

is true, since it asserts of any $x \in C$ that if $x \subseteq A$, then x belongs to $\mathcal{P}(A)$. By transfer it follows that for any $x \in {^*C}$, if $x \subseteq {^*A}$, then x belongs to $^*\mathcal{P}(A)$. But $B \in {^*C}$ and $B \subseteq {^*A}$, i.e.,

$$(\forall y \in B)(y \in {^*A}),$$

so we get $B \in {^*\mathcal{P}(A)}$ as desired. □

This result has an extremely significant consequence for statements that quantify over power sets. Since the $\mathcal{L}_\mathbb{U}$-sentence $(\forall x \in \mathcal{P}(A))\varphi$ transforms to $(\forall x \in {^*\mathcal{P}(A)})^*\varphi$, we now see from the characterisation of $^*\mathcal{P}(A)$ that a true $\mathcal{L}_\mathbb{U}$-statement of the form

for all subsets x of A, φ

gives rise to a true $\mathcal{L}_{\mathbb{U}'}$-statement of the form

for all *internal* subsets x of $^*A$, $^*\varphi$.

Consider for instance the least number principle

$$(\forall x \in \mathcal{P}(\mathbb{N}))[x \neq \emptyset \to (\exists y \in x)\,(\forall z \in x)\,(y \leq z)].$$

This transforms to show that every nonempty *internal* subset of $^*\mathbb{N}$ has a least element, which is the internal least number principle of Section 11.3.

In a similar manner we can now derive the internal induction principle by a transfer argument. Either of these principles can then be used to conclude that if $^*\mathbb{N} - \mathbb{N} \neq \emptyset$, then $\mathbb{N}$ must be external, for the reasons explained in Section 11.6.

13.13 Exercises on Internal Sets and Functions

(1) Derive in *𝕌 the principles of internal induction (Theorem 11.3.2) and internal order-completeness (Theorem 11.5.1).

(2) Show that $\mathcal{P}(A)$ is in bijective correspondence with the set of all functions $f : A \to \{0, 1\}$. Adapt this to show that $^*\mathcal{P}(A)$ is in bijective correspondence with the set of all *internal* functions $f : {^*A} \to \{0, 1\}$.

(3) Extending the previous exercise, recall that B^A denotes the set of all functions from A to B. If $A, B \in \mathbb{U}$, show that $^*(B^A)$ is the set of all *internal* functions from $^*A$ to $^*B$.

13.14 External Images Are External

Externality of $\mathbb{N}$ implies that of any set of the form $^{\text{im}}A$ with A infinite, thereby justifying the name "external image", as the following result shows.

Theorem 13.14.1 *The image* $^{\text{im}}A$ *of any infinite set* $A \in \mathbb{U}$ *is external.*

Proof. The method of proof was hinted at in Exercise 12.2(8), which is itself the special case in which A is an infinite subset of $\mathbb{R}$ (cf. also Section 11.7).

In general, if A is infinite, then there is an injection $f : \mathbb{N} \to A$. Put $X = \{f(n) : n \in \mathbb{N}\} \subseteq A$. Then $^*X$ is internal (indeed standard), and so *if* $^{\text{im}}A$ were internal, then so too would be $^{\text{im}}A \cap {^*X}$ (Section 13.11). Observe that

$$^{\text{im}}A \cap {^*X} = \{{^*a} : a \in A \text{ and } {^*a} \in {^*X}\} = \{{^*a} : a \in X\} = {^{\text{im}}X}.$$

Since the transform $^*f : {^*\mathbb{N}} \to {^*A}$ is internal, this would then imply that the inverse image of $^{\text{im}}X$ under $^*f$ is internal (Section 13.11). But by transfer $^*f$ is injective, and from this it can be shown that $^*f^{-1}({^{\text{im}}X})$ is equal to the *external* set $\mathbb{N}$.

Therefore $^{\text{im}}A$ cannot be internal. □

Exercise 13.14.2
Verify that $^*f^{-1}({^{\text{im}}X}) = \mathbb{N}$ in the proof just given. □

Since $^*A$ is internal, it follows from Theorem 13.14.1 that there are elements in $^*A - {^{\text{im}}A}$. By Theorem 13.9.2 these elements are nonstandard. All told then, it has now been established that

- *if* $^*\mathbb{N}$ *has nonstandard elements, then so does* $^*A$ *for every infinite set* $A \in \mathbb{U}$.

Since $^{\text{im}}A$ is an external subset of $^*A$, it belongs to $\mathcal{P}(^*A)$ but not to $^*\mathcal{P}(A)$. Thus $^*\mathcal{P}(A)$ is always a *proper* subset of $\mathcal{P}(^*A)$ when A is infinite.

Exercise 13.14.3
Explain why $^*\mathcal{P}(A) = \mathcal{P}(^*A)$ whenever A is finite.

13.15 Internal Set Definition Principle

In Section 11.7 we saw how internal subsets of $^*\mathbb{R}$ could be defined from properties expressible by (the transforms of) formulae from the language $\mathcal{L}_\mathfrak{R}$. We can now formulate and prove a much stronger version of this that applies to formulae of the language $\mathcal{L}_{\mathbb{U}'}$. Such a formula may refer to much more complex entities than are available in $\mathcal{L}_\mathfrak{R}$, including entities of arbitrarily high rank in the superstructure $\mathbb{U}(^*\mathbb{X})$ over $^*\mathbb{X}$.

This new version of the internal set definition principle is as follows:

If $\varphi(x)$ is an internal $\mathcal{L}_{\mathbb{U}'}$-formula (i.e., all of its constants are internal), with x the only free variable in φ, then for any internal set A, the set

$$\{a \in A : \varphi(a) \text{ is true}\}$$

is internal. In other words, if $B = \{a \in \mathbb{U}' : \varphi(a)\}$ is a subset of $\mathbb{U}'$ definable by an $\mathcal{L}_{\mathbb{U}'}$-formula with only internal constants, then for any internal set A, $A \cap B$ is internal.

Proof. Let $c_1, \ldots, c_m$ be all the (internal) constants that occur in φ. Replace each constant c_i by a new variable x_i to get a formula $\varphi(x, x_1, \ldots, x_m)$ that has no constants, hence is an $\mathcal{L}_\mathbb{U}$-formula and is equal to its own $*$-transform.

Now, the entities $A, c_1, \ldots, c_m$ all belong to some standard set $^*T$ that is transitive (Theorem 13.10.1). But the sentence

$$(\forall y, x_1, \ldots, x_m \in T)\,(\exists z \in \mathcal{P}(T))\,(\forall x \in T)$$
$$x \in z \leftrightarrow [x \in y \wedge \varphi(x, x_1, \ldots, x_m)]$$

is true, because it asserts the presence in $\mathcal{P}(T)$ of the set

$$z = \{x \in T : x \in y \text{ and } \varphi(x, x_1, \ldots, x_m)\}.$$

Applying the transfer of this sentence to $A, c_1, \ldots, c_m \in {}^*T$, we find that there is some internal set $Z \in {}^*\mathcal{P}(T)$ such that

$$Z = \{a \in {}^*T : a \in A \text{ and } \varphi(a, c_1, \ldots, c_m)\}.$$

But $A \subseteq {}^*T$ by transitivity of $^*T$, so Z is just the desired set $\{a \in A : \varphi(a)\}$. □

To illustrate this principle, let $\varphi(x)$ be the internal formula

$$(\exists k \in {}^*\mathbb{Z}) \left(0 \leq k \leq N \wedge x = a + k\tfrac{(b-a)}{N}\right).$$

Then $\{c \in {}^*\mathbb{R} : \varphi(c)\}$ is the internal partition P_N^{ab} discussed above.

13.16 Internal Function Definition Principle

Let $f : A \to B$ be a function between internal sets A and B. Suppose that there is an internal $\mathcal{L}_{\mathbb{U}'}$-term $\tau(x)$ with sole variable x such that $f(a) = \tau(a)$ for all $a \in A$. Then f is an internal function.

Proof. By the internal set definition principle, since $A \times B$ is internal and

$$f = \{c \in A \times B : (\exists x \in A)\,(\exists y \in B)\,(c = \langle x, y \rangle \wedge y = \tau(x))\}.$$

$\square$

We can illustrate this result by the internal partition example P_N^{ab} once more. Let $\tau(x)$ be the internal term $a + x\tfrac{(b-a)}{N}$. Then by the internal function definition principle the function $\{0, 1, \ldots, N\} \to {}^*\mathbb{R}$ defined by τ is internal. The image of this function is thus internal, and is just P_N^{ab}.

13.17 Hyperfiniteness

If $A \in \mathbb{U}$, let

$$\mathcal{P}_F(A) = \{B \subseteq A : B \text{ is finite}\}$$

be the set of all finite subsets of A. Then $\mathcal{P}_F(A) \in \mathbb{U}$. Under a transfer map we have

$${}^*\mathcal{P}_F(A) \subseteq {}^*\mathcal{P}(A) \subseteq \mathcal{P}({}^*A),$$

so that each member of ${}^*\mathcal{P}_F(A)$ is an internal subset of ${}^*A$. But we should not expect to have ${}^*\mathcal{P}_F(A) \subseteq \mathcal{P}_F({}^*A)$, unless A is finite. Consider for instance the true $\mathcal{L}_{\mathbb{U}}$-sentence

$$(\forall n \in \mathbb{N})\,(\exists V \in \mathcal{P}_F(\mathbb{N}))\,(\forall x \in \mathbb{N})\,[x \in V \leftrightarrow x \leq n],$$

which modifies the sentence of Section 13.9 to assert that for each natural number n the set $V = \{1, \ldots, n\}$ is actually in $\mathcal{P}_F(\mathbb{N})$. By transfer, for each $n \in {}^*\mathbb{N}$, the initial segment

$$\{1, \ldots, n\} = \{x \in {}^*\mathbb{N} : x \leq n\}$$

of ${}^*\mathbb{N}$ belongs to ${}^*\mathcal{P}_F(\mathbb{N})$, and when n is unlimited this set includes $\mathbb{N}$ and is infinite!

13.17 Hyperfiniteness

The members of $^*\mathcal{P}_F(A)$ are called *hyperfinite* subsets of $^*A$. Thus a hyperfinite set is one that belongs to $^*\mathcal{P}_F(A)$ for some $A \in \mathbb{U}$. This notion applies now to sets A of any rank, and is related by the following result to our earlier study in Chapter 12 of the case $A = \mathbb{R}$.

Theorem 13.17.1 *A set B is hyperfinite if and only if there is an internal bijection*
$$f : \{1, \ldots, n\} \to B$$
for some $n \in {^*\mathbb{N}}$.

Proof. Let $\Phi(V, n, f, Z)$ be the conjunction of the following formulae with free variables V, n, f, Z:

$(\forall x \in V)\, (x \in \mathbb{N} \wedge x \leq n)$,
$(\forall x \in \mathbb{N})\, (x \leq n \to x \in V)$,
$(\forall b \in f)\, (\exists x \in V)\, (\exists y \in Z)\, [b = \langle x, y \rangle]$,
$(\forall x \in V)\, (\forall y, z \in Z)\, [\langle x, y \rangle \in f \wedge \langle x, z \rangle \in f \to y = z]$,
$(\forall x \in V)\, (\exists y \in Z)\, [\langle x, y \rangle \in f]$,
$(\forall y \in Z)\, (\exists x \in V)\, [\langle x, y \rangle \in f]$,
$(\forall x, y \in V)\, (\forall z \in Z)\, [\langle x, z \rangle \in f \wedge \langle y, z \rangle \in f \to x = y]$.

Thus $\Phi(V, n, f, Z)$ asserts that $V = \{x \in \mathbb{N} : x \leq n\}$ and f is a function from V onto Z that is injective. Hence if we define $\varphi(n, f, Z)$ to be the formula
$$(\exists V \in \mathcal{P}(\mathbb{N}))\, \Phi(V, n, f, Z),$$
then φ asserts that f is a bijection between $\{1, \ldots, n\}$ and Z. Its transform $^*\varphi$ makes exactly the same assertion, but replaces $\mathbb{N}$ by $^*\mathbb{N}$ and so allows the possibility that $n \in {^*\mathbb{N}} - \mathbb{N}$.

Now, if $A \in \mathbb{U}$, then the $\mathcal{L}_\mathbb{U}$-sentence
$$(\forall Z \in \mathcal{P}_F(A))\, (\exists n \in \mathbb{N})\, (\exists f \in \mathcal{P}(\mathbb{N} \times A))\, \varphi(n, f, Z)$$
is true. Hence by transfer, if $B \in {^*\mathcal{P}_F(A)}$, then there is some $n \in {^*\mathbb{N}}$ and some $f \in {^*\mathcal{P}(\mathbb{N} \times A)}$, i.e., some internal $f \subseteq {^*\mathbb{N}} \times {^*A}$, such that f is a bijection from $\{1, \ldots, n\}$ to B. Moreover, f is internal, as it belongs to the standard set $^*\mathcal{P}(\mathbb{N} \times A)$.

For the converse, suppose that there is an internal bijection $f : V \to B$, where $V = \{x \in {^*\mathbb{N}} : x \leq n\}$ for some $n \in {^*\mathbb{N}}$. Thus B is internal, being the range of an internal function, and so $B \in {^*A}$ for some $A \in \mathbb{U}$, which we may take to be transitive, with $^*A$ transitive as well. Then $B \subseteq {^*A}$, so f is an internal subset of $^*\mathbb{N} \times {^*A}$, whence
$$f \in \mathcal{P}(^*\mathbb{N} \times {^*A}) \cap {^*\mathbb{U}} = \mathcal{P}(^*(\mathbb{N} \times A)) \cap {^*\mathbb{U}} = {^*\mathcal{P}(\mathbb{N} \times A)},$$

and $^*\varphi(n, f, B)$ holds.

Now consider the sentence

$$(\forall Z \in A)\big([(\exists n \in \mathbb{N})\,(\exists f \in \mathcal{P}(\mathbb{N} \times A))\,\varphi(n, f, Z)] \to Z \in \mathcal{P}_F(A)\big).$$

This sentence is true (using the fact that $Z \in A$ implies $Z \subseteq A$), since it asserts that if there is a bijection to Z from a set $\{1, \ldots, n\}$ with $n \in \mathbb{N}$, then Z is in $\mathcal{P}_F(A)$. Thus applying transfer and taking $Z = B$, we conclude that $B \in {}^*\mathcal{P}_F(A)$, so that B is a hyperfinite subset of $^*A$. □

The number n in this theorem is the *internal cardinality* of the hyperfinite set B: $n = |B|$. The map $B \mapsto |B|$ is an internal function $^*\mathcal{P}_F(A) \to {}^*\mathbb{N}$, and is the function that arises under the $*$-embedding from the corresponding function $\mathcal{P}_F(A) \to \mathbb{N}$.

13.18 Exercises on Hyperfinite Sets and Sizes

(1) If there is an internal injection $\{1, \ldots, n\} \to \{1, \ldots, m\}$, then $n \leq m$. If there is an internal *bijection* $\{1, \ldots, n\} \to \{1, \ldots, m\}$, then $n = m$.

(2) Let A be hyperfinite. Show that any internal set $B \subsetneq A$ is also hyperfinite and has $|B| < |A|$.

(3) Any hyperfinite subset of $^*\mathbb{R}$ has a greatest and a least element.

(4) If A is hyperfinite, then so is the *internal power set*

$$\mathcal{P}_I(A) = \mathcal{P}(A) \cap {}^*\mathbb{U} = \{B \subseteq A : B \text{ is internal}\}.$$

Moreover, if $|A| = n$, then $|\mathcal{P}_I(A)| = 2^n$.

(5) A set B is hyperfinite iff there is a *surjection* $f : \{1, \ldots, n\} \to B$, for some $n \in {}^*\mathbb{N}$, that is internal.

(6) Internal images of hyperfinite sets are hyperfinite: if f is internal and B is hyperfinite, then $f(B)$ is hyperfinite.

13.19 Hyperfinite Summation

In Section 12.7 the ultrapower construction was used to give meaning to the symbol $\sum_{x \in B} f(x)$ when B is a hyperfinite set and f an internal function.

In the context of a universe embedding $\mathbb{U} \xrightarrow{*} \mathbb{U}'$, summation of hyperfinite sets is obtained by applying transfer to the operation of summing finite sets. The map $B \mapsto \sum B$ from $\mathcal{P}_F(\mathbb{R})$ to $\mathbb{R}$ assigns to each finite set $B \subseteq$

ℝ the sum of its members. This map transforms to an internal function from $^*\mathcal{P}_F(\mathbb{R})$ to $^*\mathbb{R}$, thereby giving a (hyperreal) meaning to $\sum B$ for every hyperfinite set $B \subseteq {^*\mathbb{R}}$.

Now, if f is an internal $^*\mathbb{R}$-valued function whose domain includes a hyperfinite set B, then $f(B)$ will be a hyperfinite subset of $^*\mathbb{R}$ for which $\sum$ is defined. Thus we can specify

$$\sum_{x \in B} f(x) = \sum f(B).$$

13.20 Exercises on Hyperfinite Sums

(1) Explain why hyperfinite summation has the following properties: if f, g are internal functions, B, C are hyperfinite sets, and $c \in {^*\mathbb{R}}$, then

- $\sum_{x \in B} cf(x) = c \left(\sum_{x \in B} f(x) \right)$,
- $\sum_{x \in B} f(x) + g(x) = \sum_{x \in B} f(x) + \sum_{x \in B} g(x)$,
- $\sum_{x \in B \cup C} f(x) = \sum_{x \in B} f(x) + \sum_{x \in C} f(x)$ if $B \cap C = \emptyset$,
- $\sum_{x \in B} f(x) \leq \sum_{x \in B} g(x)$ if $f(x) \leq g(x)$ on B.

(2) Explain how and when it is possible to define indexed sums $\sum_1^n x_i$ with n unlimited.

14
The Existence of Nonstandard Entities

How do we know that *ℝ contains new entities in addition to the real numbers? In the ultrapower construction of *ℝ in Chapter 3, this was deduced from the fact that the nonprincipal ultrafilter $\mathcal{F}$ contains no finite sets. That allowed us to show that $^*A - A$ is nonempty whenever A is an infinite subset of ℝ.

If *ℝ is the transform of ℝ under a universe embedding, we may not be able to conclude that *ℝ ≠ ℝ (for instance, the identity function on 𝕌 is a universe embedding 𝕌 → 𝕌 making *ℝ = ℝ). The condition that *ℝ − ℝ be nonempty will have to be added as a new requirement (as was done in the previous chapter by assuming *ℕ − ℕ ≠ ∅), or else derived from some other principle.

In Section 11.11 it was shown that *countable saturation* guarantees the existence of hyperreal numbers that are characterised by countably many internal conditions. Saturation will be discussed further in Chapter 15. In this chapter we look at another principle, known as "enlargement", which is particularly convenient in the way it allows us to obtain nonstandard entities. We will then see how enlargements can be constructed by forming ultrapowers of superstructures.

14.1 Enlargements

Assume from now on that 𝕌 is a universe over some set that includes ℝ. 𝕌′ is called an *enlargement* of 𝕌 if there exists a universe embedding $\mathbb{U} \overset{*}{\to} \mathbb{U}'$ such that:

if $A \in \mathbb{U}$ is a collection of sets with the finite intersection property, then there exists an element $b \in \mathbb{U}'$ that belongs to $^*B$ for every $B \in A$, i.e.,

$$b \in \bigcap \{^*B : B \in A\} = \bigcap {}^{\mathrm{im}}A.$$

Enlargements have an abundance of nonstandard internal entities:

- Let A be the set of intervals $(0, r) \subseteq \mathbb{R}$ for all positive real r. Then $A \in \mathbb{U}_2(\mathbb{R})$ and A has the finite intersection property. If

$$b \in \bigcap \{^*(0, r) : r \in \mathbb{R}^+\},$$

then b is a positive infinitesimal member of $^*\mathbb{R}$. Indeed, in this case $\bigcap {}^{\mathrm{im}}A$ is precisely the set of positive infinitesimals, and the enlargement principle ensures that it is nonempty.

If we take A instead to consist of the intervals $(-r, r) - \{0\}$, then $\bigcap {}^{\mathrm{im}}A$ is the set of all nonzero infinitesimals.

- For $r \in \mathbb{R}^+$, let $(r, \infty) = \{x \in \mathbb{R} : r < x\}$. Then by transfer $^*(r, \infty) = \{x \in {}^*\mathbb{R} : r < x\}$. The collection of intervals (r, ∞) has the finite intersection property, and any member of

$$\bigcap \{^*(r, \infty) : r \in \mathbb{R}^+\}$$

is a positive unlimited hyperreal.

- Let $A \in \mathbb{U}$ be an infinite set. Then the collection

$$\{A - \{a\} : a \in A\}$$

has the finite intersection property. Since $^*(A - \{a\}) = {}^*A - \{^*a\}$, it follows that in any enlargement of $\mathbb{U}$ there must be an entity b that belongs to $^*A$ but is distinct from $^*a$ for all $a \in A$. Thus

$$b \in {}^*A - \{^*a : a \in A\} = {}^*A - {}^{\mathrm{im}}A.$$

Such a b will be nonstandard, because if $b = {}^*a$ for some $a \in \mathbb{U}$, then $^*a \in {}^*A$, implying $a \in A$ (recall from Theorem 13.9.2 that $^{\mathrm{im}}A$ is the set of standard members of $^*A$).

So we see that in an enlargement, any infinite standard set has nonstandard members. In particular, if A is any infinite subset of $\mathbb{R}$ (e.g., $A = \mathbb{N}, \mathbb{Z}, \mathbb{Q}$, etc.), then as $A = {}^{\mathrm{im}}A$, we deduce that A is a proper subset of $^*A$.

- If $a \in A$, let $A_a = \{Z \in \mathcal{P}_F(A) : a \in Z\}$ be the collection of finite subsets of A that contain a. Then $\{A_a : a \in A\}$ has the finite intersection property, for if $a_1, \ldots, a_n \in A$, put $Z = \{a_1, \ldots, a_n\}$ to get $Z \in A_{a_1} \cap \cdots \cap A_{a_n}$. But if

$$B \in \bigcap \{{}^*(A_a) : a \in A\},$$

then $B \in {}^*\mathcal{P}_F(A)$, since in general ${}^*(A_a) \subseteq {}^*\mathcal{P}_F(A)$, and ${}^*a \in B$ for each $a \in A$ by transfer of the sentence

$$(\forall Z \in A_a)(a \in Z).$$

Thus B is a *hyperfinite* subset of ${}^*A$ that contains $\{{}^*a : a \in A\}$, i.e., ${}^{im}A \subseteq B \subseteq {}^*A$.

The last example shows that enlargement is a stronger property than we had considered hitherto, since we were previously only able to establish the existence of such hyperfinite approximating sets in relation to countable subsets of $\mathbb{R}$ (cf. Section 12.3).

We are left then with the question of whether enlargements themselves exist. In fact, they can be obtained by applying the ultrapower construction to the superstructure $\mathbb{U}(\mathbb{X})$, as will be explained below. The outcome is this:

Enlargement Theorem. *For any set $\mathbb{X}$ there exists an enlargement of $\mathbb{U}(\mathbb{X})$ that is of the form $\mathbb{U}({}^*\mathbb{X})$.*

14.2 Concurrence and Hyperfinite Approximation

A binary relation R is called *concurrent*, or *finitely satisfiable*, if for any finite subset $\{x_1, \ldots, x_n\}$ of the domain of R there exists an element y with $x_i R y$ for all i between 1 and n.

If a concurrent relation R belongs to $\mathbb{U}$, then an enlargement will contain entities b with ${}^*x({}^*R)b$ for all $x \in \text{dom } R$. This provides another language for describing the above examples:

- Let R be the "greater than" relation on $\mathbb{R}^+$:

$$R = \{\langle r, y \rangle \in \mathbb{R}^+ \times \mathbb{R}^+ : r > y\}.$$

  R is concurrent, and by transfer ${}^*R$ is the "greater than" relation on ${}^*\mathbb{R}^+$. If $r({}^*R)b$ for all $r \in \mathbb{R}^+$, then b is a positive infinitesimal.

- Let R be the "less than" relation on $\mathbb{R}^+$:

$$R = \{\langle r, y \rangle \in \mathbb{R}^+ \times \mathbb{R}^+ : r < y\}.$$

  If $r({}^*R)b$ for all $r \in \mathbb{R}^+$, then b is a positive unlimited hyperreal.

- Let R be the nonidentity relation on a set $A \in \mathbb{U}$:
$$R = \{\langle a, y\rangle \in A \times A : a \neq y\}.$$
Then R is concurrent precisely when A is infinite, and *R is the nonidentity relation on *A. If *a(*R)b for all $a \in A$, then
$$b \in {}^*A - \{{}^*a : a \in A\},$$
and so b is a nonstandard member of *A.

- Let R be the membership relation between A and $\mathcal{P}_F(A)$:
$$R = \{\langle a, Z\rangle : a \in Z \in \mathcal{P}_F(A)\}.$$
For any $A \in \mathbb{U}$, R is a concurrent relation in $\mathbb{U}$. If *a(*R)B for all $a \in A$, then B is a *hyperfinite* subset of *A that includes ${}^{\text{im}}A$.

These notions yield alternative characterisations of the concept of enlargement:

Theorem 14.2.1 *If* $\mathbb{U} \xrightarrow{*} \mathbb{U}'$ *is a universe embedding, then the following are equivalent.*

(1) $\mathbb{U}'$ *is an enlargement of* $\mathbb{U}$ *relative to* $\xrightarrow{*}$.

(2) *For any concurrent relation* $R \in \mathbb{U}$ *there exists an entity* $b \in \mathbb{U}'$ *such that* *x(*R)b *for all* $x \in \text{dom}\, R$.

(3) *For each set* $A \in \mathbb{U}$ *there exists a hyperfinite subset* B *of* *A *that contains all the standard entities of* *A:
$$^{\text{im}}A = \{{}^*a : a \in A\} \subseteq B \in {}^*\mathcal{P}_F(A).$$

Proof. First assume (1). If R is a binary relation in $\mathbb{U}$, for each $x \in \text{dom}\, R$ let $R[x] = \{y \in \text{ran}\, R : xRy\}$. Then if R is concurrent, the collection
$$\{R[x] : x \in \text{dom}\, R\}$$
has the finite intersection property. Also, this collection is a subset of the power set $\mathcal{P}(\text{ran}\, R)$, so belongs to $\mathbb{U}$. Hence by (1) there is a $b \in \mathbb{U}'$ that belongs to every *$(R[x])$. By transfer of $(\forall y \in R[x])(xRy)$ we then have *x(*R)b for all $x \in \text{dom}\, R$, establishing (2).

The proof that (2) implies (3) was indicated in the discussion of the last example above.

To show that (3) implies (1), let $A \in \mathbb{U}$ be a collection of sets with the finite intersection property. Take a transitive $T \in \mathbb{U}$ with $A \subseteq T$. Then the sentence
$$(\forall Z \in \mathcal{P}_F(A))(\exists y \in T)(\forall z \in Z)(y \in z)$$
is true. But assuming (3), there exists a hyperfinite $B \subseteq {}^*A$ containing ${}^{\text{im}}A$. Then by transfer of this sentence, since $B \in {}^*\mathcal{P}_F(A)$, there exists some $b \in {}^*T$ with $b \in z$ for all $z \in B$, and hence $b \in {}^*C$ for all $C \in A$. Therefore (1) holds. □

14.3 Enlargements as Ultrapowers

In Chapter 3 we constructed *$\mathbb{R}$ by starting with certain real-valued sequences, i.e., functions $r, s : \mathbb{N} \to \mathbb{R}$, and identifying them when the set

$$[\![r = s]\!] = \{n \in \mathbb{N} : r_n = s_n\}$$

on which they agree belongs to an ultrafilter $\mathcal{F}$ on $\mathbb{N}$.

This method of construction can be applied much more widely by allowing $\mathcal{F}$ to be an ultrafilter on a set I other than $\mathbb{N}$. Then any two functions with domain I can be identified if the subset of I on which they agree belongs to $\mathcal{F}$. The approach can be used to build enlargements of a superstructure $\mathbb{U}(\mathbb{X})$, and we will now sketch out the way in which it works.

So, let I be an infinite set and $\mathcal{F}$ a nonprincipal ultrafilter on I. Then $\mathbb{U}(\mathbb{X})^I$ is the set of all functions from I to $\mathbb{U}(\mathbb{X})$. For $a \in \mathbb{U}(\mathbb{X})$, let $a_I \in \mathbb{U}(\mathbb{X})^I$ be the function with constant value a. For $f, g \in \mathbb{U}(\mathbb{X})^I$, put

$$\begin{aligned}
[\![f = g]\!] &= \{i \in I : f(i) = g(i)\}, \\
[\![f \in g]\!] &= \{i \in I : f(i) \in g(i)\}, \\
[\![f \in a]\!] &= [\![f \in a_I]\!] = \{i \in I : f(i) \in a\}, \\
[\![a \in f]\!] &= [\![a_I \in f]\!] = \{i \in I : a \in f(i)\}, \\
Z_n &= \{f \in \mathbb{U}(\mathbb{X})^I : [\![f \in \mathbb{U}_n(\mathbb{X})]\!] \in \mathcal{F}\}, \\
Z &= \bigcup \{Z_n : n \geq 0\}.
\end{aligned}$$

Z is then the union of the increasing chain $Z_0 \subseteq Z_1 \subseteq \cdots$. The members of Z may be viewed as functions of "bounded rank" in the sense that if $f \in Z_n$, then $f(i) \in \mathbb{U}_n(\mathbb{X})$ for $\mathcal{F}$-almost all i in I. In fact, such an f can itself be assigned a rank, because

$$[\![f \in \mathbb{U}_n(\mathbb{X})]\!] \subseteq \{i \in I : f(i) \text{ has rank } 0\} \cup \cdots \cup \{i \in I : f(i) \text{ has rank } n\},$$

and the union on the right is made up of pairwise disjoint sets, so

$$\{i \in I : f(i) \text{ has rank } k\} \in \mathcal{F}$$

for exactly one $k \leq n$, allowing us to define f to be of rank k. Then $Z_n - Z_{n-1}$ consists of the functions of rank n, and the rank of the function a_I in Z is the same as the rank of the entity a in $\mathbb{U}(\mathbb{X})$.

For $f \in Z_0$, let $[f] = \{g \in Z_0 : [\![f = g]\!] \in \mathcal{F}\}$. Put

$$\mathbb{Y} = \{[f] : f \in Z_0\}.$$

There is a map $f \mapsto [f]$ from Z to the superstructure $\mathbb{U}(\mathbb{Y})$ over $\mathbb{Y}$ that composes with the map $a \mapsto a_I$ of $\mathbb{U}(\mathbb{X})$ into Z to give a universe embedding. The definition of $[f]$ is by induction on the rank of f.

For $f \in Z_0$, $[f]$ has just been defined and is a member of $\mathbb{U}_0(\mathbb{Y})$. Thus for $n \geq 0$ we can make the inductive hypothesis that for all $f \in Z_n$, $[f]$ has been defined and is a member of $\mathbb{U}_n(\mathbb{Y})$. Then for $g \in Z_{n+1} - Z_n$ put

$$[g] = \{ [f] : f \in Z_n \text{ and } [\![f \in g]\!] \in \mathcal{F} \}.$$

This specifies $[g]$ as a subset of $\mathbb{U}_n(\mathbb{Y})$, and hence a member of $\mathbb{U}_{n+1}(\mathbb{Y})$, and completes the inductive construction.

Now, for $f, g \in Z$ it can be shown that

$$[f] \in [g] \quad \text{iff} \quad \{i \in I : f(i) \in g(i)\} \in \mathcal{F}$$

and

$$[f] = [g] \quad \text{iff} \quad \{i \in I : f(i) = g(i)\} \in \mathcal{F}.$$

For each $a \in \mathbb{U}(\mathbb{X})$ define $^*a$ to be $[a_I] \in \mathbb{U}(\mathbb{Y})$. For $a \in \mathbb{X}$, a_I is of rank 0, i.e., $a_I \in Z_0$, and we let a be identified with $^*a \in \mathbb{Y}$. For a of rank $n+1$ we find that

$$^*a = \{ [f] : f \in Z_n \text{ and } [\![f \in a]\!] \in \mathcal{F} \}.$$

Consequently,

$$^*\mathbb{U}_n(\mathbb{X}) = \{ [f] : [\![f \in \mathbb{U}_n(\mathbb{X})]\!] \in \mathcal{F} \}$$
$$= \{ [f] : f \text{ has rank } \leq n \}.$$

In particular,

$$^*\mathbb{X} = \{ [f] : [\![f \in \mathbb{X}]\!] \in \mathcal{F} \} = \mathbb{Y},$$

showing that $\mathbb{U}(\mathbb{Y})$ is just $\mathbb{U}(^*\mathbb{X})$. Also, since $[\![f \in \emptyset]\!] = \emptyset$,

$$^*\emptyset = \{ [f] : [\![f \in \emptyset]\!] \in \mathcal{F} \} = \emptyset.$$

To show that the transfer principle holds for this construction requires the demonstration of a version of Łoś's theorem. This takes the following form.

For any $\mathcal{L}_{\mathbb{U}(\mathbb{X})}$-formula $\varphi(x_1, \ldots, x_p)$ and any $f_1, \ldots, f_p \in Z$, the sentence

$$^*\varphi([f_1], \ldots, [f_p])$$

is true if and only if

$$\{i \in I : \varphi(f_1(i), \ldots, f_p(i)) \text{ is true}\} \in \mathcal{F}.$$

If the ultrafilter $\mathcal{F}$ is nonprincipal, then $^*\mathbb{R}$ will have nonstandard members, as shown in Sections 3.8 and 3.9. But to obtain $\mathbb{U}(^*\mathbb{X})$ as an enlargement of $\mathbb{U}(\mathbb{X})$ a special ultrafilter $\mathcal{F}$ has to be used. To define this, let $I = \mathcal{P}_F(\mathbb{U}(\mathbb{X}))$, the set of all finite subsets of $\mathbb{U}(\mathbb{X})$. Each member of I belongs to $\mathbb{U}(\mathbb{X})$, but I itself does not. For each $a \in I$, define

$$I_a = \{ b \in I : a \subseteq b \}.$$

Then if $a_1, \ldots, a_n \in I$, putting $b = a_1 \cup \cdots \cup a_n \in I$ gives
$$b \in I_{a_1} \cap \cdots \cap I_{a_n}.$$
This shows that the collection $\{I_a : a \in I\}$ has the finite intersection property, and so is included in some ultrafilter $\mathcal{F}$ on I (Theorem 2.6.1).

To show that the superstructure $\mathbb{U}(^*\mathbb{X})$ associated with this particular $\mathcal{F}$ is an enlargement, let $A \in \mathbb{U}(\mathbb{X})$ be a collection of sets that has the finite intersection property. For each $b \in I$, $b \cap A$ is a finite subcollection of A, so if $b \cap A$ is nonempty, it has nonempty intersection. In that case, let $f(b)$ be any member of this intersection. Hence
$$f(b) \in \bigcap(b \cap A) \in \mathcal{P}(\bigcup A).$$
If, however, $b \cap A = \emptyset$, let $f(b) = \emptyset$. The resulting function f has bounded rank, and so determines an element $[f]$ of $\mathbb{U}(^*\mathbb{X})$. It will suffice to show that
$$[f] \in \bigcap\{^*B : B \in A\}.$$
Now if $B \in A$, then $\{B\} \in I$, and so $I_{\{B\}} = \{b \in I : B \in b\} \in \mathcal{F}$. But now if $B \in b$, then $B \in b \cap A$, and so $f(b) \in B$. This shows that
$$I_{\{B\}} \subseteq \{b \in I : f(b) \in B\} = [\![f \in B]\!],$$
giving $[\![f \in B]\!] \in \mathcal{F}$, and therefore $[f] \in {^*B}$, as desired. This completes the proof of the enlargement theorem.

An alternative proof that $\mathbb{U}(^*\mathbb{X})$ is an enlargement can be given by directly proving the hyperfinite approximation property that for any set $A \in \mathbb{U}(\mathbb{X})$ there exists a B with
$$^{\text{im}}A \subseteq B \in {^*\mathcal{P}_F(A)}. \tag{i}$$
For such an A put $g(b) = b \cap A$ to define a function $g : I \to \mathcal{P}_F(A)$. Then
$$[\![g \in \mathcal{P}_F(A)]\!] = I \in \mathcal{F},$$
making $[g] \in {^*\mathcal{P}_F(A)}$. But for $a \in A$, we have $a \in b \cap A$ when $a \in b$, so
$$I_{\{a\}} = \{b \in I : a \in b \cap A\} = [\![a \in g]\!],$$
implying that $^*a \in [g]$. Thus putting $B = [g]$ fulfills (i).

14.4 Exercises on the Ultrapower Construction

(1) Verify the details of the ultrapower construction of enlargements.

(2) Suppose $F \in Z$ is such that $F(i)$ is a function for (almost all) $i \in I$. Show that $[F]$ is a function satisfying
$$[F]([h]) = [k] \quad \text{iff} \quad \{i : F(i)(h(i)) = k(i)\} \in \mathcal{F}.$$

(3) Is the ultrafilter used in the above construction nonprincipal?

15
Permanence, Comprehensiveness, Saturation

Nonstandard analysis introduces a brave new world of mathematical entities. It also has a number of distinctive structural features and principles of reasoning that can be used to explore this world. Already in the context of subsets of *$\mathbb{R}$ we have examined several of these principles: permanence, internal induction, overflow, underflow, saturation. Now we will see that in the context of a universe embedding $\mathbb{U} \xrightarrow{*} \mathbb{U}'$ they occur in a much more powerful form, since they apply to properties that may refer to any internal entities in $\mathbb{U}'$. We assume from now on that we are dealing with such an embedding for which *$\mathbb{N} - \mathbb{N} \neq \emptyset$.

15.1 Permanence Principles

Several times we have discussed situations in which a property of a certain kind that holds on a particular type of set must continue to hold on some larger set (cf. Sections 7.10 and 11.9). Here is a statement of how such situations occur in $\mathbb{U}'$:

Theorem 15.1.1 *Let $\varphi(x)$ be an internal $\mathcal{L}_{\mathbb{U}'}$-formula with only the variable x free. Then*

(1) *(Overflow) If there exists $k \in \mathbb{N}$ such that $\varphi(n)$ is true for all $n \in \mathbb{N}$ with $k \leq n$, then there exists $K \in {}^*\mathbb{N} - \mathbb{N}$ such that $\varphi(n)$ is true for all $n \in {}^*\mathbb{N}$ with $k \leq n \leq K$.*

(2) *(Underflow)* If there exists $K \in {}^*\mathbb{N} - \mathbb{N}$ such that $\varphi(n)$ is true for all $n \in {}^*\mathbb{N} - \mathbb{N}$ with $n \leq K$, then there exists $k \in \mathbb{N}$ such that $\varphi(n)$ is true for all $n \in \mathbb{N}$ with $k \leq n$.

(3) If $\varphi(a)$ is true for all hyperreals a that are infinitely close to some $b \in {}^*\mathbb{R}$, then $\varphi(a)$ is true for all hyperreals a that are within some positive real distance of b.

(4) If $\varphi(a)$ is true for arbitrarily large (resp. small) $a \in \mathbb{R}$, then it is true for some positive (resp. negative) unlimited hyperreal a.

(5) If there exists $r \in \mathbb{R}$ such that $\varphi(a)$ is true for all $a \in \mathbb{R}$ with $r \leq a$, then there exists a positive unlimited $b \in {}^*\mathbb{R}$ such that $\varphi(a)$ is true for all $a \in {}^*\mathbb{R}$ with $r \leq a \leq b$.

Proof.

(1) We adapt the proof of Theorem 11.4.1. Formula $(k < x) \land \neg\varphi(x)$ is internal, and ${}^*\mathbb{N}$ is internal, so by the version in Section 13.15 of the internal set definition principle,

$$Y = \{n \in {}^*\mathbb{N} : k < n \text{ and not } \varphi(n)\}$$

is an internal set. Now, if $\varphi(n)$ is true for all ${}^*\mathbb{N} - \mathbb{N}$, then any $K \in {}^*\mathbb{N}-\mathbb{N}$ gives the desired result (remember we assume ${}^*\mathbb{N}-\mathbb{N} \neq \emptyset$). Otherwise Y is nonempty, and so by the internal least number principle has a least element H. Then H is unlimited and $K = H - 1$ gives the result.

(2) Adapt the proof of Theorem 11.8.1, using the internal formula

$$y < K \land (\forall x \in {}^*\mathbb{N})\,(y \leq x \leq K \rightarrow \varphi(x)).$$

(3) Adapt the proof of Theorem 11.9.1.

(4) Exercise, using internal order-completeness. Likewise for (5). □

There are other permanence results in this vein, analogous to the ones appearing in Exercise 11.5.2 and Section 11.6. Formulation and proofs of these are left to the reader.

Exercise 15.1.2
Let f be an internal ${}^*\mathbb{R}$-valued function such that $f(x)$ in limited for all x in some internal set $A \subseteq \operatorname{dom} f$. Show that there is a standard $n \in \mathbb{N}$ such that $|f(x)| < n$ for all $x \in A$.

15.2 Robinson's Sequential Lemma

Overflow has the following useful consequence:

Lemma 15.2.1 *If $s : {}^*\mathbb{N} \to {}^*\mathbb{R}$ is an internal hypersequence such that s_n is infinitesimal for all standard $n \in \mathbb{N}$, then there is an unlimited $K \in {}^*\mathbb{N}$ such that s_n is infinitesimal for all hypernatural $n \leq K$.*

Proof. We cannot just apply overflow to $\{n \in {}^*\mathbb{N} : s_n \simeq 0\}$, since we do not know that this set is internal. Instead we use $\{n \in {}^*\mathbb{N} : |s_n| < \frac{1}{n}\}$. Because s is internal, the formula

$$|x \cdot s(x)| < 1$$

is internal, and for each $n \in \mathbb{N}$, since $s_n \simeq 0$ we have $n \cdot s_n \simeq 0$, and so $|n \cdot s_n| < 1$. Hence by overflow there is a $K \in {}^*\mathbb{N}_\infty$ such that if $n \leq K$, then $|n \cdot s_n| < 1$ and so $|s_n| < \frac{1}{n}$. But when such an n is unlimited, $\frac{1}{n} \simeq 0$ and so $s_n \simeq 0$. □

The argument just given can be adapted to apply to an internal hypersequence of the form $s : \{n \in {}^*\mathbb{N} : n \leq N\} \to {}^*\mathbb{R}$ with N unlimited, by using the internal formula $x \leq N \wedge |x \cdot s(x)| < 1$.

Exercise 15.2.2
If $c \in {}^*\mathbb{R}$ and $s : {}^*\mathbb{N} \to {}^*\mathbb{R}$ (or $s : \{n \in {}^*\mathbb{N} : n \leq N\} \to {}^*\mathbb{R}$) is an internal hypersequence such $s_n \simeq c$ for all standard $n \in \mathbb{N}$, then there is an unlimited $K \in {}^*\mathbb{N}$ such that $s_n \simeq c$ for all hypernatural $n \leq K$. □

Variations on the theme of Robinson's lemma can derived from other permanence principles. For instance:

Theorem 15.2.3 *If f is an internal ${}^*\mathbb{R}$-valued function and $f(x)$ is infinitesimal for all limited hyperreals x, then there is an unlimited b such that $f(x)$ is infinitesimal for all $x \in [-b, b] \subseteq {}^*\mathbb{R}$.*

Proof. As for the proof of Lemma 15.2.1, but using the internal formula $|x \cdot f(x)| < 1$ and the fact that if an internal set includes $\mathbb{L}$ then it includes $[-b, b]$ for some unlimited b (Section 11.6). □

15.3 Uniformly Converging Sequences of Functions

Robinson's sequential lemma (15.2.1) can be used to give an interesting alternative proof of the following classical result that was already discussed in Section 7.13:

- If a sequence $\langle f_n : n \in \mathbb{N} \rangle$ of continuous real-valued functions on the closed interval $[a,b] \subseteq \mathbb{R}$ converges uniformly to a function g, then g is continuous.

The sequence $\langle f_n : n \in \mathbb{N} \rangle$ is identifiable with a function of the type

$$f : \mathbb{N} \to \mathbb{R}^{[a,b]}$$

(cf. Section 7.12). This function belongs to $\mathbb{U}$ and transforms to

$${}^*f : {}^*\mathbb{N} \to {}^*\left(\mathbb{R}^{[a,b]}\right).$$

But ${}^*\left(\mathbb{R}^{[a,b]}\right)$ is the set of all internal functions from ${}^*[a,b]$ to ${}^*\mathbb{R}$, so we get

$$ {}^*f_n : {}^*[a,b] \to {}^*\mathbb{R}$$

for all $n \in {}^*\mathbb{N}$. Thus the original sequence extends to an internal hypersequence ${}^*f = \langle {}^*f_n : n \in {}^*\mathbb{N} \rangle$. For standard n and x, ${}^*f_n(x) = f_n(x)$.

Now, uniform convergence of the f_n's to g on $[a,b]$ means that for each $\varepsilon \in \mathbb{R}^+$ there exists a $k \in \mathbb{N}$ such that

$$(\forall x \in [a,b])\,(\forall n \in \mathbb{N})\,[k \leq n \to |f_n(x) - g(x)| < \varepsilon].$$

Then by transfer it follows that $|{}^*f_n(x) - {}^*g(x)| < \varepsilon$ whenever $x \in {}^*[a,b]$ and $k \leq n \in {}^*\mathbb{N}$. In particular, this will hold for any *unlimited* n, whatever positive real ε is taken here. Therefore

$$ {}^*f_n(x) \simeq {}^*g(x) \quad \text{for any } n \in {}^*\mathbb{N} - \mathbb{N} \text{ and } x \in {}^*[a,b]. \tag{i}$$

To prove that g is continuous at any real $c \in [a,b]$ we apply the theory of Section 7.1 and show that if $x \simeq c$ in ${}^*[a,b]$, then ${}^*g(x) \simeq {}^*g(c)$. But given $x \simeq c$, then for $n \in \mathbb{N}$, ${}^*f_n(x) \simeq {}^*f_n(c)$ by continuity of f_n. Thus

$$|{}^*f_n(x) - {}^*f_n(c)| \simeq 0 \quad \text{for all } n \in \mathbb{N}.$$

But the hypersequence $\langle |{}^*f_n(x) - {}^*f_n(c)| : n \in {}^*\mathbb{N} \rangle$ is internal (by the internal function definition principle 13.16), and therefore by Robinson's lemma (15.2.1) it follows that there is some unlimited K for which $|{}^*f_K(x) - {}^*f_K(c)| \simeq 0$, and so

$$ {}^*f_K(x) \simeq {}^*f_K(c).$$

Then by (i) we get that ${}^*f_K(x) \simeq {}^*g(x)$, and ${}^*f_K(c) \simeq {}^*g(c)$, so ${}^*g(x) \simeq {}^*g(c)$ follows as desired.

Exercise 15.3.1
Show that the condition (i) above is also *sufficient* for the sequence $\langle f_n : n \in \mathbb{N} \rangle$ of continuous real-valued functions on the closed interval $[a,b] \subseteq \mathbb{R}$ to converge uniformly to the function g.

15.4 Comprehensiveness

We know that a real-valued sequence $\langle s_n : n \in \mathbb{N}\rangle$ extends to a hyperreal-valued function $\langle s_n : n \in {}^*\mathbb{N}\rangle$ that is internal. But what about a hyperreal-valued sequence $\mathbb{N} \to {}^*\mathbb{R}$? Does this have an internal extension of the form ${}^*\mathbb{N} \to {}^*\mathbb{R}$?

A universe embedding is called *comprehensive* if for each set $A \in \mathbb{U}$ and each *internal* set $B \in {}^*\mathbb{U}$, any function $f : A \to B$ extends to an internal function ${}^+f : {}^*A \to B$, in the sense that ${}^+f({}^*a) = f(a)$ for all $a \in A$.

The embedding is called *countably comprehensive* if this condition holds whenever A is countable. If it holds whenever $A = \mathbb{N}$, we will call it *sequentially comprehensive*. Since ${}^*a = a$ for $a \in \mathbb{N}$, it follows that in a sequentially comprehensive embedding, any sequence $\mathbb{N} \to {}^*\mathbb{R}$ does indeed have an internal extension of the form ${}^*\mathbb{N} \to {}^*\mathbb{R}$. More generally, any sequence $\langle s_n : n \in \mathbb{N}\rangle$ of elements of any internal set B will extend to an internal hypersequence $\langle s_n : n \in {}^*\mathbb{N}\rangle$ of elements of B.

The ultrapower construction of Section 14.3 always produces a comprehensive embedding of a superstructure $\mathbb{U}(\mathbb{X})$. The reason for this, which involves some intricate details, is as follows. If B is internal, then it is equal to $[H]$, where $H \in Z$ is a function with some rank n, and we can suppose that $H(i) \in \mathbb{U}_n(\mathbb{X})$ for all $i \in I$. For each $b \in B$ choose a function $\rho_b \in Z$ to represent b, i.e., $b = [\rho_b]$. In particular, for each $a \in A$ we have $f(a) \in B$, and so $f(a) = [\rho_{f(a)}] \in [H]$. Hence we can suppose $\rho_{f(a)}(i) \in H(i)$ for all $i \in I$.

Now for each $i \in I$, define $f_i : A \to H(i)$ by putting $f_i(a) = \rho_{f(a)}(i)$ for all $a \in A$. Then $f_i \in (\mathbb{U}_n(\mathbb{X}))^A$, so putting $F(i) = f_i$ makes F a function on I of bounded rank, i.e., $F \in Z$. Let ${}^+f = [F] \in \mathbb{U}({}^*\mathbb{X})$.

Using the fact that each $F(i)$ is a function from A to $H(i)$, it can be shown (with the help of Łoś's theorem) that the internal entity $[F]$ is a function from ${}^*A$ to $[H] = B$ as desired. Its action on any $[h] \in {}^*A$ (with $h(i) \in A$ for all $i \in I$) is given by $[F]([h]) = [k]$, where k is the function $k(i) = F(i)(h(i))$. Thus in general,

$$k(i) = f_i(h(i)) = \rho_{f(h(i))}(i),$$

and so

$$ {}^+f([h]) = [\langle \rho_{f(h(i))}(i) : i \in I \rangle].$$

But for $a \in A$, we have ${}^*a = [a_I]$ where $a_I(i) = a$, and so

$$\begin{aligned}{}^+f({}^*a) = {}^+f([a_I]) &= [\langle \rho_{f(a_I(i))}(i) : i \in I\rangle] \\ &= [\langle \rho_{f(a)}(i) : i \in I\rangle] \\ &= [\rho_{f(a)}] = f(a),\end{aligned}$$

showing that ${}^+f$ extends f in the manner required to prove comprehensiveness of the embedding of $\mathbb{U}(\mathbb{X})$ into $\mathbb{U}({}^*\mathbb{X})$.

Exercise 15.4.1
For $f \mapsto {}^+f$ defined as above, prove the following.

(i) If two functions $f, g : A \to B$ differ only at finitely many elements $a_1, \ldots, a_n$ of A, then ${}^+f$ and ${}^+g$ differ only at ${}^*a_1, \ldots, {}^*a_n$.

(2) If $f : A \to {}^*C$ is the extension by $*$ of some function $h : A \to C$ in $\mathbb{U}(\mathbb{X})$, i.e., $f(a) = {}^*(h(a))$ for $a \in A$, then ${}^+f$ is just the $*$-transform ${}^*h : {}^*A \to {}^*C$ of h. (Use a suitable choice of the representatives ρ_b in this case.) □

Theorem 15.4.2 *For any universe embedding, the following are equivalent.*

(1) *The embedding is countably comprehensive with ${}^*\mathbb{N} - \mathbb{N} \neq \emptyset$.*

(2) *The embedding is sequentially comprehensive with ${}^*\mathbb{N} - \mathbb{N} \neq \emptyset$.*

(3) *(Countable Saturation) Every decreasing sequence of nonempty internal sets has nonempty intersection.*

Proof. (2) follows as a special case from (1).

To show that (2) implies (3), let $\langle A_n : n \in \mathbb{N} \rangle$ be a sequence of nonempty internal sets, with $A_n \supseteq A_{n+1}$. Because A_1 is internal, it is a subset of some standard transitive set ${}^*T$ (Theorem 13.10.1). Put $B = {}^*\mathcal{P}(T)$. Now, each A_n is an internal subset of A_1, hence of ${}^*T$, and so belongs to the internal set B. Thus by sequential comprehensiveness the sequence $\langle A_n : n \in \mathbb{N} \rangle$ extends to an internal hypersequence $\langle A_n : n \in {}^*\mathbb{N} \rangle$ of elements of B.

By hypothesis, for each standard $k \in \mathbb{N}$ we have

$$(\forall n \in {}^*\mathbb{N})\,[n \leq k \text{ implies } A_n \supseteq A_k \neq \emptyset].$$

But this is an internal statement, so if ${}^*\mathbb{N} - \mathbb{N} \neq \emptyset$, we can apply overflow to conclude that it must hold for some unlimited k. Then for such a k we have

$$\emptyset \neq A_k \subseteq \bigcap_{n \in \mathbb{N}} A_n,$$

which establishes (3).

Finally, we derive (1) from (3). First, put $A_n = \{k \in {}^*\mathbb{N} : n \leq k\}$. Then for each $n \in \mathbb{N}$, A_n is internal by the internal set definition principle 13.15, with $A_n \supseteq A_{n+1}$. Hence by (3) there exists some $K \in \bigcap \{A_n : n \in \mathbb{N}\}$, which entails $K \in {}^*\mathbb{N} - \mathbb{N}$, so ${}^*\mathbb{N} - \mathbb{N} \neq \emptyset$.

Now to prove countable comprehensiveness. Let $A = \{a_n : n \in \mathbb{N}\}$ be a countable member of $\mathbb{U}$, and $f : A \to B$ with B internal. For each $n \in \mathbb{N}$ let A_n be the set of all internal functions from ${}^*A$ to B that have $g({}^*a_k) = f(a_k)$ for all $k \leq n$. Then $A_n \supseteq A_{n+1}$, and A_n can be shown to be nonempty and internal. Hence by (3) there exists a $g \in \bigcap \{A_n : n \in \mathbb{N}\}$. Such a g is an internal function from ${}^*A$ to B that has $g({}^*a_n) = f(a_n)$ for all $n \in \mathbb{N}$, so $g({}^*a) = f(a)$ for all $a \in A$ as desired.

15.4 Comprehensiveness   197

To see that A_n is internal, take a standard set *T with *$A, B \subseteq$ *T. Let $\varphi_n(g)$ be the statement

- g *is a function from* *A *to* B *such that* $g(*a_1) = f(a_1), \cdots, g(*a_n) = f(a_n)$.

For each particular $n \in \mathbb{N}$, $\varphi_n(g)$ is expressed by an internal formula. Also, an internal function from *A to B is a subset of *$T \times$ *$T = $ *$(T \times T)$ and hence belongs to the transformed power set *$\mathcal{P}(T \times T)$. Thus

$$A_n = \{g \in {}^*\mathcal{P}(T \times T) : \varphi_n(g) \text{ is true}\},$$

and this is internal by the internal set definition principle.

Lastly, A_n is shown to be nonempty by transfer of the property that for any set $Y \in \mathcal{P}(T)$, and any $x_1, \ldots, x_n \in Y$, there is a $g \in \mathcal{P}(A \times T)$ that is a function from A to Y such that $g(a_1) = x_1, \cdots, g(a_n) = x_n$. The transferred property is applied with $Y = B \in {}^*\mathcal{P}(T)$ and $x_i = {}^*a_i$ to obtain a function from *A to B that meets the definition of A_n. □

The property expressed in Theorem 15.4.2(3) was derived in Section 11.10 for sequences of internal subsets of *$\mathbb{R}$. Now we have seen that it holds for any nested sequence of sets from *$\mathbb{U}$. From what we have learned about ultrapowers in this section it follows that the particular ultrapower construction in Section 14.3 produces an enlargement that is comprehensive, hence countably comprehensive, and so by Theorem 15.4.2 is countably saturated. The argument of Section 11.12 then applies to show that any infinite set in *$\mathbb{U}$ is uncountable.

An instance of this last property is the fact that an unlimited hypernatural number K has uncountably many hypernaturals greater than it, i.e., the internal set $\{n \in {}^*\mathbb{N} : K < n\}$ is uncountable. Under countable saturation we also have the result that K has uncountably many unlimited hypernaturals less than it: the internal set $\{n \in {}^*\mathbb{N} : n < K\}$ is uncountable, while its limited part $\mathbb{N}$ is countable. This result can instead be obtained from the following theorem, due to Abraham Robinson, which we derived using countable saturation in Section 11.11. Here now is Robinson's own proof, which uses sequential comprehensiveness and indeed was his original reason for introducing the comprehensiveness notion.

Theorem 15.4.3 *In a sequentially comprehensive enlargement, if X is any countable set of unlimited hypernaturals, then there is an unlimited hypernatural K less than every member of X.*

Proof. Write $X = \langle s_n : n \in \mathbb{N}\rangle$ and extend this to an *internal* hypersequence $\langle s_n : n \in {}^*\mathbb{N}\rangle$ by sequential comprehensiveness. Put

$$Y = \{k \in {}^*\mathbb{N} : (\forall n \in {}^*\mathbb{N})(n \leq k \to k < s_n)\}.$$

Then Y is internal, as its defining formula is internal. Moreover, Y contains each standard $k \in \mathbb{N}$, since in that case $s_1, \ldots, s_k$ are all in X, hence unlimited and so greater than k. By overflow, then, there is some unlimited K in Y. Then $K < s_n$ whenever $n \leq K$, which includes all $n \in \mathbb{N}$. Hence $K < x$ whenever $x \in X$. □

Yet another proof of this result can be obtained from Robinson's sequential lemma, by the following construction. Given an internal hypersequence $s : {}^*\mathbb{N} \to {}^*\mathbb{R}$, define the hypersequence $\underline{s} = \langle \underline{s}_n : n \in {}^*\mathbb{N}\rangle$ by putting

$$\underline{s}_n = \min\{s_m : m \leq n\}.$$

Note that $\{s_m : m \leq n\}$ is hyperfinite, being the image of $\{m \in {}^*\mathbb{N} : m \leq n\}$ under the internal function s, so $\underline{s}_n$ is indeed defined (13.18(3)). The following exercise completes the proof.

Exercise 15.4.4
Show that $\underline{s}$ is internal. If $s_n \in {}^*\mathbb{N}_\infty$ for all $n \in \mathbb{N}$, apply Robinson's sequential lemma to the reciprocals of the $\underline{s}_n$'s to obtain an unlimited hypernatural number that is less than every member of $\{s_n : n \in \mathbb{N}\}$.

15.5 Saturation

Countable saturation is itself equivalent to the assertion (cf. Corollary 11.10.2(1)):

- *Every countable collection of internal sets with the finite intersection property has nonempty intersection.*

We should not, however, expect that the corresponding statement always holds for uncountable collections of internal sets. For instance, the collection

$$\{{}^*\mathbb{R} - \{r\} : r \in {}^*\mathbb{R}\}$$

of internal sets has empty intersection, but does have the finite intersection property.

If κ is a cardinal number, then an enlargement is called κ-*saturated* when

- *any collection of fewer than κ internal sets with the finite intersection property has nonempty intersection.*

(Consequently, "countably saturated" is the same as "$\aleph_1$-saturated", where $\aleph_1$ is the least uncountable cardinal.)

By using special kinds of ultrafilters it is possible to show that for any cardinal κ, any superstructure $\mathbb{U}(\mathbb{X})$ has a κ-saturated enlargement. In such an enlargement ${}^*\mathbb{R}$ must in fact be a set of size at least κ (as indeed must every infinite internal set!).

Some theorems of nonstandard analysis require κ-saturation for large κ. For instance, in the theory of topological spaces there is the property that the shadow of an internal set is topologically closed. This was demonstrated for the topology of the real line in Section 11.13. For a general topological space it requires κ-saturation with κ larger than the number of open sets in the topology.

Exercise 15.5.1
Show that κ-saturation is equivalent to the following statement:

> *If an internal relation R is concurrent on a subset A of its domain that has cardinality less than κ (A need not be internal), then there is an entity b in the range of R such that aRb for all $a \in A$.*

Part V

Applications

16
Loeb Measure

Measure theory studies operations that assign magnitudes to sets, like measuring the length of an interval, the area of a plane region, or the volume of a solid; counting the number of elements in a set; calculating the probability of an event in a sample space or the definite integral of some function over a set; etc.

Now, the "measure spaces" on which such operations are defined are typically closed under countable set unions, and this feature is fundamental to the theory. But an *internal* collection of sets typically fails to be closed in this way. However, in 1973 Peter Loeb discovered that this very failure could be exploited to give a new way of constructing standard measure spaces out of nonstandard entities. [1] This has led to some interesting applications, particularly in probability theory and stochastic analysis. For instance, it provides a representation of Brownian motion as a "random walk with infinitesimal steps".

We will now develop Loeb's construction, elucidating the role played in it by the nonstandard principles of *countable saturation, sequential comprehensiveness,* and *overflow*. We will then apply it to show that Lebesgue measure on the real line can be represented by a weighted counting measure on hyperfinite sets, using infinitesimal weights.

But first, a review of some of the basic concepts of measure theory.

---

[1] See example 6 of Section 16.1 and example 3 of Section 16.3.

16.1 Rings and Algebras

A *ring* of sets is a nonempty collection $\mathcal{A}$ of subsets of a set S that is closed under set differences and unions:

- If $A, B \in \mathcal{A}$ then $A - B, A \cup B \in \mathcal{A}$.

It follows that $\emptyset \in \mathcal{A}$, since $A - A = \emptyset$, and that $\mathcal{A}$ is closed under symmetric differences $A \Delta B$ and intersections $A \cap B$, since

$$A \Delta B = (A - B) \cup (B - A), \quad \text{and}$$
$$A \cap B = A - (A - B).$$

An *algebra* is a ring $\mathcal{A}$ that has $S \in \mathcal{A}$ and hence (indeed equivalently) is closed under complements $A^c = S - A$. If $\mathcal{A}$ is a ring, then $\mathcal{A} \cup \{S - A : A \in \mathcal{A}\}$ is an algebra, the smallest one including $\mathcal{A}$.

A *σ-ring* is a ring that is closed under countable unions:

- If $A_n \in \mathcal{A}$ for all $n \in \mathbb{N}$, then $\bigcup_{n \in \mathbb{N}} A_n \in \mathcal{A}$.

The equation

$$\bigcap_{n \in \mathbb{N}} A_n = A_1 - \left(\bigcup_{n \in \mathbb{N}} (A_1 - A_n) \right)$$

shows that a σ-ring is also closed under countable intersections.

A *σ-algebra* is a σ-ring that is an algebra. The intersection of any family of σ-algebras is a σ-algebra. Thus for any $\mathcal{A} \subseteq \mathcal{P}(S)$ there is a smallest σ-algebra $\mathcal{S}(\mathcal{A})$ that includes $\mathcal{A}$. This $\mathcal{S}(\mathcal{A})$ is the σ-algebra *generated by* $\mathcal{A}$.

Here are some examples of these concepts:

(1) $\mathcal{P}(S)$ itself is a σ-algebra.

(2) If S is infinite, then

- the collection of all finite subsets of S is a ring that is not an algebra;
- the collection of all finite or cofinite subsets of S is an algebra that is not a σ-algebra;
- the collection of all countable subsets of S is a σ-ring that is not an algebra when S is uncountable.

(3) Let $\mathcal{C}_\mathbb{R}$ be the collection of all subsets of $\mathbb{R}$ that are finite unions of *left-open* intervals $(a, b] = \{x \in \mathbb{R} : a < x \leq b\}$ with $a, b \in \mathbb{R}$ and $a \leq b$. (Thus $\emptyset = (a, a] \in \mathcal{C}_\mathbb{R}$.) $\mathcal{C}_\mathbb{R}$ is ring in which each member is in fact a *disjoint* union of left-open intervals $(a, b]$. $\mathcal{C}_\mathbb{R}$ is not an algebra, and is not closed under countable unions: $(0, 1)$ is not in $\mathcal{C}_\mathbb{R}$, since each member of $\mathcal{C}_\mathbb{R}$ will have a greatest element, but $(0, 1)$ is the union of the intervals $(0, 1 - \frac{1}{n}]$ for $n \in \mathbb{N}$.

$\mathcal{C}_\mathbb{R}$ does, however, contain certain significant countable unions: for instance $(0, 1]$ is the union of the pairwise disjoint intervals $(\frac{1}{n+1}, \frac{1}{n}]$. Any reasonable notion of measure should thus assign to $(0, 1]$ the infinite sum of the measures of the intervals $(\frac{1}{n+1}, \frac{1}{n}]$.

(4) Let $\mathcal{B}_\mathbb{R}$ be the σ-algebra generated by $\mathcal{C}_\mathbb{R}$. Each open interval (a, b) in $\mathbb{R}$ is in $\mathcal{B}_\mathbb{R}$, being the union of the countably many left-open intervals $(a, b - \frac{1}{n}]$ for $n \in \mathbb{N}$. Hence every open subset of $\mathbb{R}$ is in $\mathcal{B}_\mathbb{R}$, being the union of countably many open intervals (take ones with rational end points).

On the other hand, if a σ-algebra contains all open intervals, it must contain any left-open $(a, b]$ as the intersection of all $(a, b + \frac{1}{n})$ for $n \in \mathbb{N}$. Thus $\mathcal{B}_\mathbb{R}$ is also the σ-algebra generated by the open intervals, as well as the σ-algebra generated by the open sets of $\mathbb{R}$.

The members of $\mathcal{B}_\mathbb{R}$ are called the *Borel* sets.

(5) Let $S = \{1, \ldots, N\}$ with N an unlimited hypernatural. Then S is hyperfinite, and the collection $\mathcal{P}_I(S)$ of all internal subsets of S is an algebra (also hyperfinite) that by transfer of the finite case will be closed under *hyperfinite* unions, i.e., unions of internal sequences $\langle A_n : n \leq K \rangle$ for $K \in {}^*\mathbb{N}$. $\mathcal{P}_I(S)$ is not, however, a σ-algebra: it contains each initial segment $\{1, \ldots, n\}$ with $n \in \mathbb{N}$, but does not contain their union because that is the *external* set $\mathbb{N}$.

This same analysis applies to the algebra of internal subsets of any nonstandard hyperfinite set $S = \{s_n : n \leq N\}$.

(6) Let $\mathcal{A}$ be an algebra in some universe $\mathbb{U}$. In any enlargement of $\mathbb{U}$, ${}^*\mathcal{A}$ will be an algebra, by transfer, but in a *countably saturated* enlargement ${}^*\mathcal{A}$ will not in general be a σ-algebra, even if $\mathcal{A}$ is. To see this, let $\langle A_n : n \in \mathbb{N} \rangle$ be a sequence of members of ${}^*\mathcal{A}$ with union A. Each A_n is internal, and if A were in $\mathcal{A}$, it would also be internal and hence by countable saturation would be equal to $\bigcup_{n \leq k} A_n$ for some $k \in \mathbb{N}$ (cf. Corollary 11.10.2). Thus if A is a genuinely infinite union of the A_n's, it cannot be in ${}^*\mathcal{A}$. This will happen, for example, if the A_n's are strictly increasing ($A_n \subsetneq A_{n+1}$) or pairwise disjoint.

For instance, in the case of the Borel algebra the internal sets ${}^*(-n, n)$ belong to ${}^*\mathcal{B}_\mathbb{R}$ for all $n \in \mathbb{N}$, but their union is not in ${}^*\mathcal{B}_\mathbb{R}$ because it is the *external* set of all limited hyperreals.

The closure condition that we do get for ${}^*\mathcal{A}$ is that the sequence $\langle A_n : n \in \mathbb{N} \rangle$ extends to an internal sequence $\langle A_n : n \in {}^*\mathbb{N} \rangle$ whose union can be shown by transfer to be in ${}^*\mathcal{A}$. In this sense ${}^*\mathcal{A}$ is a "hyper-σ-algebra", but that is not the type of structure on which a standard measure is defined.

This reasoning in fact shows that for any internal algebra of sets (not just one of the form *$\mathcal{A}$)

- *the union of a countable sequence of sets can belong to the algebra only if it is equal to the union of finitely many of its terms.*

It is this feature upon which Loeb measure is founded.

16.2 Measures

Standard measure theory employs the *extended real numbers*

$$[-\infty, +\infty] = \{-\infty\} \cup \mathbb{R} \cup \{+\infty\},$$

with $-\infty < r < +\infty$ for $r \in \mathbb{R}$, $r \pm \infty = \pm\infty$, etc. We will usually put ∞ for $+\infty$, and also make use of the set $[0, \infty] = \{r \in \mathbb{R} : r \geq 0\} \cup \{\infty\}$.

Let $\mathcal{A}$ be a ring of subsets of a set S, and μ a function from $\mathcal{A}$ to $[0, \infty]$ that has $\mu(\emptyset) = 0$. Then μ is called a *measure* if it satisfies:

(M1) *If $\langle A_n : n \in \mathbb{N}\rangle$ is a sequence of pairwise disjoint elements of $\mathcal{A}$ whose union is in $\mathcal{A}$, then*

$$\mu(\bigcup_{n\in\mathbb{N}} A_n) = \sum_{n\in\mathbb{N}} \mu(A_n).$$

This condition is called *countable additivity*. Note especially that it is not required to hold for all (pairwise disjoint) sequences $\langle A_n : n \in \mathbb{N}\rangle$, but only those whose union happens to belong to $\mathcal{A}$ (which is not guaranteed when $\mathcal{A}$ is not a σ-algebra).

The function μ is called *finitely additive* if in place of M1 it satisfies

(M2) $\mu(A \cup B) = \mu(A) + \mu(B)$ whenever $A, B \in \mathcal{A}$ with $A \cap B = \emptyset$.

Since a ring is closed under finite unions, M2 implies that

$$\mu(\bigcup_{i=1}^n A_i) = \sum_{i=1}^n (A_i)$$

whenever $A_1, \ldots, A_n$ is a finite sequence of pairwise disjoint members of $\mathcal{A}$. M2 also implies that μ is *monotonic*:

- $A \subseteq B$ implies $\mu(A) \leq \mu(B)$, for all $A, B \in \mathcal{A}$;

as well as being *subtractive*:

- $A \subseteq B$ and $\mu(B) < \infty$ implies $\mu(B - A) = \mu(B) - \mu(A)$, for all $A, B \in \mathcal{A}$.

Countable additivity implies the following important fact:

- If $\langle A_n : n \in \mathbb{N}\rangle$ is an *increasing* sequence of elements of $\mathcal{A}$ whose union is in $\mathcal{A}$, then

$$\mu(\bigcup_{n\in\mathbb{N}} A_n) = \lim_{n\to\infty} \mu(A_n).$$

An element $A \in \mathcal{A}$ is called μ-*finite* if $\mu(A) < \infty$, and μ-*null* if $\mu(A) = 0$. The function μ itself is σ-*finite* if the set S is the union of countably many μ-finite subsets.

For example:

(1) If $A \subseteq S$, put $\mu_c(A) = \begin{cases} |A|, & \text{if } A \text{ is finite,} \\ \infty, & \text{if } A \text{ is infinite.} \end{cases}$

Then μ_c is a measure on $\mathcal{P}(S)$, the *counting measure*, which is σ-finite iff S is countable. The restriction of μ_c to the ring of finite subsets of S, or to the algebra of finite or cofinite sets, is also a measure.

(2) On the ring $\mathcal{C}_\mathbb{R}$ of disjoint unions of left-open intervals $(a, b]$, put $\lambda((a, b]) = b - a$ and extend λ additively to all members of $\mathcal{C}_\mathbb{R}$. Then λ proves to be a measure on $\mathcal{C}_\mathbb{R}$, and λ is σ-finite because $\mathbb{R}$ is the union of the intervals $(-n, n]$.

Here the symbol λ may be thought of as denoting "length", but it also stands for "Lebesgue".

(3) Consider a countably saturated enlargement of a universe over a set $\mathbb{X}$ that has $[-\infty, +\infty] \subseteq \mathbb{X}$. Then the set

$$^*[0, \infty] = \{x \in {^*\mathbb{R}} : x \geq 0\} \cup \{\infty\}$$

is internal. Now let S be a hyperfinite set and $\mathcal{P}_I(S)$ the algebra of all internal subsets of S. For each $A \in \mathcal{P}_I(S)$ put

$$\mu(A) = \frac{|A|}{|S|}$$

(where $|A|$ is the internal cardinality of the hyperfinite set A). Then μ is finitely additive in the sense of M2, because $|A \cup B| = |A| + |B|$ when $A \cap B = \emptyset$ (but note that we are referring to $+$ in $^*\mathbb{R}$ rather than $\mathbb{R}$). Since $|A| \leq |S|$ whenever $A \subseteq S$, μ takes limited values between 0 and 1, i.e., $\mu : \mathcal{P}_I(S) \to {^*[0, 1]}$. Putting

$$\mu_L(A) = \text{sh}(\mu(A))$$

then defines $\mu_L : \mathcal{P}_I(S) \to [0, 1]$ as a genuinely real-valued finitely additive measure on $\mathcal{P}_I(S)$, with $\mu_L(S) = 1$.

But μ_L is in fact a measure, for the reason explained in example 6 of Section 16.1. If $\langle A_n : n \in \mathbb{N}\rangle$ is a sequence of pairwise disjoint elements of $\mathcal{P}_I(S)$ whose union A belongs to $\mathcal{P}_I(S)$, then A must be equal to $\bigcup_{n\leq k} A_n$ for some k. But then when $m > k$, $A_m = \emptyset$, since $\bigcup_{n\leq k} A_n$ and A_m are disjoint, and so $\mu_L(A_m) = 0$. Hence

$$\bigcup_{n\in\mathbb{N}} A_n = A_1 \cup \cdots \cup A_k, \text{ and}$$
$$\sum_{n\in\mathbb{N}} \mu_L(A_n) = \mu_L(A_1) + \cdots + \mu_L(A_k),$$

from which it follows that μ_L satisfies M1.

(4) Let $\mathcal{A}$ be an internal ring of subsets of some internal set S in a countably saturated enlargement, and let $\mu : \mathcal{A} \to {}^*[0,\infty]$ be a finitely additive function. Adapting the construction of (3), put

$$\mu_L(A) = \begin{cases} \operatorname{sh}(\mu(A)), & \text{if } \mu(A) \text{ is limited,} \\ \infty, & \text{if } \mu(A) \text{ is unlimited or } \infty. \end{cases}$$

Then reasoning as in (3), we show that $\mu_L : \mathcal{A} \to [0,\infty]$ is countably additive, and so is a measure on the ring $\mathcal{A}$.

(5) This last construction has (3) as a special case, and also covers other natural extensions of (3) that involve hyperfinite summation. Let $w : S \to {}^*\mathbb{R}$ be an internal "weighting" function on a hyperfinite set S. For each $A \in \mathcal{P}_I(S)$ put

$$\mu^w(A) = \sum_{s\in A} w(s)$$

(recall the definition of hyperfinite sums in Section 13.19). Then μ^w is a "weighted counting function" that is finitely additive and induces the measure μ_L^w on $\mathcal{P}_I(S)$.

In fact, every internal finitely additive function $\mu : \mathcal{P}_I(S) \to {}^*[0,\infty]$ arises in this way: put $w(s) = \mu(\{s\})$. Example (3) itself is the special case of a *uniform* weighting in which each point is assigned the same weight $w(s) = \frac{1}{|S|}$.

16.3 Outer Measures

We now review the classical procedure of Carathéodory for extending a measure μ on a ring of sets $\mathcal{A}$ to a measure on a σ-algebra including $\mathcal{A}$.

If B is an arbitrary subset of the set S on which $\mathcal{A}$ is based, put

$$\mu^+(B) = \inf\left\{\sum_{n\in\mathbb{N}} \mu(A_n) : A_n \in \mathcal{A},\ B \subseteq \bigcup_{n\in\mathbb{N}} A_n\right\}.$$

Here the infimum is taken over all sequences $\langle A_n : n \in \mathbb{N}\rangle$ of elements of $\mathcal{A}$ that cover B. The function $\mu^+ : \mathcal{P}(S) \to [0,\infty]$ is called the *outer measure* defined by μ (although it may not actually be a measure). It has the following properties:

- μ^+ agrees with μ on $\mathcal{A}$: if $B \in \mathcal{A}$, then $\mu^+(B) = \mu(B)$.
  In particular, $\mu^+(\emptyset) = 0$.

- *Monotonicity*: if $A \subseteq B$, then $\mu^+(A) \leq \mu^+(B)$.

- *Countable subadditivity*: for any sequence $\langle A_n \rangle$ of subsets of S,
$$\mu^+(\bigcup_{n \in \mathbb{N}} A_n) \leq \sum_{n \in \mathbb{N}} \mu^+(A_n).$$

- For any $B \subseteq S$ and any $\varepsilon \in \mathbb{R}^+$ there is an increasing sequence $A_1 \subseteq A_2 \subseteq \cdots$ of $\mathcal{A}$-elements that covers B and has
$$\mu^+(\bigcup_{n \in \mathbb{N}} A_n) \leq \mu^+(B) + \varepsilon.$$

A set $B \subseteq S$ is called μ^+-*measurable* if it splits every set $E \subseteq S$ μ^+-additively, in the sense that
$$\mu^+(E) = \mu^+(E \cap B) + \mu^+(E - B).$$

For this to hold it is enough that
$$\mu^+(E) \geq \mu^+(E \cap B) + \mu^+(E - B)$$

whenever $\mu^+(E) < \infty$.

The class $\mathcal{A}(\mu)$ of all μ^+-measurable sets has the following properties.

- $\mathcal{A}(\mu)$ is a σ-algebra.

- $\mathcal{A} \subseteq \mathcal{A}(\mu)$, i.e., all members of $\mathcal{A}$ are μ^+-measurable. Hence $\mathcal{A}(\mu)$ includes the σ-algebra $\mathcal{S}(\mathcal{A})$ generated by $\mathcal{A}$.

- All μ^+-null sets belong to $\mathcal{A}(\mu)$.

- μ^+ is a measure on $\mathcal{A}(\mu)$, and hence is a measure on $\mathcal{S}(\mathcal{A})$.

- If μ is σ-finite on $\mathcal{A}$, and $\mathcal{A}$ is an algebra, then μ^+ is the only extension of μ to a measure on $\mathcal{S}(\mathcal{A})$ or on $\mathcal{A}(\mu)$.

Because $\mathcal{A}(\mu)$ contains all μ^+-null sets, μ^+ is a *complete* measure on it, which means that

- if $A \subseteq B \in \mathcal{A}(\mu)$ and $\mu^+(B) = 0$, then $A \in \mathcal{A}(\mu)$.

This entails that

- if $A, B \in \mathcal{A}(\mu)$ with $A \subseteq B$ and $\mu^+(A) = \mu^+(B)$, then any subset of $B - A$ belongs to $\mathcal{A}(\mu)$ (and is μ^+-null), and hence any set C with $A \subseteq C \subseteq B$ belongs to $\mathcal{A}(\mu)$ and has $\mu^+(C) = \mu^+(A) = \mu^+(B)$.

16.4 Lebesgue Measure

Lebesgue measure is defined by the outer measure λ^+ constructed from the measure λ on $\mathcal{C}_{\mathbb{R}}$ that is determined by putting $\lambda((a,b]) = b - a$. The members of the σ-algebra $\mathcal{C}_{\mathbb{R}}(\lambda)$ are known as the *Lebesgue measurable sets* and include all members of the σ-algebra $\mathcal{B}_{\mathbb{R}}$ of Borel sets generated by $\mathcal{C}_{\mathbb{R}}$. We will write $\lambda(B)$ for $\lambda^+(B)$ whenever B is Lebesgue measurable.

Some facts about Lebesgue measure that will be needed are:

(1) λ is the only measure on $\mathcal{B}_{\mathbb{R}}$ that has $\lambda((a,b)) = b - a$: any measure on an algebra including $\mathcal{B}_{\mathbb{R}}$ that agrees with λ on open intervals must agree with λ on all Borel sets.

(2) For any Lebesgue measurable set B there exist Borel sets C, D with $C \subseteq B \subseteq D$ and $\lambda(D - C) = 0$, hence $\lambda(B) = \lambda(C) = \lambda(D)$.

(3) A set $B \subseteq \mathbb{R}$ is Lebesgue measurable if for each $\varepsilon \in \mathbb{R}^+$ there is a closed set $C_\varepsilon \subseteq B$ and an open set $D_\varepsilon \supseteq B$ such that $\lambda(D_\varepsilon - C_\varepsilon) < \varepsilon$.

By using the axiom of choice it can be shown that there is a subset of $\mathbb{R}$ that is not Lebesgue measurable.

16.5 Loeb Measures

Loeb measures are defined by applying the outer measure construction to measures of the type μ_L introduced in example 4 of Section 16.2. We work from now on in a nonstandard framework that is countably saturated, and hence sequentially comprehensive (Theorem 15.4.2). Let $(S, \mathcal{A}, \mu)$ be any "measure space" consisting of an internal finitely additive function $\mu : \mathcal{A} \to {}^*[0, \infty]$ on an internal ring $\mathcal{A}$ of subsets of an internal set S. Take $\mu_L : \mathcal{A} \to [0, \infty]$ to be the measure defined in example 16.2(4), and let μ_L^+ be its associated outer measure on $\mathcal{P}(S)$. The members of the set $\mathcal{A}(\mu_L)$ of μ_L^+-measurable subsets of S will be called the *Loeb measurable sets* determined by μ. We write $\mu_L(B)$ for $\mu_L^+(B)$ whenever B is Loeb measurable and refer to μ_L as the *Loeb measure* and $(S, \mathcal{A}(\mu_L), \mu_L)$ as the *Loeb measure space* determined by μ.

This definition of Loeb measure via the outer measure construction is the way that the notion was first arrived at. By analysing its properties we will see that $\mathcal{A}(\mu_L)$ has a characterisation that would allow it and its measure μ_L to be defined in a more direct way (cf. the comments at the end of Section 16.7).

Lemma 16.5.1 *If B is Loeb measurable with respect to μ, then*

$$\mu_L(B) = \inf\{\mu_L(A) : B \subseteq A \in \mathcal{A}\}.$$

Proof. By monotonicity, $\mu_L(B)$ is a lower bound of the $\mu_L(A)$'s for $B \subseteq A \in \mathcal{A}$. If $\mu_L(B) = \infty$, then the result follows. If, however, $\mu_L(B) < \infty$, to show that it is the greatest lower bound it suffices to show that for any $\varepsilon \in \mathbb{R}^+$ there is some set $A_\varepsilon \in \mathcal{A}$ with $B \subseteq A_\varepsilon$ and $\mu_L(A_\varepsilon) \leq \mu_L(B) + \varepsilon$.

Now, for such an ε, by properties of outer measure there is an increasing sequence $A_1 \subseteq A_2 \subseteq \cdots$ of $\mathcal{A}$-elements whose union includes B and has

$$\mu_L^+(\bigcup_{n \in \mathbb{N}} A_n) < \mu_L(B) + \varepsilon.$$

The sequence $\langle A_n : n \in \mathbb{N} \rangle$ extends *by sequential comprehensiveness* to an internal sequence $\langle A_n : n \in {}^*\mathbb{N} \rangle$ of elements of $\mathcal{A}$. Then for each $k \in \mathbb{N}$ we have

$$(\forall n \in {}^*\mathbb{N})\,(n \leq k \text{ implies } A_n \subseteq A_k \text{ and } \mu(A_n) < \mu_L(B) + \varepsilon) \quad \text{(i)}$$

(since $\mu(A_n) \simeq \mu_L(A_n) \leq \mu_L^+(\bigcup_{n \in \mathbb{N}} A_n)$). But (i) is an internal assertion, since μ and the extended sequence are internal, while k, ε, and $\mu_L(B)$ are fixed internal entities (real numbers). Therefore by *overflow* (i) must be true with some *unlimited* $K \in {}^*\mathbb{N}$ in place of k. For such a K we have $A_K \in \mathcal{A}$ and $A_n \subseteq A_K$ for all $n \in \mathbb{N}$, so that

$$B \subseteq \bigcup_{n \in \mathbb{N}} A_n \subseteq A_K,$$

while $\mu(A_K) < \mu_L(B) + \varepsilon$. Hence as $\mu(A_K) \simeq \mu_L(A_K)$,

$$\mu_L(A_K) \leq \mu_L(B) + \varepsilon,$$

establishing that A_K is the set A_ε we are looking for. □

Lemma 16.5.2 *If B is Loeb measurable and μ_L-finite, then*

$$\mu_L(B) = \sup\{\mu_L(A) : A \subseteq B \text{ and } A \in \mathcal{A}\}.$$

Proof. Given any $\varepsilon \in \mathbb{R}^+$, we will show that there is some set $A_\varepsilon \in \mathcal{A}$ such that $A_\varepsilon \subseteq B$ and $\mu_L(B) - \varepsilon < \mu(A_\varepsilon)$.

Since $\mu_L(B) < \infty$, we know from Lemma 16.5.1 that there is some $D \in \mathcal{A}$ with $B \subseteq D$ and $\mu_L(D) < \infty$. The desired result is obtained by using complementation relative to D. Firstly, $D - B$ is Loeb measurable and μ_L-finite, so by Lemma 16.5.1 there is a set C with $D - B \subseteq C \in \mathcal{A}$ and

$$\mu_L(C) < \mu_L(D - B) + \varepsilon.$$

We may assume $C \subseteq D$ (since we could replace C by $C \cap D$ here). Let $A_\varepsilon = D - C \in \mathcal{A}$. Then $A_\varepsilon \subseteq B$, and C is the disjoint union of $D - B$ and $B - A_\varepsilon$, so

$$\mu_L(D - B) + \mu_L(B - A_\varepsilon) = \mu_L(C) < \mu_L(D - B) + \varepsilon,$$

implying that $\mu_L(B - A_\varepsilon) < \varepsilon$. Therefore

$$\mu_L(B) = \mu_L(A_\varepsilon) + \mu_L(B - A_\varepsilon) < \mu_L(A_\varepsilon) + \varepsilon,$$

so $\mu_L(B) - \varepsilon < \mu(A_\varepsilon)$ as desired. □

16.6 μ-Approximability

An arbitrary subset B of S is called μ-*approximable* if for every $\varepsilon \in \mathbb{R}^+$ there exist "approximating" sets $C_\varepsilon, D_\varepsilon \in \mathcal{A}$ such that

$$C_\varepsilon \subseteq B \subseteq D_\varepsilon \quad \text{and} \quad \mu(D_\varepsilon - C_\varepsilon) < \varepsilon.$$

Equivalently, for every $\varepsilon \in \mathbb{R}^+$ there exist $C_\varepsilon, D_\varepsilon \in \mathcal{A}$ with $C_\varepsilon \subseteq B \subseteq D_\varepsilon$ and $\mu_L(D_\varepsilon - C_\varepsilon) < \varepsilon$. (The equivalence holds because $\mu(A) \simeq \mu_L(A)$ in general.)

The notion of μ-approximability will provide an alternative characterisation of Loeb measurability (Theorem 16.7.1). For this we need several preliminary results.

Lemma 16.6.1 *If B is Loeb measurable with $\mu_L(B) < \infty$, then B is μ-approximable.*

Proof. Given $\varepsilon \in \mathbb{R}^+$, by Lemmas 16.5.1 and 16.5.2 there are $C_\varepsilon, D_\varepsilon \in \mathcal{A}$ with $C_\varepsilon \subseteq B \subseteq D_\varepsilon$ and $\mu_L(D_\varepsilon) < \mu_L(B) + \frac{\varepsilon}{2}$ while $\mu_L(B) < \mu_L(C_\varepsilon) + \frac{\varepsilon}{2}$. Then

$$\mu_L(D_\varepsilon - C_\varepsilon) = \mu_L(D_\varepsilon) - \mu_L(B) + \mu_L(B) - \mu_L(C_\varepsilon) < \varepsilon.$$

□

The next lemma makes further appeal to sequential comprehensiveness and overflow.

Lemma 16.6.2 *If B is μ-approximable, then there is a set $A \in \mathcal{A}$ that is "almost equal" to B in the sense that their symmetric difference $A \triangle B$ is μ_L^+-null.*

Proof. Applying μ-approximability, construct a pair of sequences of $\mathcal{A}$-elements $\langle C_n : n \in \mathbb{N} \rangle$ and $\langle D_n : n \in \mathbb{N} \rangle$ with

$$\cdots \subseteq C_n \subseteq C_{n+1} \subseteq \cdots \subseteq B \subseteq \cdots \subseteq D_{n+1} \subseteq D_n \subseteq \cdots$$

and $\mu_L(D_n - C_n) < \frac{1}{n}$. To see how this works at the inductive stage, given C_n and D_n, take $C'_{n+1}, D'_{n+1} \in \mathcal{A}$ with

$$C'_{n+1} \subseteq B \subseteq D'_{n+1} \quad \text{and} \quad \mu_L(D'_{n+1} - C'_{n+1}) < \tfrac{1}{n+1};$$

put $C_{n+1} = C_n \cup C'_{n+1}$ and $D_{n+1} = D_n \cap D'_{n+1}$.

By sequential comprehensiveness, these sequences extend to internal hypersequences $\langle C_n : n \in {}^*\mathbb{N}\rangle$ and $\langle D_n : n \in {}^*\mathbb{N}\rangle$ of $\mathcal{A}$-elements. Then for each $k \in \mathbb{N}$,

$$(\forall n \in {}^*\mathbb{N})\, (n \leq k \text{ implies } C_n \subseteq D_k \subseteq D_n). \tag{ii}$$

This is an internal statement, so by overflow there is some unlimited $K \in {}^*\mathbb{N}$ such that for all $n \in \mathbb{N}$,

$$C_n \subseteq D_K \subseteq D_n.$$

But then for any $n \in \mathbb{N}$,

$$D_K \Delta B = (D_K - B) \cup (B - D_K) \subseteq (D_n - C_n),$$

and hence

$$\mu_L^+(D_K \Delta B) \leq \mu_L^+(D_n - C_n) < \frac{1}{n}.$$

Thus $\mu_L^+(D_K \Delta B) = 0$, and so Lemma 16.6.2 holds with $A = D_K \in \mathcal{A}$. □

Note that we cannot arrange to get $B \subseteq D_K$ in this argument, because if we included "$B \subseteq D_k$" in (ii), we would no longer have an internal statement to which we could apply overflow.

Lemma 16.6.3 *If B is μ-approximable, then B is Loeb measurable.*

Proof. We have to show that any $E \subseteq S$ is split μ_L^+-additively by B, for which it suffices that

$$\mu_L^+(E) \geq \mu_L^+(E \cap B) + \mu_L^+(E - B).$$

Now, by Lemma 16.6.2 there is an $A \in \mathcal{A}$ that is almost equal to B. The desired inequality holds with A in place of B, since A is Loeb measurable, and so as $\mu_L^+(B \Delta A) = 0$, we can show it holds for B as well. Formally, let

$$\begin{aligned} C &= (E \cap B) - A, \\ D &= (E \cap A) - B, \\ G &= E - (A \cup B), \\ H &= E \cap A \cap B \end{aligned}$$

(draw a Venn diagram!). Then C and D are included in $B \Delta A$, so $\mu_L^+(C) = \mu_L^+(D) = 0$. Thus

$$\mu_L^+(E \cap B) = \mu_L^+(C \cup H) \leq \mu_L^+(C) + \mu_L^+(H) = \mu_L^+(H),$$

and similarly

$$\mu_L^+(E - B) = \mu_L^+(D \cup G) \leq \mu_L^+(D) + \mu_L^+(G) = \mu_L^+(G).$$

Hence

$$\mu_L^+(E \cap B) + \mu_L^+(E - B) \leq \mu_L^+(H) + \mu_L^+(G)$$
$$\leq \mu_L^+(E \cap A) + \mu_L^+(E - A) \text{ by monotonicity}$$
$$= \mu_L^+(E),$$

with the last step given by the μ_L^+-measurability of A. □

Lemma 16.6.4 *If $\mu_L^+(B) < \infty$, then B is Loeb measurable with respect to μ if and only if it is μ-approximable.*

Proof. By Lemmas 16.6.1 and 16.6.3. □

16.7 Loeb Measure as Approximability

The work of the last section yields the following characterisation of Loeb measurable sets.

Theorem 16.7.1 *An arbitrary subset B of S is Loeb measurable with respect to μ if and only if $B \cap A$ is μ-approximable for all μ-finite $A \in \mathcal{A}$.*

Proof. Let B be Loeb measurable. If $A \in \mathcal{A}$ is μ-finite, then $B \cap A$ is Loeb measurable, since the Loeb measurable sets are $\cap$-closed, and μ_L^+-finite since $B \cap A \subseteq A$, so by Corollary 16.6.4 $B \cap A$ is μ-approximable.

Conversely, let $B \cap A$ be μ-approximable for all μ-finite $A \in \mathcal{A}$. To show that B is Loeb measurable, we have to show that for any μ_L^+-finite $E \subseteq S$,

$$\mu_L^+(E) \geq \mu_L^+(E \cap B) + \mu_L^+(E - B). \tag{iii}$$

But in such a case $\mu_L^+(E \cap B) < \infty$, so there must be a sequence $\langle A_n : n \in \mathbb{N} \rangle$ of $\mathcal{A}$-elements that covers $E \cap B$ with each A_n being μ-finite. Each $B \cap A_n$ is then μ-approximable, by hypothesis, and so is Loeb measurable by Lemma 16.6.3. Hence if

$$A = \bigcup_{n \in \mathbb{N}} (B \cap A_n),$$

it follows that A is Loeb measurable, and so

$$\mu_L^+(E) \geq \mu_L^+(E \cap A) + \mu_L^+(E - A). \tag{iv}$$

But

$$E \cap A = E \cap (\bigcup_{n \in \mathbb{N}} (B \cap A_n)) = E \cap B,$$

as the A_n's cover $E \cap B$. Then $E - A = E - B$, and hence (iii) follows from (iv). □

The statement of Theorem 16.7.1 could be used to provide a direct definition of the class $\mathcal{A}(\mu_L)$ of Loeb measurable sets, with Lemma 16.5.1

providing the definition of the Loeb measure μ_L itself. But then we would have work to do in proving that $\mathcal{A}(\mu_L)$ was a σ-algebra including $\mathcal{A}$, on which μ_L was a complete measure.

16.8 Lebesgue Measure via Loeb Measure

Let $N \in {}^*\mathbb{N}$ be a fixed unlimited hypernatural number, and

$$\begin{aligned} S &= \{\tfrac{k}{N} : k \in {}^*\mathbb{Z} \text{ and } |k| \leq N^2\} \\ &= \{\tfrac{-k}{N} : 1 \leq k \leq N^2\} \cup \{0\} \cup \{\tfrac{k}{N} : 1 \leq k \leq N^2\}. \end{aligned}$$

S is a hyperfinite set, of internal cardinality $2N^2+1$, forming a grid of points spread across the hyperreal line between $-N$ and N, with adjacent points being of infinitesimal distance $\tfrac{1}{N}$ apart. Each real number r is approximated infinitely closely on either side by these grid points. This follows by transfer of the statement

$$(\forall n \in \mathbb{N}) \left(|r| < n \to (\exists k \in \mathbb{Z}) \left[|k| < n^2 \text{ and } \left(\tfrac{k}{n} \leq r < \tfrac{k+1}{n}\right) \right] \right).$$

Now, let $\mathcal{A} = \mathcal{P}_I(S)$ be the set of all internal subsets of S. $\mathcal{A}$ is an algebra, is itself internal and hyperfinite, and all its members are hyperfinite. The function $\mu : \mathcal{A} \to {}^*[0, \infty)$ given by

$$\mu(A) = \frac{|A|}{N}$$

is internal and finitely additive (and similar to example 16.2(3)). μ is a weighted counting function in the sense of 16.2(5), determined by assigning the infinitesimal weight $\tfrac{1}{N}$ to each grid point. It induces the measure μ_L on $\mathcal{A}$ having

$$\mu_L(A) = \begin{cases} \operatorname{sh}\left(\tfrac{|A|}{N}\right), & \text{if } \tfrac{|A|}{N} \text{ is limited}, \\ \infty, & \text{otherwise}. \end{cases}$$

Let $(S, \mathcal{A}(\mu_L), \mu_L)$ be the associated Loeb measure space as defined in Section 16.5. Our first step is to show that the Lebesgue measure of any real interval is obtainable by using μ_L to count the weighted number of grid points between the end points of the interval.

Theorem 16.8.1 *For any $a, b \in \mathbb{R}$ with $a < b$,*

$$\mu_L(\{s \in S : a < s < b\}) = b - a.$$

Proof. Let $A = \{s \in S : a < s < b\}$. Then $A = S \cap {}^*(a, b)$, so A is internal and belongs to $\mathcal{A}$, hence is Loeb measurable. Moreover, A is hyperfinite, so has smallest and greatest elements, say s and t. Since a and b can be approximated infinitely closely by members of S, we must then have $a \simeq s$ and $b \simeq t$. Also, we can put $s = \frac{K+1}{N}$ and $t = \frac{L}{N}$ for some $K, L \in {}^*\mathbb{Z}$. Thus

$$A = \{\tfrac{K+1}{N}, \tfrac{K+2}{N}, \ldots, \tfrac{L}{N}\} = \{\tfrac{M}{N} : K < M \leq L\},$$

which is hyperfinite of cardinality $L - K$, since the internal function $f(x) = \frac{K+x}{N}$ is a bijection from $\{1, \ldots, L-K\}$ onto A. It follows that

$$\tfrac{|A|}{N} = \tfrac{L-K}{N} = \tfrac{L}{N} - \tfrac{K}{N} \simeq b - a,$$

and so $\mu_L(A) = b - a$ as desired. □

Note that the proof of Theorem 16.8.1 shows readily that μ_L assigns measure $b - a$ as well to the sets

$$S \cap {}^*(a, b], \quad S \cap {}^*[a, b), \quad S \cap {}^*[a, b].$$

Thus if B is any finite interval in $\mathbb{R}$, the Lebesgue measure of B is equal to the Loeb measure of the set $S \cap {}^*B$ of grid points that are (possibly nonstandard) members of B. One might wonder whether this equation

$$\lambda(B) = \mu_L(S \cap {}^*B)$$

holds in general, but that suggestion is quickly dispelled by the case $B = \mathbb{Q}$. Every grid point is a hyperrational number, so $S \subseteq {}^*\mathbb{Q}$ and hence $\mu_L(S \cap {}^*\mathbb{Q}) = \mu_L(S) = \infty$, while $\lambda(\mathbb{Q}) = 0$.

Rather than $S \cap {}^*B$, the appropriate set to represent B in S is the set of grid points that approximate members of B infinitely closely. This is the set

$$\begin{aligned} \mathrm{sh}^{-1}(B) &= \{s \in S : s \text{ is infinitely close to some } r \in B\} \\ &= \{s \in S : s \text{ is limited and } \mathrm{sh}(s) \in B\}, \end{aligned}$$

which may be called the *inverse shadow* of B. The definition of $\mathrm{sh}^{-1}(B)$ uses a condition that is not internal, so the set itself cannot be guaranteed to be internal, and more strongly may not be Loeb measurable, i.e., may not belong to $\mathcal{A}(\mu_L)$. One case in which it is not internal but nonetheless is Loeb measurable occurs when $B = \mathbb{R}$: since

$$\mathrm{sh}^{-1}(\mathbb{R}) = \{s \in S : s \text{ is limited}\} = \bigcup_{n \in \mathbb{N}} (S \cap {}^*(-n, n)),$$

while each set $S \cap {}^*(-n, n)$ is an internal subset of S and so belongs to $\mathcal{A}$, it follows that $\mathrm{sh}^{-1}(\mathbb{R})$ belongs to $\mathcal{A}(\mu_L)$ by closure under countable unions. But $\mathrm{sh}^{-1}(\mathbb{R})$ cannot be internal, because it is bounded in ${}^*\mathbb{R}$ but has no least upper (or greatest lower) bound.

The general situation is this:

Theorem 16.8.2 *A subset B of $\mathbb{R}$ is Lebesgue measurable if and only if $\mathrm{sh}^{-1}(B)$ is Loeb measurable. When this holds, the Lebesgue measure of B is equal to the Loeb measure of the set of grid points infinitely close to points of B:*
$$\lambda(B) = \mu_L(\mathrm{sh}^{-1}(B)).$$

Proof. Let $\mathcal{M} = \{B \subseteq \mathbb{R} : \mathrm{sh}^{-1}(B) \in \mathcal{A}(\mu_L)\}$. For $B \in \mathcal{M}$, put
$$\nu(B) = \mu_L(\mathrm{sh}^{-1}(B)).$$

Our task is to show that $\mathcal{M}$ is the class $\mathcal{C}_\mathbb{R}(\lambda)$ of Lebesgue measurable sets, and that ν is the Lebesgue measure λ.

By properties of inverse images of functions,
$$\begin{aligned}
\mathrm{sh}^{-1}(\emptyset) &= \emptyset, \\
\mathrm{sh}^{-1}(A - B) &= \mathrm{sh}^{-1}(A) - \mathrm{sh}^{-1}(B), \\
\mathrm{sh}^{-1}(\bigcup_{n \in \mathbb{N}} A_n) &= \bigcup_{n \in \mathbb{N}} \mathrm{sh}^{-1}(A_n).
\end{aligned}$$

Since $\mathcal{A}(\mu_L)$ contains $\emptyset$ and is closed under set differences and countable unions, these facts imply that $\mathcal{M}$ has the same closure properties. Since $\mathrm{sh}^{-1}(\mathbb{R}) \in \mathcal{A}(\mu_L)$, as was shown above, we also have $\mathbb{R} \in \mathcal{M}$. Altogether then $\mathcal{M}$ is a σ-algebra, on which ν proves to be a measure.

At this point we need the the following lemma.

Lemma 16.8.3 *$\mathcal{M}$ includes the Borel algebra $\mathcal{B}_\mathbb{R}$, and ν agrees with Lebesgue measure on all Borel sets.*

Proof. Each open interval $(a,b) \subseteq \mathbb{R}$ belongs to $\mathcal{M}$, since $\mathrm{sh}^{-1}((a,b))$ is the union of the sequence $\langle A_n : n \in \mathbb{N} \rangle$, where
$$A_n = S \cap {}^*(a + \tfrac{1}{n}, b - \tfrac{1}{n}) \in \mathcal{A}.$$

But $\mathcal{B}_\mathbb{R}$ is the smallest σ-algebra containing all open intervals (a,b), so this implies that $\mathcal{B}_\mathbb{R} \subseteq \mathcal{M}$. Also, by Theorem 16.8.1,
$$\mu_L(A_n) = (b - \tfrac{1}{n}) - (a + \tfrac{1}{n}) = b - a - \tfrac{2}{n},$$
and hence as the A_n's form an *increasing* sequence,
$$\nu((a,b)) = \mu_L(\mathrm{sh}^{-1}((a,b))) = \lim_{n \to \infty} \mu_L(A_n) = b - a.$$

Thus ν is a measure on $\mathcal{B}_\mathbb{R}$ that agrees with λ on all open intervals. But any such measure must agree with λ on all Borel sets (16.4(1)). □

Now, if $B \subseteq \mathbb{R}$ is Lebesgue measurable, then (16.4(2)) there are Borel sets C, D with $C \subseteq B \subseteq D$ and $\lambda(C) = \lambda(B) = \lambda(D)$. Then
$$\mathrm{sh}^{-1}(C) \subseteq \mathrm{sh}^{-1}(B) \subseteq \mathrm{sh}^{-1}(D),$$

and by Lemma 16.8.3, $C, D \in \mathcal{M}$, whence $\mathrm{sh}^{-1}(C), \mathrm{sh}^{-1}(D) \in \mathcal{A}(\mu_L)$, and

$$\mu_L(\mathrm{sh}^{-1}(C)) = \nu(C) = \lambda(C) = \lambda(D) = \nu(D) = \mu_L(\mathrm{sh}^{-1}(D)).$$

Since μ_L is a complete measure on $\mathcal{A}(\mu_L)$ (by the general theory of outer measures), it follows that $\mathrm{sh}^{-1}(B) \in \mathcal{A}(\mu_L)$, and hence $B \in \mathcal{M}$, with

$$\nu(B) = \mu_L(\mathrm{sh}^{-1}(B)) = \mu_L(\mathrm{sh}^{-1}(C)) = \lambda(C) = \lambda(B).$$

This establishes that every Lebesgue measurable set is in $\mathcal{M}$ (i.e., $\mathcal{C}_\mathbb{R}(\lambda) \subseteq \mathcal{M}$) and that ν agrees with λ on all Lebesgue measurable sets.

It remains now to show that $\mathcal{M} \subseteq \mathcal{C}_\mathbb{R}(\lambda)$, and for this we need the result from Section 11.13 that the shadow of any internal subset of *$\mathbb{R}$ is topologically closed as a subset of $\mathbb{R}$, and so is a Borel set.

Let $B \in \mathcal{M}$, i.e., $\mathrm{sh}^{-1}(B) \in \mathcal{A}(\mu_L)$. First we consider the case that $\mathrm{sh}^{-1}(B)$ is μ_L-finite and show that B is Lebesgue measurable by the criterion 16.4(3). But if $\mu_L(\mathrm{sh}^{-1}(B)) < \infty$, then by Lemma 16.6.1, $\mathrm{sh}^{-1}(B)$ is μ-approximable, so for any given $\varepsilon \in \mathbb{R}^+$ there exist sets $C, D \in \mathcal{A}$ with

$$C \subseteq \mathrm{sh}^{-1}(B) \subseteq D \quad \text{and} \quad \mu_L(D) - \mu_L(C) < \varepsilon.$$

Let $C_\varepsilon = \mathrm{sh}(C) = \{\mathrm{sh}(s) : s \in C \cap \mathbb{L}\}$. Now, C is internal, being a member of $\mathcal{A}$, and so by Theorem 11.13.1 C_ε is closed, hence Borel, and therefore $C_\varepsilon \in \mathcal{M}$ by Lemma 16.8.3. But since $C \subseteq \mathrm{sh}^{-1}(B)$, every member of C has a shadow, so

$$C \subseteq \mathrm{sh}^{-1}(\mathrm{sh}(C)) = \mathrm{sh}^{-1}(C_\varepsilon) \in \mathcal{A}(\mu_L).$$

Hence
$$\mu_L(C) \leq \mu_L(\mathrm{sh}^{-1}(C_\varepsilon)). \tag{v}$$

Similarly, $\mathrm{sh}(S-D)$ is closed, and is disjoint from B because $\mathrm{sh}^{-1}(B) \subseteq D$. Thus $D_\varepsilon = \mathbb{R} - \mathrm{sh}(S-D)$ is open in $\mathbb{R}$, hence Borel, and has $B \subseteq D_\varepsilon$. Moreover, $\mathrm{sh}^{-1}(D_\varepsilon) \subseteq D$, and so

$$\mu_L(\mathrm{sh}^{-1}(D_\varepsilon)) \leq \mu_L(D). \tag{vi}$$

Thus we have $C_\varepsilon \subseteq B \subseteq D_\varepsilon$, with C_ε closed, D_ε open, and

$$\begin{aligned}\lambda(D_\varepsilon) - \lambda(C_\varepsilon) &= \mu_L(\mathrm{sh}^{-1}(D_\varepsilon)) - \mu_L(\mathrm{sh}^{-1}(C_\varepsilon)) \quad \text{by Lemma 16.8.3}\\ &\leq \mu_L(D) - \mu_L(C) \quad \text{by (v) and (vi)}\\ &< \varepsilon.\end{aligned}$$

According to 16.4(3), this is enough to ensure that B is Lebesgue measurable.

For the general case, we use the fact that the set B of reals is the countable union

$$B = \bigcup_{n \in \mathbb{N}} (B \cap (-n, n)),$$

so that it is enough to show that each $B \cap (-n, n)$ is Lebesgue measurable in order to deduce that B itself is Lebesgue measurable.

Now, $\operatorname{sh}^{-1}((-n,n))$ belongs to $\mathcal{A}(\mu_L)$, since $(-n,n)$ belongs to $\mathcal{M}$, and so $\mathcal{A}(\mu_L)$ contains the set

$$\operatorname{sh}^{-1}(B) \cap \operatorname{sh}^{-1}((-n,n)) = \operatorname{sh}^{-1}(B \cap (-n,n)).$$

But this set is μ_L-finite, because

$$\mu_L(\operatorname{sh}^{-1}(B \cap (-n,n))) \leq \mu_L(\operatorname{sh}^{-1}((-n,n))) = 2n,$$

so the case just considered proves that $B \cap (-n, n)$ is Lebesgue measurable.

This finishes the proof of Theorem 16.8.2, completing our demonstration that the Lebesgue measure $\lambda(B)$ of a subset B of $\mathbb{R}$ can be obtained as the Loeb measure $\mu_L(\operatorname{sh}^{-1}(B))$.

□

17
Ramsey Theory

So far, the nonstandard methodology has been applied to calculus, analysis, and topology. This is to be expected, since the notions of infinitely small and large numbers, infinitely close approximation, limiting concepts, etc. belong to those subjects. But there are other areas of mathematics that can be illuminated by the methodology of enlargement, and we will look at one such now: the *combinatorics* of infinite sets.

17.1 Colourings and Monochromatic Sets

Let $C_1, \ldots, C_r$ be a finite sequence of pairwise disjoint sets and A a set satisfying
$$A \subseteq C_1 \cup \cdots \cup C_r. \qquad (\text{i})$$
If each of the sets C_i is finite, then so too is A. To put this another way:

- If A is infinite, then at least one of the C_i's must be infinite.

This observation can be reformulated in the language of *colourings*. We regard (i) as inducing a partition of the set A given by an assignment of r different colours to the members of A. Then $B_i = A \cap C_i$ is the set of elements of A that are assigned colour i. In these terms we have we have:

- For any colouring of an infinite set A using finitely many colours, there must be an infinite subset B of A that is *monochromatic*, i.e., all members of B get the same colour.

This is the simplest case of a powerful principle known as *Ramsey's theorem*, which has significant combinatorial applications and which forms the basis of a subject called *Ramsey theory*. To formulate the general case we introduce the notation $[A]^k$ for the set of all k-element subsets of A (where $k \in \mathbb{N}$):
$$[A]^k = \{B \subseteq A : |B| = k\}.$$
Notice that if $B \subseteq A$, then $[B]^k \subseteq [A]^k$. Given a finite colouring
$$[A]^k \subseteq C_1 \cup \cdots \cup C_r$$
of the k-element subsets of A, a set $B \subseteq A$ is called *monochromatic* if all its k-element subsets get the same colour, i.e., $[B]^k \subseteq C_i$ for some i.

Ramsey's Theorem. *If A is infinite, then for any finite colouring of $[A]^k$ there exists an infinite monochromatic subset of A.*

Here is an illustration of the use of this principle.

Theorem 17.1.1 *If $(P, \leq)$ is an infinite partially ordered set, then P contains a sequence $\langle p_n : n \in \mathbb{N}\rangle$ that is*

(1) *strictly increasing: $p_1 < p_2 < \cdots$ or*

(2) *strictly decreasing: $p_1 > p_2 > \cdots$ or*

(3) *an antichain, i.e., p_n and p_m are incomparable under the ordering $\leq$ for all $n \neq m$.*

Proof. Take a colouring
$$[P]^2 = C_b \cup C_w$$
of the 2-element subsets of P in which
$$C_b = \{\{p, q\} : p \leq q \text{ or } q \leq p\}$$
and
$$C_w = [P]^2 - C_b.$$
Thus C_b (the black sets) consists of pairs of elements that are comparable, and C_w (white) consists of the incomparable pairs.

By Ramsey's theorem there is an infinite set $Q \subseteq P$ that is monochromatic for this colouring. If $[Q]^2 \subseteq C_w$, then any sequence of distinct points in Q will be an antichain fulfilling (3). If, however, $[Q]^2 \subseteq C_b$, then Q is an infinite chain, from which we can extract a sequence satisfying (1) or (2). Indeed, if Q is not well-ordered by $\leq$, then it must contain an infinite decreasing sequence (2), and if it is well-ordered, then each of its nonempty subsets has a least element, and we can use this fact to construct an increasing sequence (1) inductively. □

17.2 A Nonstandard Approach

The proof of Theorem 17.1.1 reveals nothing about the structure of the poset $(P, \leq)$, and leaves an air of mystery: we simply invoke this marvelous new principle called Ramsey's theorem, and it delivers the set Q we need. We do not see whether $[Q]^2$ turned out to be black or white.

Here now is a nonstandard argument that does analyse the structure of the partial ordering. We work in an enlargement of a universe $\mathbb{U}$ that contains P and $\leq$ and in which members of P are individuals. The extension of the relation $\leq$ to $^*P$ will also be denoted by $\leq$. It will be assumed that P has a least element 0 and a greatest element 1 (we can always add such elements and then take them away at the end to get the desired result).

Let π be a nonstandard member of $^*P$ in this enlargement (recall from Section 14.1 that infinite sets from $\mathbb{U}$ always have such members). Put

$$L = \{p \in P : p < \pi\},$$
$$U = \{p \in P : p > \pi\}.$$

These sets are nonempty, since they contain 0 and 1 respectively. Several cases are considered.

First, if L has no maximal member, then it must contain an increasing sequence of the type (1) (p is maximal in L if there is no $x \in L$ with $p < x$). Likewise, if U has no minimal member, it yields a decreasing sequence (2). If neither of these cases holds, then L has a maximal element q_l and U has a minimal element q_u. Since $q_l < \pi < q_r$, the statement

$$(\exists y \in {}^*P)\,(q_l < y < q_u)$$

is then true, and so by transfer it follows that the set

$$M = \{p \in P : q_l < p < q_u\}$$

is nonempty, and contains an element p_1. We will now inductively define an antichain in M to complete the proof. The entity π belongs to $^*M$ and is used to guide the construction of the antichain.

Assume that $p_1, \ldots p_n \in M$ have been defined and are pairwise incomparable. Now, M is disjoint from L, since all members of M are greater than q_l, which is maximal in L. Similarly, M is disjoint from U. But then no member of M can be comparable with π, because an element of P that is comparable with π must belong to L or U. Hence the statement

$$(\exists y \in {}^*M)\,(p_1 \not\leq y \not\leq p_1 \wedge \cdots \wedge p_n \not\leq y \not\leq p_n)$$

is true (when $y = \pi$). By transfer it follows that there is some $p_{n+1} \in M$ that is incomparable with each of $p_1, \ldots, p_n$. This completes the inductive step, showing that we can indeed obtain $\langle p_n : n \in \mathbb{N} \rangle$ as an antichain in M.

17.3 Proving Ramsey's Theorem

The essence of the inductive construction just given is that at each stage the nonstandard element π of $^*M$ has the desired property (incomparability with $p_1,\ldots,p_n$), and so by existential transfer gives rise to a (standard) element of M having the same property.

By an argument similar to this we can prove Ramsey's theorem itself, using a nonstandard entity to guide the construction of a monochromatic set. But first we observe that it is enough to obtain the result for 2-colourings, because the rest can be done by standard induction. Fixing the parameter k, assume inductively that any r-colouring of a set of the form $[B]^k$ with B infinite has an infinite monochromatic subset of B. Then given an $r+1$-colouring

$$[A]^k \subseteq C_1 \cup \cdots \cup C_{r+1},$$

put $C_b = C_1 \cup \cdots \cup C_r$ and $C_w = C_{r+1}$ to turn it into a 2-colouring. If the 2-colouring case holds, then it yields an infinite monochromatic set $B \subseteq A$ for this 2-colouring. If $[B]^k \subseteq C_b$, then $C_1,\ldots,C_r$ induce an r-colouring on $[B]^k$ that by the induction hypothesis on r has an infinite monochromatic set $B' \subseteq B$. Then B', being a subset of A, is an infinite monochromatic set for the original $(r+1)$-colouring of $[A]^k$. If on the other hand $[B]^k \subseteq C_w = C_{r+1}$, then B itself is a monochromatic set for the $(r+1)$-colouring.

Thus we are reduced to proving Ramsey's theorem for 2-colourings

$$[A]^k \subseteq C_b \cup C_w$$

for infinite sets A. Now we proceed by induction on the parameter k. The case $k=1$ is just the special case described at the beginning: $[A]^1$ can be identified with A, via the correspondence $\{a\} \leftrightarrow a$, and if $A \subseteq C_b \cup C_w$, then one of C_b and C_w must be infinite. For the inductive case, assume that Ramsey's theorem holds for any 2-colouring of any set of the form $[D]^k$. Let

$$[A]^{k+1} \subseteq C_b \cup C_w \qquad (ii)$$

be a 2-colouring of $[A]^{k+1}$, with A infinite. Then the enlarged sets $^*C_b$ and $^*C_w$ give a 2-colouring of $[^*A]^{k+1}$, i.e., the partition of $(k+1)$-element subsets of A into "black" and "white" extends to all $(k+1)$-element subsets of $^*A$. To see this, observe that the fact that every $(k+1)$-element subset of A belongs to $[A]^{k+1}$ can be expressed by the sentence

$$(\forall a_1,\ldots,a_{k+1} \in A)$$
$$\left(\bigwedge_{1 \leq i \neq j \leq k+1} a_i \neq a_j\right) \to (\exists z \in [A]^{k+1})\, z = \{a_1,\ldots,a_{k+1}\},$$

where "$z = \{a_1,\ldots,a_{k+1}\}$" abbreviates the formula

$$a_1 \in z \wedge \cdots \wedge a_{k+1} \in z \wedge (\forall x \in z)(x = a_1 \vee \cdots \vee x = a_{k+1}).$$

17.3 Proving Ramsey's Theorem

By transfer of this sentence it follows that every $(k+1)$-element subset of $^*A$ belongs to $^*([A]^{k+1})$:

$$[^*A]^{k+1} \subseteq {}^*([A]^{k+1}).$$

Exercise 17.3.1 Show that in fact, $[^*A]^{k+1} = {}^*([A]^{k+1})$. □

Now by transfer of (ii),

$$^*([A]^{k+1}) \subseteq {}^*(C_b \cup C_w) = {}^*C_b \cup {}^*C_w,$$

and $^*C_b$ and $^*C_w$ are disjoint, by transfer of the fact that $C_b \cap C_w = \emptyset$. Altogether then we get a 2-colouring

$$[^*A]^{k+1} \subseteq {}^*C_b \cup {}^*C_w$$

of the claimed type.

Because A is infinite, it includes a denumerable subset, so we may as well assume $\mathbb{N} \subseteq A$. Now, let $\pi \in {}^*\mathbb{N} - \mathbb{N}$ be an unlimited hypernatural number. Then $\pi \in {}^*A$, and we use π to control the construction of an increasing sequence $s_1 < s_2 < \cdots$ of standard positive integers with the property that whenever $n_1 < \cdots < n_{k+1}$, then

$$\{s_{n_1}, \ldots, s_{n_k}, s_{n_{k+1}}\} \in C_b \quad \text{iff} \quad \{s_{n_1}, \ldots, s_{n_k}, \pi\} \in {}^*C_b. \quad \text{(iii)}$$

This property asserts that each member of the sequence behaves in relation to its predecessors just like π.

If such a sequence exists, then putting $D = \{s_n : n \in \mathbb{N}\} \subseteq A$, a 2-colouring

$$[D]^k = C'_b \cup C'_w \quad \text{(iv)}$$

of $[D]^k$ may be defined by

$$\{s_{n_1}, \ldots, s_{n_k}\} \in C'_b \quad \text{iff} \quad \{s_{n_1}, \ldots, s_{n_k}, \pi\} \in {}^*C_b, \quad \text{(v)}$$

and $C'_w = [D]^k - C'_b$. (Here we list the members of any k-element subset of D in increasing order.)

By the inductive assumption that Ramsey's theorem holds for k, it follows that there is an infinite $B \subseteq D$ that is monochromatic for the colouring (iv). But then B will also be our desired monochromatic set for (ii). For if $[B]^k \subseteq C'_b$, then $[B]^{k+1} \subseteq C_b$, because if

$$\{s_{n_1}, \ldots, s_{n_k}, s_{n_{k+1}}\} \in [B]^{k+1}$$

(with $n_1 < \cdots < n_{k+1}$), then

$$\{s_{n_1}, \ldots, s_{n_k}\} \in [B]^k \subseteq C'_b,$$

and so by (v),

$$\{s_{n_1}, \ldots, s_{n_k}, \pi\} \in {}^*C_b,$$

and hence by (iii),
$$\{s_{n_1},\ldots,s_{n_k},s_{n_{k+1}}\} \in C_b.$$
Similarly, if $[B]^k \subseteq C'_w$, then $[B]^{k+1} \subseteq C_w$, because (iii) and (v) also imply
$$\{s_{n_1},\ldots,s_{n_k}\} \in C'_w \quad \text{iff} \quad \{s_{n_1},\ldots,s_{n_k},s_{n_{k+1}}\} \in C_w.$$

It remains now to construct a sequence of standard integers satisfying (iii). Choose $s_1 < \cdots < s_k$ arbitrarily. For $n \geq k$ suppose inductively that $s_1,\ldots,s_n$ have been defined in such a way that (iii) holds whenever $n_1 < \cdots < n_{k+1} \leq n$. We have to define s_{n+1} so that it behaves in relation to its predecessors just like π. Thus s_{n+1} must satisfy the condition

$$\{s_{n_1},\ldots,s_{n_k},s_{n+1}\} \in C_b \tag{vi}$$

whenever $n_1 < \cdots < n_k \leq n$ and

$$\{s_{n_1},\ldots,s_{n_k},\pi\} \in {}^*C_b. \tag{vii}$$

Likewise, s_{n+1} must satisfy the condition

$$\{s_{n_1},\ldots,s_{n_k},s_{n+1}\} \notin C_b \tag{viii}$$

whenever $n_1 < \cdots < n_k \leq n$ and

$$\{s_{n_1},\ldots,s_{n_k},\pi\} \notin {}^*C_b. \tag{ix}$$

But there are only finitely many such conditions to be fulfilled, and so our requirements can be expressed in a sentence of the formal language associated with $\mathbb{U}$.

To facilitate this we will list the members of any $(k+1)$-element subset of $\{s_1,\ldots,s_n\}$ in increasing order, so that such $(k+1)$-element subsets can be identified with $(k+1)$-tuples $\langle s_{n_1},\ldots,s_{n_{k+1}}\rangle$ having $n_1 < \cdots < n_{k+1}$. Now, let $\varphi(x)$ be the conjunction of all (atomic) formulae

$$\langle s_{n_1},\ldots,s_{n_k},x\rangle \in C_b$$

such that (vii) holds and all formulae

$$\langle s_{n_1},\ldots,s_{n_k},x\rangle \notin C_b$$

such that (ix) holds (where $n_1 < \cdots < n_k \leq n$).

The transformed formula ${}^*\varphi(x)$ is a conjunction of formulae of the form $\langle s_{n_1},\ldots,s_{n_k},x\rangle \in {}^*C_b$ or $\langle s_{n_1},\ldots,s_{n_k},x\rangle \notin {}^*C_b$. It makes an assertion about x that is true of the unlimited hypernatural number π, i.e., ${}^*\varphi(\pi)$ is true. Since s_n is standard, it follows that the sentence

$$(\exists x \in {}^*\mathbb{N})\,(s_n < x \wedge {}^*\varphi(x))$$

is true. By existential transfer this implies that there is an $s_{n+1} \in \mathbb{N}$ with $s_n < s_{n+1}$ and $\varphi(s_{n+1})$ true. The definition of φ then ensures that (vi) and (viii) hold according as (vii) and (ix) hold. Together with the induction hypothesis on n, this guarantees that (iii) holds whenever $n_1 < \cdots < n_{k+1} \leq n+1$.

This completes the construction of a sequence fulfilling (iii) in general, and hence completes the proof of Ramsey's theorem.

17.4 The Finite Ramsey Theorem

If a set A is finite, then of course no colouring of $[A]^k$ can produce an infinite monochromatic subset of A. But we can ask for a *large* monochromatic set, by specifying in advance a minimum size for it. In this situation there is a finitary version of Ramsey's theorem, which can be deduced in an interesting way from the infinite version by further nonstandard reasoning. In essence the new version says that if we specify the required size m for a monochromatic subset of A, then such a monochromatic subset will exist if the size n of A itself is great enough.

To simplify the formulation, we will confine the discussion of finite sets to initial segments $\{1,\ldots,n\}$ of $\mathbb{N}$.

Finite Ramsey Theorem. *For any given numbers $k, r, m \in \mathbb{N}$ there exists a number $n \in \mathbb{N}$ such that the following holds:*

$\psi(n)$: *for any r-colouring*

$$[\{1,\ldots,n\}]^k \subseteq C_1 \cup \cdots \cup C_r$$

of $[\{1,\ldots,n\}]^k$ there is a subset B of $\{1,\ldots,n\}$ that is monochromatic, i.e., $[B]^k \subseteq C_i$ for some i, and has at least m elements, i.e., $|B| \geq m$.

Proof. Fix the numbers k, r, m. Then the condition $\psi(n)$ can be written out as a formula of the language for $\mathbb{U}$, with n as a variable. This formula begins with the quantifier form

$$\forall C_1, \ldots, C_r \in \mathcal{P}([\mathbb{N}]^k)$$

and has B as a variable bound by the quantifier $(\exists B \in \mathcal{P}_F(\mathbb{N}))$. The transformed formula $^*\psi(n)$ asserts that

> for any r-colouring of $[\{1,\ldots,n\}]^k$ by sets in $^*\mathcal{P}([\mathbb{N}]^k)$ (i.e., by internal subsets of $[^*\mathbb{N}]^k$) there exists a hyperfinite subset of $\{1,\ldots,n\}$ that is monochromatic and has at least m elements.

Now, if n is interpreted as some *unlimited* hypernatural number N, then the sentence $^*\psi(N)$ is readily shown to be true by using the infinite Ramsey theorem. To see this, observe that any r-colouring

$$[\{1, \ldots, N\}]^k \subseteq C_1 \cup \cdots \cup C_r$$

of $[\{1, \ldots, N\}]^k$ induces an r-colouring of $[\mathbb{N}]^k$, because $\mathbb{N} \subseteq \{1, \ldots, N\}$. But for the latter colouring we know that there is an infinite set $D \subseteq \mathbb{N}$ that is monochromatic: $[D]^k \subseteq C_i$ for some i. Since D is infinite, we can then take any m-element subset B of D to get $[B]^k \subseteq C_i$ and $|B| = m$, giving a B that fulfills the condition $^*\psi(N)$.

Thus by interpreting n as an unlimited hypernatural we establish that

$$(\exists n \in {}^*\mathbb{N})\, {}^*\psi(n).$$

By transfer it then follows that there is some $n \in \mathbb{N}$ such that $\psi(n)$, and so the finite Ramsey theorem is proved. □

Exercise 17.4.1
Verify in detail that $\psi(n)$ can be written out as a formula of the language for $\mathbb{U}$.

17.5 The Paris–Harrington Version

The proof just given of the finite Ramsey theorem leaves room for its conclusion to be strengthened, since the desired monochromatic set B was chosen arbitrarily from the infinitely many m-element subsets of the infinite monochromatic set D. This gives us scope to impose further properties on B. For instance, consider the requirement

$$|B| \geq \min(B)$$

that the size of B be at least as great as the smallest element of B. This is sometimes expressed by saying that B is *relatively large*. From an infinite set $D \subseteq \mathbb{N}$ we can select relatively large finite subsets of unbounded size: given a lower bound m for $|B|$, choose any number $j \in D$ that is at least as big as m, and add to j another $j-1$ elements of D that are bigger than j to form a B having $|B| = \min(B) = j \geq m$.

Applying these observations to the above arguments leads to a proof that

> for any given numbers $k, r, m \in \mathbb{N}$ there exists a number $n \in \mathbb{N}$ such that for any r-colouring of $[\{1, \ldots, n\}]^k$ there exists a relatively large subset B of $\{1, \ldots, n\}$ that is monochromatic and has at least m elements.

This statement may appear to be a rather innocuous strengthening of the finite Ramsey theorem, but in fact it is a remarkable one. It was shown by Paris and Harrington that the statement cannot be proved in the first-order axiom system of Peano arithmetic. This was the first *mathematically significant* example of the Gödel incompleteness phenomenon: a true statement about the structure of the natural numbers that cannot be proven from an appropriate set of axioms. Gödel's own argument constructs such sentences that are designed specifically to make his proof work and do not have independent status. The Paris–Harrington version of Ramsey's theorem, however, involves natural combinatorial ideas that arise quite independently of considerations of the proof theory of statements about $\mathbb{N}$.

17.6 Reference

The nonstandard approach to Ramsey theory presented in this chapter is due to Joram Hirshfeld, and was presented in the following article:

> JORAM HIRSHFELD. *Nonstandard Combinatorics.* Studia Logica 47 (1980), 221–232.

18
Completion by Enlargement

There are many mathematical structures that are "incomplete" because they lack certain elements, such as the limit of a Cauchy sequence, the sum of infinite series, the least upper bound of a set of elements, a point "at infinity", and so on. A variety of standard techniques exist for *completing* such structures by adding the missing elements.

Now, the enlargement of a structure in a nonstandard framework is a kind of completion, and we are going to explore ways in which enlargements give an alternative approach to standard completions. From this perspective there is some redundancy in the enlargement process because in a sense it "saturates" a structure with all the elements one could ever imagine adjoining to it. Some of these new elements are irrelevant to completion, while others may be distinct but indistinguishable in terms of their role in completing the original structure. Thus we need to factor out such redundancy, and as we shall see, standard completions can typically be obtained as quotients of certain kinds of enlargement.

18.1 Completing the Rationals

The set *$\mathbb{Q}$ of hyperrationals contains infinitely close approximations of all real numbers. For if $r \in \mathbb{R}$, then by transfer

$$(\forall x \in {}^*\mathbb{R})\, [r < x \to (\exists q \in {}^*\mathbb{Q})\, (r < q < x)].$$

So, putting $x = r + \varepsilon$ with ε a positive infinitesimal implies that there is some $q \in {}^*\mathbb{Q}$ with $r < q < r + \varepsilon$ and hence $q \simeq r$. Thus $q \in {}^*\mathbb{Q} \cap \text{hal}(r)$ and

232   18. Completion by Enlargement

r is the shadow of q.

There are two points to immediately note here:

- By this argument, for each $x \in \mathrm{hal}(r)$ there is a member of *$\mathbb{Q}$ between x and r, and these members are all equally good as infinitely close hyperrational approximations to r.

- There are many members of *$\mathbb{Q}$, namely all the unlimited ones, that are "infinitely far away" and hence irrelevant to the issue of approximating reals.

This suggests that we confine attention to the set

$$\text{*}\mathbb{Q}^{\mathrm{lim}} = \{x \in \text{*}\mathbb{Q} : x \text{ is limited}\} = \text{*}\mathbb{Q} \cap \mathbb{L}$$

of limited hyperrationals, and that we identify those that are in the same halo. The way to make this work is to use the shadow map

$$\mathrm{sh} : \text{*}\mathbb{Q}^{\mathrm{lim}} \to \mathbb{R}$$

introduced in Section 5.6. We have just seen in effect that this map is a surjection from *$\mathbb{Q}^{\mathrm{lim}}$ to $\mathbb{R}$: for each $r \in \mathbb{R}$ there is an element $q \in \text{*}\mathbb{Q}^{\mathrm{lim}}$ with $\mathrm{sh}(q) = r$. But *$\mathbb{Q}^{\mathrm{lim}}$ is closed under addition and multiplication, so forms a subring of *$\mathbb{R}$, and the shadow map preserves addition and multiplication (Theorem 5.6.2) so is a ring homomorphism from *$\mathbb{Q}^{\mathrm{lim}}$ onto $\mathbb{R}$. Thus by the fundamental homomorphism theorem for rings, $\mathbb{R}$ is isomorphic to the quotient ring of *$\mathbb{Q}^{\mathrm{lim}}$ factored by the *kernel*

$$\{x \in \text{*}\mathbb{Q}^{\mathrm{lim}} : \mathrm{sh}(x) = 0\}$$

of the shadow map. But this kernel is just the set

$$\text{*}\mathbb{Q}^{\mathrm{inf}} = \{x \in \text{*}\mathbb{Q} : x \simeq 0\} = \text{*}\mathbb{Q} \cap \mathbb{I}$$

of infinitesimal hyperrationals. So we have an isomorphism

$$\text{*}\mathbb{Q}^{\mathrm{lim}}/\text{*}\mathbb{Q}^{\mathrm{inf}} \cong \mathbb{R}$$

(cf. Exercise 5.7(4)). The members of *$\mathbb{Q}^{\mathrm{lim}}/\text{*}\mathbb{Q}^{\mathrm{inf}}$ are the *cosets*

$$\text{*}\mathbb{Q}^{\mathrm{inf}} + x = \{q + x : q \in \text{*}\mathbb{Q}^{\mathrm{inf}}\}$$

of elements $x \in \text{*}\mathbb{Q}^{\mathrm{lim}}$. These are the same as the equivalence classes of *$\mathbb{Q}^{\mathrm{lim}}$ under the relation $\simeq$ of infinite closeness, because the following conditions are all equivalent:

$$\begin{aligned} x &\simeq y, \\ \mathrm{sh}(x) &= \mathrm{sh}(y), \\ \mathrm{sh}(x - y) &= 0, \\ (x - y) &\in \text{*}\mathbb{Q}^{\mathrm{inf}}, \\ \text{*}\mathbb{Q}^{\mathrm{inf}} + x &= \text{*}\mathbb{Q}^{\mathrm{inf}} + y. \end{aligned}$$

Hence
$$^*\mathbb{Q}^{\inf} + x = \{y \in {}^*\mathbb{Q}^{\lim} : x \simeq y\} = \mathrm{hal}(x) \cap {}^*\mathbb{Q},$$
and $^*\mathbb{Q}^{\lim}/{}^*\mathbb{Q}^{\inf}$ can also be described as the quotient set $^*\mathbb{Q}^{\lim}/{\simeq}$. Its isomorphism with $\mathbb{R}$ is given by the map $^*\mathbb{Q}^{\inf} + x \mapsto \mathrm{sh}(x)$.

This construction can be viewed as providing an alternative way of building the reals out of the rationals. As it stands, we have assumed the existence of $\mathbb{R}$ in the above analysis and used its Dedekind completeness in obtaining the shadow map on which the discussion was based. But we could try to prove directly that $^*\mathbb{Q}^{\lim}/{}^*\mathbb{Q}^{\inf}$ is a complete ordered field: since all complete ordered fields are isomorphic, this would show that the construction was independent of the choice of nonstandard framework in which $^*\mathbb{Q}$, $^*\mathbb{Q}^{\lim}$, and $^*\mathbb{Q}^{\inf}$ reside. The question of completeness of quotient structures like $^*\mathbb{Q}^{\lim}/{}^*\mathbb{Q}^{\inf}$ will be addressed in the next two sections. We will see that there are many structures $\mathbb{X}$ for which $^*\mathbb{X}^{\lim}/{\simeq}$ is a "completion" of $\mathbb{X}$.

18.2 Metric Space Completion

A *metric* on a set $\mathbb{X}$ is a function $d : \mathbb{X} \times \mathbb{X} \to \mathbb{R}^{\geq} = \mathbb{R}^+ \cup \{0\}$ satisfying the axioms

- $d(x,y) = 0$ iff $x = y$;

- $d(x,y) = d(y,x)$;

- $d(x,y) \leq d(x,z) + d(z,y)$ (*triangle inequality*).

The pair $(\mathbb{X}, d)$ is a *metric space*, in which the number $d(x,y)$ is to be thought of as the distance from x to y. The *Euclidean* metric on $\mathbb{R}$ is given by $d(x,y) = |x-y|$.

When $\mathbb{X}$ carries a commutative ring structure, a metric sometimes comes from a *norm*, which is a function $x \mapsto \|x\| \in \mathbb{R}^{\geq}$ satisfying

- $\|x\| = 0$ iff $x = 0$;

- $\|x \cdot y\| = \|x\| \cdot \|y\|$;

- $\|x + y\| \leq \|x\| + \|y\|$.

Then putting $d(x,y) = \|x-y\|$ induces a metric on $\mathbb{X}$. The absolute value function $|x|$ is a norm on $\mathbb{R}$ that induces the Euclidean metric.

A sequence $\langle x_n : n \in \mathbb{N} \rangle$ in a metric space $(\mathbb{X}, d)$ is *Cauchy* if

$$(\forall \varepsilon \in \mathbb{R}^+)\,(\exists k_\varepsilon \in \mathbb{N})\,(\forall m, n \in \mathbb{N})\,[m, n \geq k_\varepsilon \to d(x_m, x_n) < \varepsilon\,].$$

This is just like the definition of a Cauchy sequence in $\mathbb{R}$, but with d in place of the Euclidean metric on $\mathbb{R}$. In fact, many of the ideas and results about convergence etc. of sequences can be lifted to an abstract metric space in this way. For instance, the sequence $\langle x_n : n \in \mathbb{N} \rangle$ *converges to x* in $(\mathbb{X}, d)$ if

$$(\forall \varepsilon \in \mathbb{R}^+)\,(\exists k_\varepsilon \in \mathbb{N})\,(\forall n \in \mathbb{N})\,[n \geq k_\varepsilon \to d(x_n, x) < \varepsilon].$$

Then the metric space can be defined to be *complete* if every Cauchy sequence in the space converges to a point in the space.

A *completion* of a metric space $(\mathbb{X}, d)$ is another space $(\mathbb{X}', d')$ such that

- $\mathbb{X} \subseteq \mathbb{X}'$ and d is the restriction of d' to $\mathbb{X}$, i.e., $(\mathbb{X}, d)$ is a *subspace* of $(\mathbb{X}', d')$;

- $(\mathbb{X}', d')$ is complete;

- $\mathbb{X}$ is *dense in* $\mathbb{X}'$.

The last condition means that any point x' in $\mathbb{X}'$ can be approximated arbitrarily closely by points of $\mathbb{X}$, i.e., there is a sequence $\langle x_n : n \in \mathbb{N} \rangle$ of points $x_n \in \mathbb{X}$ that converges to x'. For this it suffices that for each $\varepsilon \in \mathbb{R}^+$ there is some $x_\varepsilon \in \mathbb{X}$ with $d'(x', x_\varepsilon) < \varepsilon$. Thus $\mathbb{R}$ is a completion of $\mathbb{Q}$ under the Euclidean metric.

It can be shown that any two completions of a metric space are *isometric*, meaning that there is a bijection between them that preserves their metrics and leaves the original space fixed. In this sense a completion of a metric space is unique. In particular, $\mathbb{R}$ is *the* completion of $\mathbb{Q}$.

18.3 Nonstandard Hulls

Consider a nonstandard framework for a set $\mathbb{X}$ that carries a metric d. We will take this framework to be a sequentially comprehensive enlargement.

The extended function *d on *$\mathbb{X}$ is not a metric, because it takes values in *$\mathbb{R}^{\geq}$ rather than $\mathbb{R}^{\geq}$. But it does satisfy the axioms of a metric, by transfer, and this is enough to ensure that we can define equivalence relations $\simeq$ and $\sim$ of infinitesimal and limited proximity in *$\mathbb{X}$. Put

$$x \simeq y \quad \text{iff} \quad {}^*d(x, y) \simeq 0,$$
$$x \sim y \quad \text{iff} \quad {}^*d(x, y) \text{ is limited},$$

for all $x, y \in {}^*\mathbb{X}$. The equivalence classes under $\simeq$ are the *halos*

$$\mathrm{hal}(x) = \{y \in {}^*\mathbb{X} : x \simeq y\},$$

while the equivalence classes under $\sim$ are the *galaxies*

$$\mathrm{gal}(x) = \{y \in {}^*\mathbb{X} : x \sim y\}.$$

18.3 Nonstandard Hulls

A member x of *$\mathbb{X}$ is *limited* if it is of limited distance from some member of $\mathbb{X}$, i.e., if $x \sim y$ for some $y \in \mathbb{X}$. Let

$$^*\mathbb{X}^{\lim} = \{x \in {}^*\mathbb{X} : x \text{ is limited}\}.$$

*$\mathbb{X}^{\lim}$ proves to be a galaxy including $\mathbb{X}$, and is sometimes called the *principal* galaxy. At first sight it might be thought that the metric d could be extended to *$\mathbb{X}^{\lim}$ by taking the distance between limited points x, y to be the real number $\text{sh}(^*d(x,y))$. But this number will be 0 whenever $x \simeq y$, so the first axiom for a metric is not satisfied. What we must do therefore is identify points that are infinitely close, by passing to the quotient set

$$\widehat{\mathbb{X}} = (^*\mathbb{X}^{\lim}/\simeq) = \{\text{hal}(x) : x \text{ is limited}\}$$

(note that if $x \in {}^*\mathbb{X}^{\lim}$, then $\text{hal}(x) \subseteq {}^*\mathbb{X}^{\lim}$, so *$\mathbb{X}^{\lim}$ is partitioned by the halos of its points). A metric is then defined on $\widehat{\mathbb{X}}$ by

$$d(\text{hal}(x), \text{hal}(y)) = \text{sh}(^*d(x,y)).$$

This is well-defined, since if $\text{hal}(x) = \text{hal}(x')$ and $\text{hal}(y) = \text{hal}(y')$, then $^*d(x,y) \simeq {}^*d(x',y')$.

The pair $(\widehat{\mathbb{X}}, d)$ is called the *nonstandard hull* of $(\mathbb{X}, d)$. Observe that if x, y are distinct members of $\mathbb{X}$, then $\text{hal}(x)$ and $\text{hal}(y)$ are distinct (indeed, they are disjoint), so the mapping $x \mapsto \text{hal}(x)$ is an injection of $\mathbb{X}$ into $\widehat{\mathbb{X}}$, allowing us to identify $\mathbb{X}$ with a subset of its nonstandard hull. Moreover, when $x, y \in \mathbb{X}$,

$$d(x,y) = {}^*d(x,y) = \text{sh}(^*d(x,y)) = d(\text{hal}(x), \text{hal}(y)),$$

so under this identification $(\mathbb{X}, d)$ becomes a subspace of $(\widehat{\mathbb{X}}, d)$, and this is what justifies us continuing to use the symbol "d" for the metric on $\widehat{\mathbb{X}}$.

Theorem 18.3.1 *The nonstandard hull $(\widehat{\mathbb{X}}, d)$ is complete.*

Proof. Let $\langle \text{hal}(x_n) : n \in \mathbb{N} \rangle$ be a Cauchy sequence in $\widehat{\mathbb{X}}$. The sequence $\langle x_n : n \in \mathbb{N} \rangle$ of points in *$\mathbb{X}^{\lim}$ extends to an *internal* hypersequence $\langle x_n : n \in {}^*\mathbb{N} \rangle$ in *$\mathbb{X}$, by sequential comprehensiveness. We will show that $\langle \text{hal}(x_n) : n \in \mathbb{N} \rangle$ converges to $\text{hal}(x_K)$ for some $K \in {}^*\mathbb{N}_\infty$.

Now, for each $n \in \mathbb{N}$, by the Cauchy property there exists $k_n \in \mathbb{N}$ such that for all standard $m \geq k_n$,

$$d(\text{hal}(x_m), \text{hal}(x_{k_n})) < \frac{1}{2n},$$

and hence

$$^*d(x_m, x_{k_n}) < \frac{1}{2n}. \tag{i}$$

But the set $\{m \in {}^*\mathbb{N} : {}^*d(x_m, x_{k_n}) < 1/(2n)\}$ is internal, so by overflow we conclude that there is some unlimited $K_n \in {}^*\mathbb{N}$ such that (i) holds for all $m \in {}^*\mathbb{N}$ with $k_n \leq m \leq K_n$.

Invoking sequential comprehensiveness again, there is some unlimited $K \in {}^*\mathbb{N}$ that is smaller than every K_n (cf. Theorem 15.4.3). Then x_K is limited (e.g., ${}^*d(x_K, x_{k_1}) < \frac{1}{2}$ and x_{k_1} is limited), so $\mathrm{hal}(x_K) \in \widehat{\mathbb{X}}$. To show that $\langle \mathrm{hal}(x_n) : n \in \mathbb{N} \rangle$ converges to $\mathrm{hal}(x_K)$ it is enough to show that for each $n \in \mathbb{N}$ we get

$$^*d(x_m, x_K) < \frac{1}{n}$$

whenever $m \in \mathbb{N}$ and $k_n \leq m$. But for such m we have $k_n \leq m, K < K_n$, so by two applications of (i),

$$^*d(x_m, x_K) \leq {}^*d(x_m, x_{k_n}) + {}^*d(x_{k_n}, x_K) < \frac{1}{2n} + \frac{1}{2n} = \frac{1}{n}.$$

□

The nonstandard hull $\widehat{\mathbb{X}}$ need not be a completion of $(\mathbb{X}, d)$. It may contain points that cannot be approximated arbitrarily closely (in the real sense) by points of $\mathbb{X}$. To clarify this situation we introduce the concept of a point $x \in {}^*\mathbb{X}$ being *approachable from* $\mathbb{X}$, meaning that for each $\varepsilon \in \mathbb{R}^+$ there is some (standard) x_ε in $\mathbb{X}$ such that ${}^*d(x, x_\varepsilon) < \varepsilon$. Let

$$^*\mathbb{X}^{\mathrm{ap}} = \{x \in {}^*\mathbb{X} : x \text{ is approachable from } \mathbb{X}\} \subseteq {}^*\mathbb{X}^{\mathrm{lim}}.$$

Now, if x is approachable from $\mathbb{X}$, then so is any point infinitely close to x, i.e., $\mathrm{hal}(x) \subseteq {}^*\mathbb{X}^{\mathrm{ap}}$. Thus

$$({}^*\mathbb{X}^{\mathrm{ap}}/\simeq) = \{\mathrm{hal}(x) : x \in {}^*\mathbb{X}^{\mathrm{ap}}\} \subseteq \widehat{\mathbb{X}},$$

and so ${}^*\mathbb{X}^{\mathrm{ap}}/\simeq$ is a subspace of $\widehat{\mathbb{X}}$ in which $\mathbb{X}$ is dense, as follows readily from the definition of "approachable". Moreover, in the completeness proof of Theorem 18.3.1, if the points $x_n \in {}^*\mathbb{X}^{\mathrm{lim}}$ are all approachable from $\mathbb{X}$, then so is the point x_K. This implies that Cauchy sequences in ${}^*\mathbb{X}^{\mathrm{ap}}/\simeq$ converge in ${}^*\mathbb{X}^{\mathrm{ap}}/\simeq$, and so

Theorem 18.3.2 $({}^*\mathbb{X}^{\mathrm{ap}}/\simeq, d)$ *is a completion of* $(\mathbb{X}, d)$. □

A point $x \in {}^*\mathbb{X}$ is *near to* $\mathbb{X}$ if it is infinitely close to some $y \in \mathbb{X}$, i.e., if x has the same halo as a (standard) point from $\mathbb{X}$. Thus if a limited point x is not near to $\mathbb{X}$, then its halo is distinct from the halos of all points of $\mathbb{X}$, so $\mathrm{hal}(x)$ is a point of $\widehat{\mathbb{X}}$ that is not (identifiable with) a point of $\mathbb{X}$.

All points near to $\mathbb{X}$ are approachable from $\mathbb{X}$. If, conversely, every member of ${}^*\mathbb{X}^{\mathrm{ap}}$ is near to $\mathbb{X}$, then

$$({}^*\mathbb{X}^{\mathrm{ap}}/\simeq) = \{\mathrm{hal}(x) : x \in \mathbb{X}\},$$

which is the set we identify with $\mathbb{X}$ itself. Thus:

Corollary 18.3.3 *If in *$\mathbb{X}$ *every point approachable from* $\mathbb{X}$ *is actually near to* $\mathbb{X}$*, then* $(\mathbb{X}, d)$ *is a complete metric space.* □

Exercise 18.3.4
Prove, conversely to Corollary 18.3.3, that if a metric space $(\mathbb{X}, d)$ is complete, then in *$\mathbb{X}$ every point approachable from $\mathbb{X}$ is near to $\mathbb{X}$. □

Consider for example a hyperrational $x \in$ *$\mathbb{Q}$ that is infinitely close to $\sqrt{2}$ (recall that every real number is the shadow of some hyperrational). Then x is not near to $\mathbb{Q}$, because it is not infinitely close to any standard *rational* number, but x is approachable from $\mathbb{Q}$, since there is a sequence in $\mathbb{Q}$ converging to $\sqrt{2}$. This is a manifestation of the fact that $\mathbb{Q}$ is not complete under the Euclidean metric.

A point near $\mathbb{X}$ is often called *nearstandard*, since it is infinitely close to a standard point. (Points approachable from $\mathbb{X}$ are sometimes referred to by the less evocative term *pre-nearstandard*.) The hyperrational x just considered as not being near to $\mathbb{Q}$ is, on the other hand, near to $\mathbb{R}$, hence nearstandard, because $x \simeq \sqrt{2} \in \mathbb{R}$. This underlines the point that nearness is always specified in relation to a particular set.

In the case of the rationals it turns out that

$$*\mathbb{Q}^{\mathrm{ap}} = *\mathbb{Q}^{\mathrm{lim}},$$

and so the nonstandard hull $\widehat{\mathbb{Q}}$ is equal to the completion *$\mathbb{Q}^{\mathrm{ap}}/\simeq$ of $\mathbb{Q}$ (which is isomorphic to $\mathbb{R}$). This is because if $x \in$ *$\mathbb{Q}^{\mathrm{lim}}$, then the shadow $\mathrm{sh}(x)$ is a real number that can be approximated arbitrarily closely by rational numbers. Therefore, x can be approximated by rationals in this way too and so is approachable from $\mathbb{Q}$.

In Section 18.6 we will see an example, involving power series, of a metric space $(\mathbb{X}, d)$ whose enlargement has limited points that are not approachable from $\mathbb{X}$. In that case the nonstandard hull $\widehat{\mathbb{X}}$ is strictly larger than the completion *$\mathbb{X}^{\mathrm{ap}}/\simeq$.

18.4 *p*-adic Integers

Integers Modulo *m*

Recall that for $x, y \in \mathbb{Z}$, $x \equiv y \pmod{m}$ means that x is congruent to y modulo m, i.e., the difference $(x - y)$ is divisible by m, or equivalently, x and y have the same remainder upon division by m.

$$\mathbb{Z}/m = \{0, 1, \ldots, m-1\}$$

is the set of residues modulo m, i.e., remainders for division by m. Each integer z is congruent modulo m to exactly one member of $\mathbb{Z}/m$, which

p-adic Distance

Fix a prime number $p \in \mathbb{N}$. Each positive integer z has an expansion in base p of the form

$$z = z_0 + z_1 p + z_2 p^2 + \cdots + z_n p^n$$

with each z_i belonging to $\mathbb{Z}/p$, i.e., $0 \leq z_i < p$. Let

$$\begin{aligned}
a_1 &= z_0 & &\in \mathbb{Z}/p, \\
a_2 &= z_0 + z_1 p & &\in \mathbb{Z}/p^2, \\
a_3 &= z_0 + z_1 p + z_2 p^2 & &\in \mathbb{Z}/p^3, \\
&\;\;\vdots \\
a_{n+1} &= z_0 + z_1 p + z_2 p^2 + \cdots + z_n p^n & &\in \mathbb{Z}/p^{n+1}.
\end{aligned}$$

Consider the notion that the list $a_1, a_2, \ldots, a_{n+1} = z$ provides a sequence of successively better approximations to z. This would require that successive a_i's get closer to each other, which is evidently not the case if the Euclidean metric is used to measure proximity. But notice that

$$a_2 \equiv a_1 \pmod{p}, \quad a_3 \equiv a_2 \pmod{p^2}, \quad a_4 \equiv a_3 \pmod{p^3}, \ldots,$$

which suggests that we should view two numbers as being "close" if their difference is divisible by a power of p: the higher the power of p, the closer together are the numbers in question.

Now, every nonzero integer $z \in \mathbb{Z}$ has a highest power of p dividing it, i.e., there is a largest $n \geq 0$ such that $z \equiv 0 \pmod{p^n}$. In other words, z can be written uniquely in the form

$$z = p^n y$$

with y an integer that does not have p as a factor. This unique n will be denoted by $o_p(n)$ and called the *p-adic order* of z. (It is also known as the *p*-adic *valuation* of z, denoted by $v_p(z)$, and could equally naturally be thought of as the *p*-adic "exponent" or "logarithm"—important concepts often have more than one name.) Thus if $\mathbb{Z}^{\geq} = \mathbb{N} \cup \{0\}$, we have

$$o_p : \mathbb{Z} - \{0\} \longrightarrow \mathbb{Z}^{\geq}$$

(sometimes $o_p(0)$ is set equal to ∞, but we will avoid this). The *p*-adic order satisfies the laws

$$\begin{aligned}
o_p(zw) &= o_p(z) + o_p(w), \\
o_p(z+w) &\geq \min\{o_p(z), o_p(w)\}.
\end{aligned}$$

Now put

$$|z|_p = \begin{cases} p^{-o_p(z)} = \frac{1}{p^{o_p(z)}} & \text{if } z \neq 0, \\ 0 & \text{if } z = 0. \end{cases}$$

The function $|\ |_p$ is a norm on $\mathbb{Z}$ (sometimes called the p-adic *absolute value*), and satisfies

$$|zw|_p = |z|_p |w|_p,$$
$$|z+w|_p \leq \max\{|z|_p, |w|_p\}.$$

It gives rise to the p-*adic metric*

$$d_p(x, y) = |x - y|_p.$$

Observe that $|p^n|_p = p^{-n}$, so the sequence $p, p^2, p^3, \ldots$ converges to zero in the sense of this p-adic size of its terms. Thus an expression like

$$z_0 + z_1 p + z_2 p^2 + \cdots + z_n p^n + \cdots,$$

with $0 \leq z_i < p$, can be seen as analogous to the decimal representation of certain real numbers in the form

$$r_0 + r_1 \left(\tfrac{1}{10}\right) + r_2 \left(\tfrac{1}{10}\right)^2 + \cdots + r_n \left(\tfrac{1}{10}\right)^n + \cdots$$

with $0 \leq r_i < 10$.

p-adic Integers

A p-*adic integer* is a sequence $a = \langle a_n : n \in \mathbb{N} \rangle$ such that for each $n \in \mathbb{N}$,

(1) $a_n \in \mathbb{Z}/p^n$, and

(2) $a_{n+1} \equiv a_n \pmod{p^n}$.

This implies that

(3) $a_m \equiv a_n \pmod{p^n}$ whenever $m \geq n$

(hence if $a_m = 0$, then $a_n = 0$ for all $n < m$).

The set $\mathbb{Z}_p$ of all p-adic integers is a subset of the direct product

$$\mathbb{Z}/p \times \mathbb{Z}/p^2 \times \cdots \times \mathbb{Z}/p^n \times \cdots$$

and inherits operations of addition and multiplication from this direct product. Thus if a and b are p-adic integers, then

$$a + b = \langle a_n \oplus_n b_n : n \in \mathbb{N} \rangle,$$
$$ab = \langle a_n \otimes_n b_n : n \in \mathbb{N} \rangle,$$

where $a_n \oplus_n b_n$ and $a_n \otimes_n b_n$ are the sum and product modulo n. $\mathbb{Z}_p$ proves to be an integral domain under these operations, and we now briefly review its basic structure.

The divisibility relation $|$ is defined in $\mathbb{Z}_p$ just as it is in $\mathbb{Z}$: for $x, y \in \mathbb{Z}_p$,

$$x | y \quad \text{iff} \quad (\exists z \in \mathbb{Z}_p)\, y = xz.$$

Each integer $z \in \mathbb{Z}$ (positive or negative) can be identified with the p-adic integer

$$a_z = \langle z \bmod p,\ z \bmod p^2,\ \ldots,\ z \bmod p^n,\ \ldots \rangle.$$

For instance, -1 corresponds to the p-adic integer

$$\langle p - 1,\ p^2 - 1,\ \ldots,\ p^n - 1,\ \ldots \rangle.$$

The map $z \mapsto a_z$ is an injection of $\mathbb{Z}$ into $\mathbb{Z}_p$ that preserves addition and multiplication and allows us to identify $\mathbb{Z}$ with a subring of $\mathbb{Z}_p$.

A p-adic integer a has a multiplicative inverse in $\mathbb{Z}_p$ iff $a_1 \neq 0$. When $a_1 \neq 0$, then in general $a_n \not\equiv 0 \pmod{p}$ and a_n has an inverse b_n in $\mathbb{Z}/p^n$, i.e., $a_n b_n \equiv 1 \pmod{p^n}$. Then $\langle b_n : n \in \mathbb{N} \rangle$ is the inverse of a in $\mathbb{Z}_p$. Invertible elements of $\mathbb{Z}_p$ are called p-adic *units*, and can also be characterised as those elements that divide 1 in $\mathbb{Z}_p$.

Let a be a unit. Then any factor of a is also a unit. But p itself is not a unit, since it corresponds to the sequence $a_p = \langle 0, p, p, \ldots \rangle$, so p is not a factor of a, and therefore the only way to express a in the form $p^n b$ is to put $n = 0$ and $b = a$.

On the other hand, if a is a nonunit, and nonzero, then taking the least $n \geq 0$ such that $a_{n+1} \neq 0$, we have $n \geq 1$ and $a_n = 0$, so $a_{n+m} \equiv 0 \pmod{p^n}$ for all $m \in \mathbb{N}$. Then

$$b = \langle a_{n+1}/p^n \bmod p,\ a_{n+2}/p^n \bmod p^2,\ \ldots \rangle$$

proves to be a unit of $\mathbb{Z}_p$ with $a = p^n b$.

This shows that the p-adic units are precisely those members of $\mathbb{Z}_p$ that are not divisible by p. The representation of any nonzero a in the form $p^n b$ with b not divisible by p is unique, and this allows us to put $o_p(a) = n$. Then defining $|a|_p = p^{-o_p(a)}$ and $|0|_p = 0$ gives a norm on $\mathbb{Z}_p$ extending the p-adic norm on $\mathbb{Z}$ and inducing the associated extended metric d_p on $\mathbb{Z}_p$. Note that a is a p-adic unit iff $|a|_p = 1$.

Any p-adic integer a has

$$a \equiv a_n \pmod{p^n}$$

in $\mathbb{Z}_p$, i.e., the difference $a - a_n$ is equal to $p^n b$ for some $b \in \mathbb{Z}_p$. This implies $o_p(a - a_n) \geq n$, hence $|a - a_n|_p \leq p^{-n}$. Thus the sequence $a_1, a_2, a_3, \ldots$ converges to a in the p-adic metric, showing that $\mathbb{Z}$ is dense in the metric

space $(\mathbb{Z}_p, d_p)$. But in fact, using (1) and (2) it can be shown that a has the form
$$\langle z_0, z_0 + z_1 p, z_0 + z_1 p + z_2 p^2, \ldots \rangle$$
with $0 \le z_n < p$, and so we can also write
$$a = z_0 + z_1 p + z_2 p^2 + \cdots + z_n p^n + \cdots.$$

For instance, when $p = 5$,
$$\begin{aligned}
-1 &= \langle -1 \bmod 5,\ -1 \bmod 25,\ \ldots,\ -1 \bmod 5^n,\ \ldots \rangle \\
&= \langle 4,\ 24,\ 124,\ 624,\ \ldots \rangle \\
&= 4 + 4 \cdot 5 + 4 \cdot 5^2 + 4 \cdot 5^3 + \cdots + 4 \cdot 5^n + \cdots.
\end{aligned}$$

The Nonstandard Analysis

$\mathbb{Z}_p$ is complete under the p-adic metric d_p, and is a completion of $(\mathbb{Z}, d_p)$. We are going to demonstrate this fact, not by appealing to any of the convergence results just claimed for $\mathbb{Z}_p$, but by showing:

- $(\mathbb{Z}_p, d_p)$ is isometric to the nonstandard hull of $(\mathbb{Z}, d_p)$.

- In $({}^*\mathbb{Z}, {}^*d_p)$, every limited point is approachable from $\mathbb{Z}$.

This implies that the nonstandard hull of $(\mathbb{Z}, d_p)$ is equal to the completion of $(\mathbb{Z}, d_p)$ based on approachable points, as given by Theorem 18.3.2.

The symbols $|\ |_p$ and o_p will continue to be used for the extension of these functions from $\mathbb{Z}$ to the commutative ring ${}^*\mathbb{Z}$, as provided by the nonstandard framework. In general, $o_p(x)$ is a nonnegative hyperinteger, i.e., o_p takes values in the set ${}^*\mathbb{Z}^{\ge} = {}^*\mathbb{N} \cup \{0\}$. The basic properties and relationships of $|\ |_p$ and o_p are preserved by transfer. In particular,

$$|x|_p = \frac{1}{p^{o_p(x)}}$$

for all nonzero hyperintegers x, so $|\ |_p$ takes hyperreal values in the set $\{p^{-n} : n \in {}^*\mathbb{Z}^{\ge}\}$. These values consist of positive infinitesimals (when n in unlimited) and real numbers ≤ 1 (when n is standard).

These functions can then be used to define the sets of limited and infinitesimal elements in the p-adic sense as

$$\begin{aligned}
{}^*\mathbb{Z}^{\lim_p} &= \{x \in {}^*\mathbb{Z} : |x - z|_p \text{ is limited for some } z \in \mathbb{Z}\}, \\
{}^*\mathbb{Z}^{\inf_p} &= \{x \in {}^*\mathbb{Z} : |x|_p \simeq 0 \text{ in } {}^*\mathbb{R}\}.
\end{aligned}$$

Now, $|x - 0|_p = |x|_p \le 1$ holds for all standard integers $x \in \mathbb{Z}$, and hence for all hyperintegers $x \in {}^*\mathbb{Z}$ by transfer. This means that every member of ${}^*\mathbb{Z}$ is p-adically limited, so ${}^*\mathbb{Z}^{\lim_p} = {}^*\mathbb{Z}$ and the nonstandard hull here is just $^*\mathbb{Z}/\simeq$ with the metric induced by the shadows of $^*d_p$. To explain what the p-adic infinitesimals are, observe first that p^N will be infinitesimal whenever N is unlimited because then by transfer $|p^N|_p = p^{-N} \simeq 0$. The idea that a p-adically "small" number is one that is divisible by a "large" power of p finds its ultimate expression in the following characterisation.

Theorem 18.4.1 *For any nonzero hyperinteger $x \in {}^*\mathbb{Z}$, the following are equivalent.*

(1) $x \in {}^*\mathbb{Z}^{\inf_p}$.

(2) $o_p(x)$ is unlimited.

(3) x is divisible by p^n in $^*\mathbb{Z}$ for all $n \in \mathbb{N}$.

(4) x is divisible by p^N for some unlimited $N \in {}^*\mathbb{N}$.

Proof. $|x|_p$ is the reciprocal of $p^{o_p(x)}$, so is infinitesimal iff $p^{o_p(x)}$ is unlimited, which holds iff $o_p(x)$ is unlimited, as p is standard. Thus (1) and (2) are equivalent.

Since the divisibility relation $|$ is defined in $\mathbb{Z}$ by

$$x|y \quad \text{iff} \quad (\exists z \in \mathbb{Z})\, y = xz,$$

it follows by transfer that $x|y$ for hyperintegers in $^*\mathbb{Z}$ iff $y = xz$ for some $z \in {}^*\mathbb{Z}$. Now, the statement

$$p^n|x \quad \text{iff} \quad n \le o_p(x) \tag{ii}$$

holds for all $x \in {}^*\mathbb{Z}$ and $n \in {}^*\mathbb{Z}^{\ge}$, again by transfer. Hence if (2) holds, then for every $n \in \mathbb{N}$ we have $n \le o_p(x)$, as $o_p(x)$ is unlimited, and so $p^n|x$ by (ii). Thus (2) implies (3).

Next, observe that for each $x \in {}^*\mathbb{Z}$ the set

$$\{n \in {}^*\mathbb{N} : p^n|x\}$$

is internal, by the internal set definition principle, so if (3) holds, this set contains all members of $\mathbb{N}$, and hence by overflow it contains some unlimited N, establishing (4).

Finally, if $p^N|x$ with N unlimited, then (ii) gives $N \le o_p(x)$, so $o_p(x)$ is also unlimited. Thus (4) implies (2). □

This result shows that

$$^*\mathbb{Z}^{\inf_p} = \{p^N q : N \text{ is unlimited and } q \in {}^*\mathbb{Z}\}.$$

The main properties of congruence relations lift to $^*\mathbb{Z}$ by transfer. In particular, if $m \in \mathbb{N}$, then each $x \in {}^*\mathbb{Z}$ is congruent modulo m to a unique element $r \in \mathbb{Z}/m$, i.e., $x - r$ is divisible by m in $^*\mathbb{Z}$. We continue to denote this unique element r by $x \bmod m$. The map $x \mapsto x \bmod m$, which is a ring

18.4 p-adic Integers

homomorphism from $\mathbb{Z}$ onto $\mathbb{Z}/m$, thereby lifts to a ring homomorphism from $^*\mathbb{Z}$ onto $^*(\mathbb{Z}/m) = \mathbb{Z}/m$. This allows us to define a homomorphism

$$\theta_p : {}^*\mathbb{Z} \to \mathbb{Z}_p$$

by putting

$$\theta_p(x) = \langle x \bmod p,\ x \bmod p^2,\ \ldots,\ x \bmod p^n,\ \ldots \rangle .$$

It is left as an instructive exercise to check that $\theta_p(x) \in \mathbb{Z}_p$, i.e.,

$$x \bmod p^{n+1} \equiv x \bmod p^n \pmod{p^n},$$

and that θ_p preserves addition and multiplication. Notice also that since we identify the standard integer z with the p-adic integer

$$\langle z \bmod p,\ z \bmod p^2,\ \ldots,\ z \bmod p^n,\ \ldots \rangle ,$$

it follows that θ_p leaves all members of $\mathbb{Z}$ fixed. The kernel

$$\{x \in {}^*\mathbb{Z} : \theta_p(x) = 0\}$$

of θ_p consists of those $x \in {}^*\mathbb{Z}$ such that for all $n \in \mathbb{N}$ we have $x \bmod p^n = 0$, which means that p^n divides x. By Theorem 18.4.1 this holds precisely when $|x|_p \simeq 0$. Thus the kernel is exactly the set $^*\mathbb{Z}^{\inf_p}$ of p-adic infinitesimals, which is therefore an ideal of the ring $^*\mathbb{Z}$. Then the coset

$$^*\mathbb{Z}^{\inf_p} + x = \{p^N q + x : N \text{ is unlimited and } q \in {}^*\mathbb{Z}\}$$

is the set of all hyperintegers that are infinitely close to x in the p-adic metric, because $|y - x|_p$ is infinitesimal if and only if $y - x$ is of the form $p^N q$ with N unlimited.

If we can show that θ_p maps *onto* $\mathbb{Z}_p$, then by the homomorphism theorem we will have a ring isomorphism

$$^*\mathbb{Z}/{}^*\mathbb{Z}^{\inf_p} \cong \mathbb{Z}_p.$$

To prove that θ_p is onto $\mathbb{Z}_p$ requires us to invoke the concurrence version of enlargement (Theorem 14.2.1). If $a = \langle a_n : n \in \mathbb{N} \rangle \in \mathbb{Z}_p$, define a relation $R^a \subseteq \mathbb{N} \times \mathbb{Z}$ by putting

$$n R^a y \quad \text{iff} \quad y \equiv a_n \pmod{p^n}.$$

R^a has domain $\mathbb{N}$ and is concurrent: given integers $n_1, \ldots, n_k \in \mathbb{N}$, take any $m \in \mathbb{N}$ with $m \geq n_1, \ldots, n_k$. Then since $a_m \equiv a_n \pmod{p^n}$ whenever $m \geq n$ (condition (3) of the definition of p-adic integer), it follows that $n_1 R^a a_m, \ldots, n_k R^a a_m$. Hence as R^a is concurrent, there must be an $x \in {}^*\mathbb{Z}$ such that $n({}^*R^a)x$ for all $n \in \mathbb{N}$. Transferring the definition of R^a then shows that for all such n, $x \equiv a_n (\bmod\ p^n)$, and hence $x \bmod p^n = a_n$. Thus $\theta_p(x) = a$, and the proof that θ_p is onto is complete.

Note that in proving the concurrence of R^a here we can choose m such that $a_m > 0$, so the proof shows that R^a is concurrent as a relation from N to N, and hence will produce a *positive* x (i.e., a member of *N) with $\theta_p(x) = a$. This fact will be used at the end of this chapter to derive a description of p-adic integers as certain hyperfinite formal sums. Another explanation of why such a positive x can be found is that the hyperintegers whose θ_p-image is equal to a form a coset *$\mathbb{Z}^{\inf_p} + y$, and any coset must contain positive elements. Indeed, for a given y, $p^N + y$ (which belongs to *$\mathbb{Z}^{\inf_p} + y$) will be positive for large enough unlimited N.

Preserving the Metric

We have observed that the coset *$\mathbb{Z}^{\inf_p} + x$ is just the $\simeq$-equivalence class

$$\mathrm{hal}_p(x) = \{y \in {}^*\mathbb{Z} : |x - y|_p \simeq 0\}$$

of x in *$\mathbb{Z}$. Hence

$$({}^*\mathbb{Z}/\simeq) = {}^*\mathbb{Z}/{}^*\mathbb{Z}^{\inf_p} \cong \mathbb{Z}_p.$$

This bijection between the nonstandard hull *$\mathbb{Z}/\simeq$ and $\mathbb{Z}_p$ is given by $\mathrm{hal}_p(x) \mapsto \theta_p(x)$. If we can show that it preserves the metrics, we will have our desired demonstration that the nonstandard hull is isometric to $(\mathbb{Z}_p, d_p)$. But the metric on *$\mathbb{Z}/\simeq$ is induced by the norm function

$$|\mathrm{hal}_p(x)|_p = \mathrm{sh}|x|_p,$$

so we want $\mathrm{sh}|x|_p = |\theta_p(x)|_p$, or equivalently, $|x|_p \simeq |\theta_p(x)|_p$. There are two cases:

(1) $\theta_p(x) = 0$. Then $x \in {}^*\mathbb{Z}^{\inf_p}$ and $|x|_p \simeq 0 = |0|_p = |\theta_p(x)|_p$.

(2) $\theta_p(x) \ne 0$. Then by definition of $\theta_p(x)$ there must be some standard $n \ge 0$ such that $x \bmod p^{n+1} \ne 0$, and $o_p(\theta_p(x))$ is the least such n by definition of the p-adic norm on $\mathbb{Z}_p$. Now,

$$p^n | x \quad \text{iff} \quad n \le o_p(x)$$

for all $n \ge 0$ (cf. (ii) in the proof of Theorem 18.4.1), and $o_p(x)$ is a standard integer because $|x|_p \not\simeq 0$, so $o_p(x)$ is the least standard n for which $p^{n+1} \nmid x$, i.e., the least standard n for which $x \bmod p^{n+1} \ne 0$. Thus in this case $o_p(x) = o_p(\theta_p(x))$ and $|x|_p = |\theta_p(x)|_p$.

Having now shown that the nonstandard hull of $(\mathbb{Z}, d_p)$ is isometric to $(\mathbb{Z}_p, d_p)$, it remains to show that this hull is a completion of $(\mathbb{Z}, d_p)$ by showing that any point $x \in {}^*\mathbb{Z}$ is approachable from $\mathbb{Z}$. But for each $n \in \mathbb{N}$, p^n divides $x - (x \bmod p^n)$, so

$$o_p(x - x \bmod p^n) \geq n \quad \text{and} \quad |x - x \bmod p^n|_p \leq p^{-n}.$$

Then for any $\varepsilon \in \mathbb{R}^+$, by choosing a standard n large enough that $p^{-n} < \varepsilon$ we get the *standard* integer $x \bmod p^n$ that is within ε of x in the p-adic metric. This shows that x is approachable from $\mathbb{Z}$. Thus

$$ {}^*\mathbb{Z} = {}^*\mathbb{Z}^{\lim_p} = {}^*\mathbb{Z}^{\mathrm{ap}}.$$

Nearstandardness

To round out this discussion, consider the points in ${}^*\mathbb{Z}$ that are near to $\mathbb{Z}$. If $|x - z|_p \simeq 0$, then p^N divides $x - z$ for some unlimited $N \in {}^*\mathbb{N}$. So the nearstandard points of ${}^*\mathbb{Z}$ are precisely those of the form $x = p^N y + z$ with N unlimited and z standard.

Since $(\mathbb{Z}, d_p)$ is incomplete, there must be points in ${}^*\mathbb{Z}$ that are approachable from $\mathbb{Z}$ but not near to $\mathbb{Z}$ (Corollary 18.3.3). Indeed, if $|x - z|_p \simeq 0$ with z standard, then $\theta_p(x - z) = 0$, and so $\theta_p(x) = \theta_p(z) = z$. This shows that the nearstandard points in ${}^*\mathbb{Z}$ are just the θ_p-preimages of members of $\mathbb{Z}$. Any $x \in {}^*\mathbb{Z}$ with $\theta_p(x) \in (\mathbb{Z}_p - \mathbb{Z})$ fails to be near to $\mathbb{Z}$.

18.5 p-adic Numbers

The ring $\mathbb{Z}_p$ has a *field of fractions*

$$\mathbb{Q}_p = \left\{ \frac{a}{b} : a, b \in \mathbb{Z}_p \text{ and } b \neq 0 \right\}.$$

Members of $\mathbb{Q}_p$ are called *p-adic numbers*, and equality between them is given by

$$\frac{a}{b} = \frac{c}{d} \quad \text{iff} \quad ad = bc \text{ in } \mathbb{Z}_p.$$

The field operations are given by the familiar formulae from rational arithmetic:

$$\frac{a}{b} + \frac{c}{d} = \frac{ad + bc}{bd},$$
$$\frac{a}{b} \cdot \frac{c}{d} = \frac{ac}{bd},$$
$$-\left(\frac{a}{b}\right) = \left(\frac{-a}{b}\right),$$
$$\left(\frac{a}{b}\right)^{-1} = \left(\frac{b}{a}\right).$$

Thus $\mathbb{Q}_p$ stands in the same relation to $\mathbb{Z}_p$ that $\mathbb{Q}$ stands to $\mathbb{Z}$. Moreover, since $\mathbb{Z} \subseteq \mathbb{Z}_p$, it follows that $\mathbb{Q} \subseteq \mathbb{Q}_p$.

The p-adic order function extends to $\mathbb{Q}_p$ by putting
$$o_p(a/b) = o_p(a) - o_p(b).$$
This is well-defined, because if $a/b = c/d$, then $o_p(a) - o_p(b) = o_p(c) - o_p(d)$ by the "logarithmic" law $o_p(ad) = o_p(a) + o_p(d)$ etc. Then we put $|x|_p = p^{-o_p(x)}$ and $d_p(x, y) = |x - y|_p$ as before, but now for $x, y \in \mathbb{Q}_p$. In particular, this gives a p-adic order function and norm on $\mathbb{Q}$. To analyse this further, recall that nonzero p-adic integers a, b can be written uniquely in the form $a = p^n c$ and $b = p^m d$ with $n, m \geq 0$, c, d units in $\mathbb{Z}_p$, and p not a factor of c or d. Then
$$\frac{a}{b} = \frac{p^n c}{p^m d} = p^{n-m} \frac{c}{d}.$$
Here $o_p(a/b) = n - m \in \mathbb{Z}$, and c/d is a unit in $\mathbb{Z}_p$.

In general, then, any nonzero p-adic number has a representation in the form $p^m b$ with m a standard integer and b a unit in $\mathbb{Z}_p$, and this representation is unique (the case $m \geq 0$ giving the p-adic integers). Moreover, in view of the representation in Section 18.4 of p-adic units as power series with nonzero initial term, each p-adic number $x \neq 0$ is uniquely expressible in the form
$$\begin{aligned} x &= p^m(z_0 + z_1 p + z_2 p^2 + \cdots + z_n p^n + \cdots) \\ &= z_0 p^m + z_1 p^{m+1} + z_2 p^{m+2} + \cdots, \end{aligned}$$
where m is the integer $o_p(x)$, $0 \leq z_n < p$, and $z_0 \geq 1$. Since m can be negative here, it follows that a p-adic number can be written in the general form
$$z_{-k} p^{-k} + \cdots + z_{-1} p^{-1} + z_0 + z_1 p + z_2 p^2 + \cdots + z_n p^n + \cdots,$$
with $0 \leq z_i < p$, showing that it is the sum of a standard rational number $z_{-k} p^{-k} + \cdots + z_{-1} p^{-1}$ and a p-adic integer. Note the analogy with the fact that any real number can be represented as an infinite decimal expression
$$r_{-k} \left(\tfrac{1}{10}\right)^{-k} + \cdots + r_{-1} \left(\tfrac{1}{10}\right)^{-1} + r_0 + r_1 \left(\tfrac{1}{10}\right) + r_2 \left(\tfrac{1}{10}\right)^2 \cdots + r_n \left(\tfrac{1}{10}\right)^n + \cdots.$$
Under the metric d_p, $\mathbb{Q}_p$ is a completion of $\mathbb{Q}$.

Limited p-adics

In a nonstandard framework, the functions o_p and $|\ |_p$ extend from $\mathbb{Q}$ to $*\mathbb{Q}$ by the transfer map and continue to satisfy the usual properties, including
$$|x|_p = p^{-o_p(x)},$$
$$\left|\frac{x}{y}\right|_p = \frac{|x|_p}{|y|_p}$$

for $x, y \in {}^*\mathbb{Q}$. The sets of hyperrationals that are *limited* or *infinitesimal* in the p-adic sense are given by

$${}^*\mathbb{Q}^{\lim_p} = \{x \in {}^*\mathbb{Q} : |x - q|_p \text{ is limited for some } q \in \mathbb{Q}\}$$
$$= \{x \in {}^*\mathbb{Q} : |x|_p \text{ is limited}\}$$

and
$${}^*\mathbb{Q}^{\inf_p} = \{x \in {}^*\mathbb{Q} : |x|_p \simeq 0 \text{ in } {}^*\mathbb{R}\}.$$

The p-adic order $o_p(x)$ of a nonzero hyperrational x is itself a hyperinteger, so falls under one of three cases:

(1) $o_p(x)$ is limited, and hence is a standard integer. Then $|x|_p$ is a nonnegative real number, equal to 0 when $x = 0$, and otherwise of the form p^m with $m \in \mathbb{Z}$.

(2) $o_p(x)$ is positive unlimited (i.e., in ${}^*\mathbb{N}_\infty$). Then $p^{o_p(x)}$ is positive unlimited, and $|x|_p$ is a positive infinitesimal: $|x|_p \simeq 0$.

(3) $o_p(x)$ is negative unlimited. Then $-o_p(x) \in {}^*\mathbb{N}_\infty$, and so $|x|_p$ is positive unlimited.

This shows that x can fail to be p-adically limited only when case (3) occurs, so

$${}^*\mathbb{Q}^{\lim_p} = \{x \in {}^*\mathbb{Q} : o_p(x) \text{ is not negative unlimited}\}$$
$$= \{x \in {}^*\mathbb{Q} : o_p(x) \in \mathbb{Z} \cup {}^*\mathbb{N}_\infty\}.$$

This characterisation gives rise to a more useful one: p-adic limitedness of a hyperrational depends on the size of the denominator, as the next result indicates.

Theorem 18.5.1 *Let $y, z \in {}^*\mathbb{Z}$. If $|z|_p$ is not infinitesimal, then y/z is p-adically limited.*

Proof. $o_p(z)$ is a nonnegative hyperinteger, so if $|z|_p = p^{-o_p(z)} \not\simeq 0$, then $o_p(z)$ must be limited, i.e., $o_p(z) \in \mathbb{N} \cup \{0\}$. But then since $o_p(y) \geq 0$,

$$o_p(y/z) = o_p(y) - o_p(z)$$

cannot be negative unlimited. Hence as above, $y/z \in {}^*\mathbb{Q}^{\lim_p}$. □

The converse of this can fail. If $|z|_p \simeq 0$, then $|y/z|_p$ will still be limited if $|y|_p \leq n|z|_p$ for some $n \in \mathbb{N}$ (in which case $|y|_p$ is also infinitesimal). For instance, this happens when $y = 2p^N$ and $z = p^N$ with N unlimited.

Now, we can express any hyperrational as a ratio of hyperintegers that have no factors on common. It is the presence of unlimited powers p^N of p as factors that makes a hyperinteger p-adically infinitesimal, and it turns out that the absence of common factors of this particular kind is enough to give the converse to Theorem 18.5.1.

Theorem 18.5.2 *Let y, z be hyperintegers that have no common factors of the form p^N with $N \in {}^*\mathbb{N}_\infty$. If $|y/z|_p$ is limited, then $|z|_p$ is not infinitesimal.*

Proof. Suppose that $|y/z|_p$ is limited, but $|z|_p \simeq 0$. Then $o_p(z)$ is positive unlimited (Theorem 18.4.1), while $o_p(y/z) = o_p(y) - o_p(z)$ is not negative unlimited. But this can be so only if $o_p(y)$ is also positive unlimited. Then if N is the smaller of $o_p(y)$ and $o_p(z)$, we have $N \in {}^*\mathbb{N}_\infty$ and p^N a factor of both y and z. However, this contradicts the hypothesis. □

The Completion

We are going to show that $\mathbb{Q}_p$ is a completion of $\mathbb{Q}$ under the p-adic metric by exhibiting an isomorphism

$$^*\mathbb{Q}^{\lim_p}/^*\mathbb{Q}^{\inf_p} \cong \mathbb{Q}_p \qquad \text{(iii)}$$

and demonstrating that all elements of $^*\mathbb{Q}^{\lim_p}$ are approachable from $\mathbb{Q}$. For the isomorphism we need a homomorphism from $^*\mathbb{Q}^{\lim_p}$ onto $\mathbb{Q}_p$. We already have a homomorphism $\theta_p : {}^*\mathbb{Z} \to \mathbb{Z}_p$, and the relationships between $^*\mathbb{Q}$ and $^*\mathbb{Z}$ and $\mathbb{Q}_p$ and $\mathbb{Z}_p$ suggest that we extend θ_p to hyperrationals by putting

$$\theta_p^+\left(\frac{x}{y}\right) = \frac{\theta_p(x)}{\theta_p(y)}. \qquad \text{(iv)}$$

Of course for this to be defined we need $\theta_p(y) \neq 0$, but that is exactly where limitedness comes in. We apply the definition (iv) only when x/y is in *reduced form*, i.e., x and y have no proper factors in common. In particular, they have no common factors p^N with N unlimited, so by Theorem 18.5.2 if $x/y \in {}^*\mathbb{Q}^{\lim_p}$, then $|y|_p \not\simeq 0$, and so $\theta_p(y) \neq 0$. Thus (iv) is well-defined for all members of $^*\mathbb{Q}^{\lim_p}$.

The fact that $\theta_p : {}^*\mathbb{Z} \to \mathbb{Z}_p$ preserves addition and multiplication and maps $^*\mathbb{Z}$ onto $\mathbb{Z}_p$ can be used to show:

- θ_p^+ is a ring homomorphism from $^*\mathbb{Q}^{\lim_p}$ onto $\mathbb{Q}_p$ that extends θ_p.

Thus to obtain the isomorphism (iii) we have only to show that $^*\mathbb{Q}^{\inf_p}$ is the kernel of θ_p^+. But for $x/y \in {}^*\mathbb{Q}^{\lim_p}$ in reduced form, $|x/y|_p = |x|_p/|y|_p$ with $|y|_p \not\simeq 0$, and hence $|y|_p$ is a standard real number. Therefore

$$|x/y|_p \simeq 0 \quad \text{iff} \quad |x|_p \simeq 0 \quad \text{iff} \quad \theta_p(x) = 0 \quad \text{iff} \quad \theta_p^+(x/y) = 0,$$

so indeed x/y belongs to $^*\mathbb{Q}^{\inf_p}$ iff it is in the kernel of θ_p^+.

Preserving the Metric

In order to show that the isomorphism (iii) preserves the metric of the space $^*\mathbb{Q}^{\lim_p}/^*\mathbb{Q}^{\inf_p}$, and hence show that $\mathbb{Q}_p$ is isomorphic to the nonstandard hull of $(\mathbb{Q}, d_p)$, we need to show that $\text{sh}|v|_p = |\theta_p^+(v)|_p$, or equivalently, $|v|_p \simeq |\theta_p^+(v)|_p$, for all $v \in {}^*\mathbb{Q}^{\lim_p}$. As with the integer case in Section 18.4, there are two parts to this:

(1) $\theta_p^+(v) = 0$. Then $v \in {}^*\mathbb{Q}^{\inf_p}$ and $|v|_p \simeq 0 = |0|_p = |\theta_p^+(v)|_p$.

(2) $\theta_p^+(v) \neq 0$. Then in fact, $|\theta_p^+(v)|_p = |v|_p$, because if $v = x/y$ in reduced form with $x, y \in {}^*\mathbb{Z}$, then $\theta_p(x), \theta_p(y) \neq 0$, so $|x|_p = |\theta_p(x)|_p$ and $|y|_p = |\theta_p(y)|_p$ by the integer case, and therefore
$$|x/y|_p = |x|_p/|y|_p = |\theta_p(x)|_p/|\theta_p(y)|_p = |\theta_p(x)/\theta_p(y)|_p = |\theta_p^+(x/y)|_p.$$

It remains now to prove that each $v \in {}^*\mathbb{Q}^{\lim_p}$ is approachable from $\mathbb{Q}$. Again there are two parts:

(1) If $\theta_p^+(v) = 0$, then $|v - 0|_p \simeq 0$, so v is actually near to $\mathbb{Q}$.

(2) If $\theta_p^+(v) \neq 0$, then as just shown, $|v|_p = |\theta_p^+(v)|_p$. Since $\theta_p^+(v)$ is a p-adic number, it is equal to $p^m b$ for some $m \in \mathbb{Z}$ and some $b \in \mathbb{Z}_p$ with $p \nmid b$, and so $|b|_p = 1$. Now choose an $x \in {}^*\mathbb{Z}$ with $\theta_p(x) = b$. Then θ_p^+ leaves p^m fixed because it is a standard integer, and
$$\theta_p^+(v - p^m x) = \theta_p^+(v) - \theta_p^+(p^m x) = \theta_p^+(v) - (p^m b) = 0,$$
so $|v - p^m x|_p \simeq 0$.

But given any $\varepsilon \in \mathbb{R}^+$, since $x \in {}^*\mathbb{Z}$ is approachable from $\mathbb{Z}$, there must be a standard $z \in \mathbb{Z}$ such that
$$|x - z|_p < \frac{\varepsilon p^m}{2}.$$
Then $p^m z$ is a standard rational number that is p-adically within ε of v, since
$$\begin{aligned}|v - p^m z|_p &\leq |v - p^m x|_p + |p^m x - p^m z|_p \\ &= |v - p^m x|_p + |p^m|_p |x - z|_p \\ &< |v - p^m x|_p + p^{-m}(\varepsilon p^m)/2 \\ &< \varepsilon\end{aligned}$$
because $|v - p^m x|_p$ is infinitesimal.

18.6 Power Series

Polynomials

Let $\langle R, +, -, \cdot, 0, 1 \rangle$ be a commutative ring. A *polynomial in x of degree n over R* is a "finite formal sum"
$$a_0 + a_1 x + a_2 x^2 + \cdots + a_n x^n,$$

where $a_0, \ldots, a_n$ are elements of R, called the *coefficients* of the polynomial. Coefficient a_i is of *degree* i. The *leading coefficient* is a_n, which is required to be nonzero if $n \neq 0$. The set of all polynomials in x over R of all possible degrees $n \in \mathbb{Z}^{\geq}$ is denoted by $R[x]$.

When $n = 0$, a single element a_0 of R is regarded as a polynomial, and has degree 0 (unless $a_0 = 0$: the *zero polynomial* 0 will not be assigned a degree). Thus we have $R \subseteq R[x]$. Members of R are *constant* polynomials.

Two polynomials are equal if they have the same degree and corresponding coefficents (i.e., those of the same degree) are identical. Thus a polynomial is uniquely determined by its list of coefficents, and this suggests that a more explicit way to define a polynomial is to view it as a sequence $a = \langle a_0, \ldots, a_n, \ldots \rangle$ of elements of R, or equivalently, a function $a : \mathbb{Z}^{\geq} \to R$, that is *ultimately zero* in the sense that

$$(\exists n \in \mathbb{Z}^{\geq})(\forall m \in \mathbb{N})(m > n \to a_m = 0).$$

The least such n is the degree of a. The inclusion of R in $R[x]$ arises by identifying each $r \in R$ with the sequence $\langle r, 0, 0, \ldots, 0, \ldots \rangle$.

The set $R[x]$ of polynomials over R forms a commutative ring under the operations

$$\begin{aligned} a + b &= \langle a_0 + b_0, \ldots, a_n + b_n, \ldots \rangle, \\ -a &= \langle -a_0, \ldots, -a_n, \ldots \rangle, \\ ab &= \langle a_0 b_0, a_0 b_1 + a_1 b_0, \ldots, a_0 b_n + a_1 b_{n-1} + \cdots + a_n b_0, \ldots \rangle. \end{aligned}$$

Power Series

A *power series over* R is an "infinite formal sum"

$$a = a_0 + a_1 x + a_2 x^2 + \cdots + a_n x^n + \cdots$$

with coefficients from R. Thus we may simply say that a power series is any sequence $a = \langle a_0, \ldots, a_n, \ldots \rangle$ of elements of R, or equivalently, any function $a : \mathbb{Z}^{\geq} \to R$. The set of all power series over R will be denoted by $R[\![x]\!]$. It forms a ring under the operations defined as for $R[x]$ and has $R[x]$ as a subring. Altogether now we have

$$R \subseteq R[x] \subseteq R[\![x]\!].$$

If a power series a is nonzero, then it must have a nonzero coefficient. The least n such that $a_n \neq 0$ is called the *order* of a, denoted by $o(a)$. Put

$$|a| = \begin{cases} 2^{-o(a)} & \text{if } a \neq 0, \\ 0 & \text{if } a = 0. \end{cases}$$

Then $d(a, b) = |a - b|$ defines a metric on $R[\![x]\!]$. Note that $|a| \leq 1$ in general.

A power series a as above determines the sequence

$$a_0,\ a_0 + a_1x,\ a_0 + a_1x + a_2x^2,\ \ldots,\ a_0 + a_1x + \cdots + a_nx^n,\ \ldots$$

of partial sums, which are polynomials. The order of

$$a - (a_0 + a_1x + \cdots + a_nx^n)$$

is at least $n+1$, and so

$$|a - (a_0 + a_1x + \cdots + a_nx^n)| < 2^{-n}.$$

It follows that the sequence of partial sums converges to a in the metric just defined. This implies that $R[x]$ is dense in $R[\![x]\!]$. In fact, $R[\![x]\!]$ is a complete metric space, hence a completion of $R[x]$, as we will now show by invoking the nonstandard hull construction again.

Enlargement

Let $^*R[x]$ abbreviate the enlargement $^*(R[x])$ of $R[x]$ in a nonstandard framework for R. Since members of $R[x]$ are functions from $\mathbb{Z}^\geq$ to R, the members of $^*R[x]$ are *internal* functions from $^*\mathbb{Z}^\geq$ to $^*R$ (Exercise 13.13(3)), or alternatively internal hypersequences $a = \langle a_n : n \in {}^*\mathbb{Z}^\geq \rangle$. Since polynomials are ultimately zero, so too are members of $^*R[x]$. This is because

$$(\exists n \in \mathbb{Z}^\geq)\,(\forall m \in \mathbb{N})\,(m > n \to a_m = 0)$$

is true for all $a \in R[x]$, so

$$(\exists n \in {}^*\mathbb{Z}^\geq)\,(\forall m \in {}^*\mathbb{N})\,(m > n \to a_m = 0)$$

is true for all $a \in {}^*R[x]$. But now the largest n for which $a_n \neq 0$ may be unlimited, so in general a member of $^*R[x]$ may be thought of as a *hyperfinite* formal sum

$$a_0 + a_1x + a_2x^2 + \cdots + a_N x^N$$

with its degree $N \in {}^*\mathbb{N}$ possibly being unlimited. The coefficients a_n can be nonstandard here, even when n is standard. Thus a member of $^*R[x]$ is an *internal hyperpolynomial* with coefficients from $^*R$ (note that $^*R$ is a commutative ring, by transfer of the fact that R is).

$^*R[x]$ is not the same thing as $(^*R)[x]$. The latter is the ring of (finite) polynomials $a_0 + a_1x + a_2x^2 + \cdots + a_nx^n$ with coefficients from $^*R$. Of course we can view a polynomial as a special case of a hyperpolynomial, and so identify each member of $(^*R)[x]$ with a member of $^*R[x]$. To be precise this requires a use of transfer: for a fixed $n \in \mathbb{Z}^\geq$, the statement

$$(\forall a_0, \ldots, a_n \in R)\,(\exists b \in R[x])$$
$$[b_0 = a_0 \wedge \cdots \wedge b_n = a_n \wedge (\forall m \in \mathbb{N})\,(m > n \to b_m = 0)]$$

asserts (correctly) that for any list $a_0, \ldots, a_n$ of elements of R there is a polynomial b in $R[x]$ having this list as its coefficients. By transfer then, for any list $a_0, \ldots, a_n$ of elements of $^*R$ (possibly including nonstandard elements) there is a hyperpolynomial b in $^*R[x]$ with $a_0, \ldots, a_n$ as its coefficients.

In particular, $^*R[x]$ includes all members of $^*R$ as constant hyperpolynomials. Also, if $a \in R[x]$, then a is regarded as being in $^*R[x]$ by identifying it with its extension to $^*\mathbb{Z}^{\geq}$ having $a_n = 0$ for all unlimited n. The functions $o(a)$, $|a|$, $a+b$, ab all extend from $R[x]$ to $^*R[x]$, preserving many of their properties by transfer.

Theorem 18.6.1 $a \in {}^*R[x]$ *is approachable from* $R[x]$ *if and only if the coefficient* a_n *belongs to* R *for all* **standard** n.

Proof. Fix a standard $n \in \mathbb{Z}^{\geq}$. Then if two polynomials $a, b \in R[x]$ are closer than 2^{-n} to each other (i.e., $|a - b| < 2^{-n}$), the order of $a - b$ must be at least $n + 1$, so $(a - b)_n = 0$ and hence $a_n = b_n$. Thus the statement

$$|a - b| < 2^{-n} \;\rightarrow\; a_n = b_n$$

holds for all $a, b \in R[x]$, and so by transfer holds for all $a, b \in {}^*R[x]$.

Now suppose that a is in $^*R[x]^{\mathrm{ap}}$, the set of all members of $^*R[x]$ approachable from $R[x]$. Then for each standard n there must be some polynomial $b \in R[x]$ with $|a - b| < 2^{-n}$. From the previous paragraph it then follows that $a_n = b_n \in R$. Thus the coefficient a_n is in R for each standard n.

Conversely, suppose $a \in {}^*R[x]$ has $a_n \in R$ for all standard n. For each such n, the polynomial

$$a \upharpoonright n = a_0 + a_1 x + a_2 x^2 + \cdots + a_n x^n$$

belongs to $R[x]$. But $a \upharpoonright n$ is within 2^{-n} of a, because the statement

$$|a - a \upharpoonright n| < 2^{-n}$$

holds for all $a \in R[x]$ (see above) so holds for all $a \in {}^*R[x]$ by transfer. This shows that a is approachable from $R[x]$. $\square$

At the end of Section 18.3 we promised to provide an example of a metric space having limited entities that are not approachable. The theorem just proved furnishes many examples. All members of $^*R[x]$ are limited, and indeed satisfy $|a| \leq 1$ by transfer. But if R is infinite then $^*R[x]$ will have members that have some coefficents of standard degree that are nonstandard, i.e., belong to $^*R - R$. Such hyperpolynomials are not approachable from $R[x]$, as Theorem 18.6.1 shows.

Infinitesimals

The infinitesimal members of $^*R[x]$ can be characterised as those internal hyperpolynomials whose coefficients of standard degree all vanish:

Theorem 18.6.2 *For any nonzero $a \in {}^*R[x]$, the following are equivalent.*

(1) $|a| \simeq 0$.

(2) $o(a)$ *is unlimited.*

(3) *There is an unlimited* $N \in {}^*\mathbb{N}$ *such that* $a_n = 0$ *for all* $n < N$.

(4) $a_n = 0$ *for all standard n.*

Proof. In general, $|a| = 2^{-o(a)}$ and $o(a)$ is a nonnegative hyperinteger, so $|a|$ will be appreciable iff $o(a)$ is limited, or equivalently, $|a|$ will be infinitesimal iff $o(a)$ is unlimited. Thus (1) and (2) are equivalent.

Now, by transfer we have that for any nonzero $a \in {}^*R[x]$,

$$(\forall m \in {}^*\mathbb{Z}^{\geq})\, [m < o(a) \leftrightarrow (\forall n \in {}^*\mathbb{Z}^{\geq})\,(n \leq m \rightarrow a_n = 0)].$$

From this, (2) implies (3) by putting $N = o(a)$. It is immediate that (3) implies (4). Finally, if (4) holds, then the above transferred sentence ensures that each standard m is smaller than $o(a)$, so (2) follows. □

Corollary 18.6.3 *In $^*R[x]$, two hyperpolynomials are infinitely close precisely when their coefficients of standard degree are identical: $a \simeq b$ if and only if $a_n = b_n$ for all standard n.* □

The Completion

Let $\theta : {}^*R[x]^{\text{ap}} \to R[\![x]\!]$ be the restriction map

$$a = \langle a_n : n \in {}^*\mathbb{Z}^{\geq}\rangle \longmapsto \langle a_n : n \in \mathbb{Z}^{\geq}\rangle,$$

i.e., $\theta(a)$ is the standard power series defined by putting $\theta(a)_n = a_n$ for all standard n. By Theorem 18.6.1, $\theta(a)$ is indeed a member of $R[\![x]\!]$ whenever $a \in {}^*R[x]^{\text{ap}}$.

The map θ is a ring homomorphism. To see that it preserves addition, notice that

$$(a+b)_n = a_n + b_n$$

holds for all $n \in {}^*\mathbb{Z}^{\geq}$ and all $a, b \in {}^*R[x]$, by transfer, and this is more than enough to guarantee

$$\theta(a+b) = \theta(a) + \theta(b).$$

For multiplication, observe that for any fixed standard n,

$$(ab)_n = a_0 b_n + a_1 b_{n-1} + \cdots + a_n b_0$$

for all $a, b \in R[x]$, and hence for all $a, b \in {}^*R[x]$. But this equation asserts that $(\theta(ab))_n = (\theta(a)\theta(b))_n$. As this holds for all standard n, we conclude that

$$\theta(ab) = \theta(a)\theta(b).$$

Next we want to establish that θ maps *onto* $R[\![x]\!]$. Given a power series $a \in R[\![x]\!]$, then a is a function from $\mathbb{Z}^{\geq}$ to R, and so it transforms to a function ${}^*a : {}^*\mathbb{Z}^{\geq} \to {}^*R$ that has ${}^*a_n = a_n \in R$ for all standard n. In spite of Theorem 18.6.1 we cannot conclude from this that $\theta({}^*a) = a$, because we do not know whether ${}^*a$ is in the domain ${}^*R[x]^{\mathrm{ap}}$ of θ at all. Indeed, ${}^*a$ will not even be in ${}^*R[x]$ unless it is ultimately zero, and if all the coefficients of a are nonzero, then we will have ${}^*a_m \neq 0$ for all $m \in {}^*\mathbb{Z}^{\geq}$ by transfer. To overcome this, consider the statement

- for any function $c \in R^{\mathbb{Z}^{\geq}}$ and any $n \in \mathbb{Z}^{\geq}$ there is a polynomial $b \in R[x]$ that agrees with c up to n, i.e., $c_m = b_m$ for all $m \leq n$.

Since this is manifestly true, so is its $*$-transform. But ${}^*a$ is a standard, hence internal, function from ${}^*\mathbb{Z}^{\geq}$ to ${}^*R$, so belongs to $*(R^{\mathbb{Z}^{\geq}})$. Therefore if we take an unlimited $N \in {}^*\mathbb{N}$, by this $*$-transform we deduce that there is some $b \in {}^*R[x]$ that agrees with ${}^*a$ up to N. Hence b agrees with ${}^*a$ on all standard n, so that $b_n = a_n \in R$ for all such n, implying both that $b \in {}^*R[x]^{\mathrm{ap}}$ (Theorem 18.6.1) and $\theta(b) = a$. Thus θ maps onto $R[\![x]\!]$.

From the definition of θ we have that

$$\theta(a) = 0 \quad \text{iff} \quad a_n = 0 \quad \text{for all } n \in \mathbb{Z}^{\geq}.$$

Theorem 18.6.2 then gives

$$\theta(a) = 0 \quad \text{iff} \quad |a| \simeq 0,$$

so the kernel of θ is the set ${}^*R[x]^{\mathrm{inf}}$ of infinitesimal elements of ${}^*R[x]$. As with previous cases, the cosets of the kernel are the equivalence classes under the infinite closeness relation $\simeq$, and we conclude that $R[\![x]\!]$ is isomorphic to the corresponding quotient:

$$({}^*R[x]^{\mathrm{ap}}/\simeq) \cong R[\![x]\!].$$

It remains only to show that this isomorphism preserves metrics, in the sense that $\mathrm{sh}|a| = |\theta(a)|$, to conclude that the space $R[\![x]\!]$ of power series over R is isometric to the completion $({}^*R[x]^{\mathrm{ap}}/\simeq, d)$ of $(R[x], d)$. This is left as an exercise:

Exercise 18.6.4

If a is a nonzero member of ${}^*R[x]^{\mathrm{ap}}$, show:

(1) If $o(a)$ is limited, then $|a| = |\theta(a)|$.

(2) If $o(a)$ is unlimited, then $|a| \simeq |\theta(a)|$.

18.7 Hyperfinite Expansions in Base p

Any p-adic integer can be represented as an infinite sum

$$\sum_{n=0}^{\infty} z_n p^n = z_0 + z_1 p + z_2 p^2 + \cdots + z_n p^n + \cdots$$

with coefficients z_n from $\mathbb{Z}/p$. In view of our discussion of power series in the last section, this suggests that we could view it instead as a hyperfinite sum.

To see how this works, recall that each standard positive integer $z \in \mathbb{N}$ has a unique expansion in base p of the form

$$z = z_0 + z_1 p + z_2 p^2 + \cdots + z_n p^n$$

and so is represented in this base by the sequence $\langle z_i : 0 \leq i \leq n \rangle$ of numbers that are between 0 and $p-1$. The representation gives a bijection between $\mathbb{N}$ and the set $Seq(p)$ of all finite sequences of elements of $\mathbb{Z}/p$. This bijection is provided by the operator

$$\sum : Seq(p) \to \mathbb{N}$$

taking $\langle z_i : 0 \leq i \leq n \rangle$ to the number $\sum_{i=0}^{n} z_i p^i$. In a nonstandard framework $\sum$ will lift to a bijection

$$\sum : {}^*Seq(p) \to {}^*\mathbb{N}.$$

By appropriate transfer arguments we can see that $^*Seq(p)$ is the set of all *internal hyperfinite sequences of elements of* $^*(\mathbb{Z}/p) = \mathbb{Z}/p$. A typical member of $^*Seq(p)$ is an internal function of the form

$$\langle z_i : i \in {}^*\mathbb{Z}^{\geq} \text{ and } i \leq n \rangle,$$

with $0 \leq z_i < p$ and n possibly unlimited. The operator $\sum$ takes this hypersequence to an element of $^*\mathbb{N}$ that we denote by $\sum_{i=0}^{n} z_i p^i$. Every member of $^*\mathbb{N}$ is represented in this way as a hyperfinite sum determined by a unique member of $^*Seq(p)$, and so has an expansion in base p.

Now, within $\mathbb{Z}^{\geq}$, if $n < m$, then the difference $\left(\sum_{i=0}^{m} z_i p^i\right) - \left(\sum_{i=0}^{n} z_i p^i\right)$ is divisible by p^{n+1}, so

$$\sum_{i=0}^{m} z_i p^i \equiv \sum_{i=0}^{n} z_i p^i \pmod{p^{n+1}}. \tag{v}$$

By transfer, (v) holds for all $n, m \in {}^*\mathbb{Z}^{\geq}$ with $n < m$ when these sums are defined. This property can be used to analyse the relation of infinite closeness of hyperintegers in terms of the behaviour of the coefficients of their base p expansions.

Consider two hyperintegers that have base p expansions

$$z = \sum_{i=0}^{N} z_i p^i, \qquad w = \sum_{i=0}^{M} w_i p^i$$

256   18. Completion by Enlargement

with N, M unlimited. If z and w are infinitely close in the p-adic metric, i.e., $|z-w|_p \simeq 0$, then for any *standard* $n \geq 0$, p^{n+1} divides $z - w$ (Theorem 18.4.1), so
$$\textstyle\sum_{i=0}^{N} z_i p^i \equiv \sum_{i=0}^{M} w_i p^i \pmod{p^{n+1}}.$$
But $n < N, M$, so applying result (v) gives
$$\textstyle\sum_{i=0}^{N} z_i p^i \equiv \sum_{i=0}^{n} z_i p^i \pmod{p^{n+1}}$$
and likewise
$$\textstyle\sum_{i=0}^{M} w_i p^i \equiv \sum_{i=0}^{n} w_i p^i \pmod{p^{n+1}}.$$
Consequently,
$$\textstyle\sum_{i=0}^{n} z_i p^i \equiv \sum_{i=0}^{n} w_i p^i \pmod{p^{n+1}}.$$
But then
$$\textstyle\sum_{i=0}^{n} z_i p^i = \sum_{i=0}^{n} w_i p^i \pmod{p^{n+1}}$$
because both sums belong to $\mathbb{Z}/p^{n+1}$, and so the uniqueness of the base p expansion of standard integers implies that $z_i = w_i$ for $i \leq n$, and in particular, $z_n = w_n$.

This argument can be worked in reverse, to establish the following analogue of Corollary 18.6.3.

Theorem 18.7.1 *Two positive hyperintegers are p-adically infinitely close precisely when their base p expansions have identical coefficients of standard degree:*
$$\textstyle\sum_{i=0}^{N} z_i p^i \simeq \sum_{i=0}^{M} w_i p^i \quad \text{iff} \quad z_i = w_i \text{ for all standard } i.$$

□

Now, we saw in Section 18.4 that if
$$a = z_0 + z_1 p + z_2 p^2 + \cdots + z_n p^n + \cdots$$
is a p-adic integer, then there exists a *positive* hyperinteger x with
$$a = \theta_p(x) = \langle x \bmod p,\ x \bmod p^2, \ldots, x \bmod p^n, \ldots \rangle.$$
Hence $x \bmod p^{n+1} = z_0 + z_1 p + z_2 p^2 + \cdots + z_n p^n$ for all $n \in \mathbb{Z}^{\geq}$.
But x has a base-p expansion
$$x = x_0 + x_1 p + \cdots + x_N p^N$$
for some $N \in {}^*\mathbb{Z}^{\geq}$, and for each standard $n \geq 0$ we get by result (v) that
$$x \equiv \textstyle\sum_{i=0}^{n} x_i p^i \pmod{p^{n+1}},$$
so
$$\textstyle\sum_{i=0}^{n} x_i p^i = x \bmod p^{n+1} = \sum_{i=0}^{n} z_i p^i,$$

and therefore $x_i = z_i$ for all $i \leq n$. Thus the p-adic integer a and the hyperinteger x have the same coefficients of standard degree in these base-p expansions.

Of course for any given $a \in \mathbb{Z}_p$ there will be more than one $x \in {}^*\mathbb{N}$ representing a in this way, i.e., having $\theta_p(x) = a$, but all such x's will be infinitely close in the p-adic metric. Altogether, this discussion shows that we can view any p-adic integer as a hyperfinite base-p expansion

$$z_0 + z_1 p + \cdots + z_N p^N$$

with $0 \leq z_i < p$, provided that we identify any two such expansions that differ only at coefficients of unlimited degree.

18.8 Exercises

(1) Write out in full the transfer arguments showing that members of $^*Seq(p)$ are internal hyperfinite sequences of members of $\mathbb{Z}/p$ (cf. the proof of Theorem 13.17.1 for guidance).

(2) Complete the proof of Theorem 18.7.1 by showing that if two positive hyperintegers have identical coefficients of standard degree in their hyperfinite base p expansions, then they are p-adically infinitely close.

19
Hyperfinite Approximation

In some nonstandard frameworks there are infinite sets that can be "approximated" by hyperfinite sets. From the discussion in Chapter 14 we know that in an enlargement of a universe over a set A there will be a hyperfinite set B with

$$A \subseteq B \subseteq {}^*A.$$

This phenomenon suggests a new methodology for analysing infinite structures by "lifting" a corresponding analysis that is known for finite ones. The steps involved are as follows:

(1) Obtain information about finite structures by standard reasoning.

(2) Use transfer to lift this to hyperfinite structures, including the set B above.

(3) Find some way of "pushing down" the results from B to the infinite set A.

This procedure will now be illustrated with three applications: colouring of graphs, representation of Boolean algebras, and the Hahn–Banach theorem about extensions of linear functionals on vector spaces. For the first two of these, the pushing-down step (3) is immediate. For the Hahn–Banach theorem, however, it requires a little further nonstandard analysis in the form of an appeal to the shadow map.

19.1 Colourings and Graphs

An *r-colouring* of a set G is a sequence $C_1, \ldots, C_r$ of *pairwise disjoint* sets satisfying
$$G \subseteq C_1 \cup \cdots \cup C_r.$$

Thus each member of G belongs to exactly one of the sets C_i. This situation induces a partition of G, which we regard as being given by an assignment of r different colours to the members of G. $G \cap C_i$ is the set of elements of G that are assigned colour i (cf. Section 17.1).

Notice that the way we have defined this notion ensures that any r-colouring of G is also an r-colouring of any subset of G.

In any nonstandard framework for G we will get
$$^*G \subseteq {^*C_1} \cup \cdots \cup {^*C_r},$$

with $^*C_i \cap {^*C_j} = \emptyset$ whenever $i \neq j$. Hence the enlarged sets $^*C_1, \ldots, {^*C_r}$ form an r-colouring of $^*G$, which moreover agrees with the original colouring when restricted to G.

A *graph* is a structure $\langle G, E \rangle$ comprising a nonempty set G with a binary relation E on G that is irreflexive and symmetric:

$$(\forall x \in G)\, \langle x, x \rangle \notin E,$$
$$(\forall x, y \in G)\, (\langle x, y \rangle \in E \to \langle y, x \rangle \in E).$$

This is visualised as a collection of *nodes*, or *vertices*, labelled by the members of G, with a line connecting the nodes labelled by x and y whenever $\langle x, y \rangle \in E$. A pair $\langle x, y \rangle$ that belongs to E is called an *edge with vertices x and y*.

Any subset H of G defines the *subgraph* of $\langle G, E \rangle$ determined by retaining just those edges whose vertices both belong to H. In other words, there is a line connecting x and y in this subgraph precisely when $x, y \in H$ and $\langle x, y \rangle \in E$.

In a nonstandard framework for G the structure $\langle {^*G}, {^*E} \rangle$ is a graph, since transfer ensures that $^*E$ is irreflexive and symmetric. Moreover, the subgraph of this enlarged graph defined by G is just the original graph $\langle G, E \rangle$, because
$$\langle x, y \rangle \in E \quad \text{iff} \quad \langle x, y \rangle \in {^*E}$$
for each $x, y \in G$ by transfer (note here that $^*\langle x, y \rangle = \langle {^*x}, {^*y} \rangle = \langle x, y \rangle$, since x and y are individuals in a nonstandard framework for G).

An *r-colouring of a graph* is an r-colouring of its set of nodes with the additional property that the vertices of any edge have different colours, i.e., there is no line connecting two nodes of the same colour. This requires that each of the statements

$$(\forall x, y \in G)(\langle x, y \rangle \in E \wedge x \in C_i \to y \notin C_i)$$

must hold true for $i = 1, \ldots, r$. An r-colouring of any graph is automatically an r-colouring of any of its subgraphs.

Now, the assertion

"$C_1, \ldots, C_r$ is an r-colouring of $\langle G, E \rangle$"

can be expressed by a formula that we abbreviate as

$$Colour(C_1, \ldots, C_r, G, E).$$

If $C_1, \ldots, C_r, G, E$ are all taken to be constants naming particular entities in a universe over G, then this formula is a sentence whose $*$-transform asserts that $^*C_1, \ldots, {}^*C_r$ is an r-colouring of the enlarged graph $\langle {}^*G, {}^*E \rangle$. If, however, we take $C_1, \ldots, C_r$ to be variables, then we can form the sentence

$$(\exists C_1, \ldots, C_r \in \mathcal{P}(G))\, Colour(C_1, \ldots, C_r, G, E),$$

which states that there exists an r-colouring of $\langle G, E \rangle$. This can be modified further to express the assertion that every finite subgraph of $\langle G, E \rangle$ has an r-colouring: replace G by a variable H and form

$$(\forall H \in \mathcal{P}_F(G))(\exists C_1, \ldots, C_r \in \mathcal{P}(G))\, Colour(C_1, \ldots, C_r, H, E). \quad \text{(i)}$$

Note that for (i) to be true it is required only that each finite subgraph have its own r-colouring. The colourings of different finite subgraphs need not agree with each other, so it is not obvious from (i) that the whole graph $\langle G, E \rangle$ can itself be r-coloured. Nonetheless, it is true that

- *if every finite subgraph of a graph has an r-colouring, then the graph itself has an r-colouring.*

A proof of this result will now be given by a simple application of the hyperfinite approximation methodology. We work in an enlargement of a universe over G, and observe first that by applying transfer to (i) we can infer that each member of $^*\mathcal{P}_F(G)$ defines a subgraph of $\langle {}^*G, {}^*E \rangle$ that has an r-colouring. Thus the fact that every finite subgraph of $\langle G, E \rangle$ has an r-colouring can be lifted to the conclusion that

every hyperfinite subgraph of $\langle {}^*G, {}^*E \rangle$ has an r-colouring.

But in the enlargement there is a hyperfinite approximant of G, i.e., a set $H \in {}^*\mathcal{P}_F(G)$ with $G \subseteq H \subseteq {}^*G$. Then the subgraph of $\langle {}^*G, {}^*E \rangle$ defined by H has an r-colouring, and it remains only to push this situation down to $\langle G, E \rangle$ itself. But this is immediate, since $\langle G, E \rangle$ is a subgraph of the graph defined by H, so the r-colouring of H is also an r-colouring of $\langle G, E \rangle$.

Exercise 19.1.1
Write out explicitly the formula $Colour(C_1, \ldots, C_r, G, E)$.

19.2 Boolean Algebras

George Boole (1815–1864) was a pioneer of *mathematical* logic. He showed that the study of logical connectives, and the validity of inferences, could be carried out by a mathematical analysis of equations involving operations that constitute what we now refer to as *Boolean algebra*.

To explain this, consider first the collection $\mathcal{P}(A)$ of all subsets of a set A. This is closed under the binary operations $\cap$ and $\cup$ of intersection and union of sets, and under the unary operation $-$ of complementation relative to A. The structure

$$\langle \mathcal{P}(A); \cap, \cup, -, \emptyset, A \rangle$$

is the *power set algebra* of A. More generally, a *field of sets* is any nonempty collection B of subsets of a set A (i.e., $B \subseteq \mathcal{P}(A)$) that is closed under $\cap$, $\cup$, and $-$. Then

$$\langle B; \cap, \cup, -, \emptyset, A \rangle$$

is a *subalgebra* of the power set algebra of A.

These are concrete examples of the abstract notion of a *Boolean algebra*, which can be defined as any structure

$$\langle B; \sqcap, \sqcup, ', 0, 1 \rangle$$

in which B is a nonempty set that contains elements 0 (the *zero*) and 1 (the *unit*) and carries binary operations $\sqcap$ and $\sqcup$ and a unary operation $'$ such that the following equations hold for all x, y, z in B:

$$\begin{align*}
x \sqcap y &= y \sqcap x, \\
x \sqcup y &= y \sqcup x, \\
x \sqcap (y \sqcup z) &= (x \sqcap y) \sqcup (x \sqcap z), \\
x \sqcup (y \sqcap z) &= (x \sqcup y) \sqcap (x \sqcup z), \\
x \sqcap 1 &= x, \\
x \sqcup 0 &= x, \\
x \sqcap x' &= 0, \\
x \sqcup x' &= 1.
\end{align*}$$

From these many other properties are deducible, including

$$\begin{align*}
x \sqcap x &= x, \\
x \sqcup x &= x, \\
x \sqcap (y \sqcap z) &= (x \sqcap y) \sqcap z, \\
x \sqcup (y \sqcup z) &= (x \sqcup y) \sqcup z, \\
x \sqcup (y \sqcap x) &= x,
\end{align*}$$

19.2 Boolean Algebras

$$x \sqcap (y \sqcup x) = x,$$
$$(x \sqcap y)' = x' \sqcup y',$$
$$(x \sqcup y)' = x' \sqcap y',$$
$$x \sqcup 1 = 1,$$
$$x \sqcap 0 = 0,$$
$$0' = 1,$$
$$1' = 0,$$
$$x \sqcap y' = 0 \quad \text{iff} \quad x \sqcap y = x.$$

The element $x \sqcap y$ is called the *meet* of x and y, while $x \sqcup y$ is their *join*. A relation $\leq$ is defined in any Boolean algebra by

$$x \leq y \quad \text{iff} \quad x \sqcap y = x \quad (\text{iff} \quad x \sqcup y = y \quad \text{iff} \quad x \sqcap y' = 0).$$

Then $\leq$ proves to be a *partial ordering* (reflexive, transitive, antisymmetric) in which the meet $x \sqcap y$ is the greatest lower bound of x and y, and the join $x \sqcup y$ is the least upper bound. In a field of sets, $\leq$ is the relation $\subseteq$ of set inclusion.

Notice that in any nonstandard framework for B, the operations $\sqcap, \sqcup, '$ will extend to corresponding operations on $^*B$, for which we continue to use the same symbols. The Boolean algebra axioms hold for these operations by transfer, and so $^*B$ becomes a Boolean algebra having B as a subalgebra.

Of singular importance is the two-element algebra based on the set $\{0,1\}$, in which 1 is identified with "true" and 0 with "false", and $\sqcap, \sqcup, '$ are the operations specified by the *truth tables* that give the usual meanings of the logical connectives $\land, \lor, \neg$. This algebra is denoted by **2**. It is isomorphic to the power set algebra of any one-element set $\{a\}$, identifying 1 with $\{a\}$ and 0 with $\emptyset$. The algebra **2** is a fundamental building block: from the representation to be discussed below it can be shown that any Boolean algebra is isomorphic to a subalgebra of an algebra that is constructed as a direct product of copies of **2**. This implies that any equation satisfied by **2** will be satisfied by every Boolean algebra.

Now, the properties of the set operations $\cap, \cup, -$ follow from their definitions,

$$C \cap D = \{x : x \in C \text{ and } x \in D\},$$
$$C \cup D = \{x : x \in C \text{ or } x \in D\},$$
$$-C = \{x \in A : \text{not } x \in C\},$$

and hence depend on the meaning of the words *and*, *or*, *not*. Thus it is the behaviour of these logical connectives that dictate that $\mathcal{P}(A)$ should be a Boolean algebra, indicating a natural connection between the algebra of sets and the algebra of connectives. This is further exemplified by a construction that builds Boolean algebras out of formulae. Let Π be a nonempty set,

whose members will be called *sentence letters*, and consider the class of
Π-*formulae* generated inductively from members of Π by the connectives
$\wedge, \vee, \neg, \rightarrow, \leftrightarrow$. A Π-*valuation* is a function $v : \Pi \rightarrow \{0,1\}$ assigning a truth
value to each sentence letter. Any valuation extends in a unique way to all
formulae by the usual truth conditions:

$$v(\neg \varphi) = 1 \quad \text{iff} \quad v(\varphi) = 0,$$
$$v(\varphi \wedge \psi) = 1 \quad \text{iff} \quad v(\varphi) = v(\psi) = 1,$$
$$v(\varphi \vee \psi) = 1 \quad \text{iff} \quad v(\varphi) = 1 \text{ or } v(\psi) = 1,$$
$$v(\varphi \rightarrow \psi) = 1 \quad \text{iff} \quad v(\varphi) = 0 \text{ or } v(\psi) = 1,$$
$$v(\varphi \leftrightarrow \psi) = 1 \quad \text{iff} \quad v(\varphi) = v(\psi).$$

A Π-formula is a *tautology* if it is assigned the value 1 by every Π-valuation.
Examples of tautologies are $\varphi \vee \neg\varphi$, $\varphi \rightarrow \varphi$, and $\varphi \wedge \psi \rightarrow \varphi$, where ϕ and
ψ are any formulae. An *inconsistent* formula is one that always takes value
0, such as $\varphi \wedge \neg\varphi$.

An equivalence relation $\sim$ on the set of all Π-formulae is defined by

$$\varphi \sim \psi \quad \text{iff} \quad \varphi \leftrightarrow \psi \text{ is a tautology}$$
$$\text{iff} \quad v(\varphi) = v(\psi) \text{ for all } \Pi\text{-valuations } v.$$

The purpose of this relation is to identify formulae that are indistinguishable by any truth-value assignment. Thus $\varphi \wedge \psi$ and $\psi \wedge \varphi$, while being distinct formulae, will belong to the same equivalence class.

The equivalence class of a formula φ will be denoted by $[\varphi]$. The set B_Π of all such equivalence classes becomes a Boolean algebra under the operations

$$[\varphi] \sqcap [\psi] = [\varphi \wedge \psi],$$
$$[\varphi] \sqcup [\psi] = [\varphi \vee \psi],$$
$$[\varphi]' = [\neg\varphi],$$
$$1 = [\varphi \vee \neg\varphi],$$
$$0 = [\varphi \wedge \neg\varphi].$$

In this algebra we get

$$[\varphi] \leq [\psi] \quad \text{iff} \quad \varphi \rightarrow \psi \text{ is a tautology},$$
$$[\varphi] = 1 \quad \text{iff} \quad \varphi \text{ is a tautology},$$
$$[\varphi] = 0 \quad \text{iff} \quad \varphi \text{ is inconsistent}.$$

If Π is infinite, then so too is B_Π, since no two sentence letters are equivalent. On the other hand, if Π is finite, then the algebra B_Π will be finite, even though there are infinitely many Π-formulae. Indeed, if Π consists of n sentence letters, then there are only 2^n valuations $\Pi \rightarrow \{0,1\}$ for distinguishing formulae. Each equivalence class can be identified with the set of valuations $\{v : v(\varphi) = 1\}$, and there are no more than 2^{2^n} such sets. Further investigation establishes that B_Π has exactly this number 2^{2^n} of members:

Exercise 19.2.1
Show that if Π is finite, then for any set $\mathcal{V}$ of Π-valuations there is a Π-formula φ such that in general, $v(\varphi) = 1$ iff $v \in \mathcal{V}$. Hence verify in detail that B_Π has exactly 2^{2^n} members if Π has n.

19.3 Atomic Algebras

We are going to prove (in Section 19.4) that any Boolean algebra has an isomorphic representation as a field of sets. The key to this is the notion of an "atom". To see what this means, observe that in the power set algebra of a finite set each nonzero element of the algebra is itself a finite set and so can be decomposed as the union/join

$$\{a_1\} \cup \cdots \cup \{a_n\}$$

of finitely many *singleton* sets. These singletons themselves cannot be further decomposed into smaller elements (so this is analogous to the decomposition of integers into primes). One way to characterise the singletons is by the fact that the only element smaller than them is $\emptyset$.

In an abstract Boolean algebra we define an *atom* to be a nonzero element that has no nonzero element smaller than itself in the partial ordering. This can be symbolized by the formula $atom(a, B)$, defined as

$$a \neq 0 \wedge (\forall x \in B)\,(x \leq a \rightarrow x = 0 \vee x = a),$$

which expresses "a is an atom of B". Note that reference to the ambient algebra B is crucial here, since an atom of one algebra may fail to be an atom within a larger algebra. This is the case with the atom $\{1, 2\}$ of the field of sets

$$\{\emptyset, \{0\}, \{1, 2\}, \{0, 1, 2\}\},$$

which is not an atom in $\mathcal{P}(\{0, 1, 2\})$.

Now let
$$B_x = \{a \in B : atom(a, B) \wedge a \leq x\}$$

be the set of atoms in B that are "below" x. Thus B_1 is the set of *all* atoms of B, while $B_0 = \emptyset$. In general, it can be shown for any atom a that

$$a \leq x \sqcap y \quad \text{iff} \quad a \leq x \text{ and } a \leq y,$$
$$a \leq x \sqcup y \quad \text{iff} \quad a \leq x \text{ or } a \leq y,$$
$$a \leq x' \quad \text{iff} \quad a \not\leq x,$$

implying that

$$B_{x \sqcap y} = B_x \cap B_y,$$
$$B_{x \sqcup y} = B_x \cup B_y,$$
$$B_{x'} = -B_x.$$

These conditions state that the map $x \mapsto B_x$ is a homomorphism from B into the power set algebra $\mathcal{P}(B_1)$ of all subsets of the set of atoms of B.

A Boolean algebra is *atomic* if each of its nonzero elements has an atom below it:

$$(\forall x \in B)\,[x \neq 0 \to (\exists a \in B)\,(atom(a, B) \wedge a \leq x)].$$

This means that

$$x \neq 0 \quad \text{implies} \quad B_x \neq \emptyset,$$

which can be used to show that

$$x \neq y \quad \text{implies} \quad B_x \neq B_y,$$

and so altogether the map $x \mapsto B_x$ is an *injective* homomorphism from B into $\mathcal{P}(B_1)$. The image of B under this injection will be a subalgebra of $\mathcal{P}(B_1)$ isomorphic to B itself. This establishes that

- *any atomic Boolean algebra is isomorphic to a field of sets.*

If B is finite, the injection $x \mapsto B_x$ maps *onto* $\mathcal{P}(B_1)$ (and so B has exactly 2^n elements, where n is its number of atoms). This is because each nonzero element $\{a_1, \ldots a_k\}$ of $\mathcal{P}(B_1)$ is equal to B_x, where

$$x = a_1 \sqcup \cdots \sqcup a_k.$$

For infinite B the injection need not be onto, so we cannot conclude that B is isomorphic to $\mathcal{P}(B_1)$. For example, if Π is countably infinite, then so is the formula algebra B_Π, and hence this algebra cannot be isomorphic to *any* power set. The point is that there is no such thing as a countably infinite power set: if A is finite, then so is $\mathcal{P}(A)$ (since $|\mathcal{P}(A)| = 2^{|A|}$), while if A is infinite, then $\mathcal{P}(A)$ is *uncountable* (by a famous diagonalisation argument of Cantor, showing that there is no map from A onto $\mathcal{P}(A)$).

This whole theory fails to apply to an algebra that is not atomic, i.e., has at least one element that lacks atoms below it. An extreme example of this is provided by the formula algebra B_Π whenever Π is infinite. This has no atoms at all! For if $[\varphi]$ is a nonzero element of B_Π and Π is infinite, we can choose a sentence letter $p \in \Pi$ that does not occur in φ. Then

$$0 \neq [\varphi \wedge p] < [\varphi],$$

showing that $[\varphi]$ is not an atom in B_Π. We have $[\varphi \wedge p] \leq [\varphi]$ because $\varphi \wedge p \to \varphi$ is a tautology. As $[\varphi] \neq 0$, there is a valuation v with $v(\varphi) = 1$. Since p does not occur in φ, we can define a valuation v_1 that agrees with v on the letters occurring in φ and has $v_1(p) = 1$. Then $v_1(\varphi) = v(\varphi) = 1$, so $v_1(\varphi \wedge p) = 1$, showing $[\varphi \wedge p] \neq 0$. Similarly, there is a v_2 with $v_2(\varphi) = 1$ but $v_2(p) = 0$, so $v_2(\varphi \wedge p) \neq v_2(\varphi)$, implying $[\varphi \wedge p] \neq [\varphi]$.

On the other hand, the theory does always apply to a finite Boolean algebra, since a finite one is always atomic. To see this, observe that if a nonzero element a is not an atom, then there must be some nonzero a_1 below it. If a_1 is not an atom, then there must be an $a_2 \neq 0$ with $a_2 < a_1$, and so on. If this argument could be repeated infinitely often, it would generate an infinite chain

$$\cdots < a_n < \cdots < a_2 < a_1 < a$$

of distinct elements of B. Thus if B is finite, the argument must stop at some atom a_n below a.

To sum up:

- *Every finite Boolean algebra is atomic and is isomorphic to the power set algebra of its set of atoms.*

19.4 Hyperfinite Approximating Algebras

The discussion of Boolean algebras so far has been entirely concerned with their standard theory. We now bring in some nonstandard ideas to prove that for any Boolean algebra B there is an *atomic* Boolean algebra B^+ that has B as a subalgebra. From the results described above we know that B^+ is isomorphically embeddable into the power set algebra of its set of atoms, and so B is likewise embeddable into that power set by the map

$$(x \in B) \mapsto B_x^+$$

that takes each member of B to the set of atoms of B^+ that are below it. Thus the representation of B^+ as a field of sets immediately pushes down to B itself. This establishes the fundamental representation result that

- *every Boolean algebra is isomorphic to a field of sets.*

Our desired algebra B^+ is hyperfinite, and will be obtained as an approximation to B in the sense that

$$B \subseteq B^+ \subseteq {}^*B \tag{ii}$$

in a nonstandard framework for B. The essential reason why B^+ turns out to be atomic is that all finite Boolean algebras are atomic, and this property is preserved by transfer to the hyperfinite B^+.

We already know how to realise B^+ as a hyperfinite *set* fulfilling (ii), but now we want it to be a Boolean algebra. For this it suffices that it be closed under the Boolean operations $\sqcap$, $\sqcup$, and $'$, so that it is a subalgebra of the Boolean algebra $^*B$ and hence is a Boolean algebra in its own right. There is one more piece of standard theory that we need to make this work:

- *Every finitely generated Boolean algebra is finite.*

To explain this, let C be a subset of Boolean algebra B. Then there is a *smallest* subalgebra of B that includes C. This is known as the subalgebra *generated by C*. It may be defined in a "top-down" way as the intersection of all subalgebras of B that include C, or given a "bottom-up" construction by starting with the elements of C and repeatedly applying $\sqcap$, $\sqcup$, and $'$ until a set of elements is produced that is closed under these operations. A *finitely generated* algebra is one that is generated in this way by a finite set C.

Now, if $C = \{b_1, \ldots, b_n\}$, then because of the particular equations satisfied by Boolean algebras it can be shown that any member of the subalgebra of B generated by C is equal to the join of finitely many elements of the form

$$b_1^\circ \sqcap \cdots \sqcap b_i^\circ \sqcap \cdots \sqcap b_n^\circ,$$

where b_i° is either b_i or b_i'. But there are at most 2^n elements of this form in B, and hence at most 2^{2^n} joins of sets of such elements. In other words,

- *a Boolean algebra with n generators has at most 2^{2^n} elements*

(cf. Exercise 19.5(2) below for an alternative proof of this).

Now let us work in a nonstandard framework for B that is an *enlargement*. Define a binary relation $R \subseteq B \times \mathcal{P}_F(B)$ by letting

$$bRA \quad \text{iff} \quad b \in A \in \mathcal{P}_F(B) \text{ and } A \text{ is a subalgebra of } B.$$

Then R is concurrent, for if $b_1, \ldots, b_k \in B$, and A is the subalgebra of B generated by $\{b_1, \ldots, b_k\}$, then (as above) A is finite, and so $b_i R A$ for all $1 \leq i \leq k$. In particular, this shows (when $k = 1$) that the domain of R is B itself.

It follows (Theorem 14.2.1) that in the enlargement there exists an entity B^+ such that $b(^*R)B^+$ for all $b \in B$. This B^+ is a hyperfinite subset of $^*B$ including B, since by transfer $^*R \subseteq {^*B} \times {^*\mathcal{P}_F(B)}$ and

$$b(^*R)A \rightarrow b \in A$$

in general. Hence (ii) holds.

Next we show that B^+ is a subalgebra of $^*B$, and so is a Boolean algebra. Every member of the range of R is a subalgebra of B, hence is closed under the Boolean operations:

$$(\forall A \in \operatorname{ran} R)\,(\forall x, y \in B)\,(x, y \in A \rightarrow x \sqcap y \in A \wedge x \sqcup y \in A \wedge x' \in A).$$

By transfer it follows that every member of the range of *R is likewise closed (recall that *(ran R) = ran (*R)). In particular, this applies to B^+, which is thereby a subalgebra of *B.

Finally, we want B^+ to be atomic. But every member of the range of R is a finite Boolean algebra, so is atomic:

$$(\forall A \in \text{ran } R)\,(\forall x \in A)\,[x \neq 0 \to (\exists a \in A)\,(atom(a,A) \wedge a \leq x)],$$

where $atom(a, A)$, as defined earlier, is the formula

$$a \neq 0 \wedge (\forall x \in A)\,(x \leq a \to x = 0 \vee x = a).$$

Transfer of this asserts that all members of the range of *R, and in particular B^+, are atomic.

This completes the proof of the existence of a B^+ with the desired properties.

19.5 Exercises on Generation of Algebras

(1) Let C be a hyperfinite subset of *B. Prove that there exists a smallest *internal* subalgebra of *B including C, and that this subalgebra is hyperfinite.

(2) If B is a Boolean algebra with an n-element generating set, show that there is a homomorphism from B_Π onto B where Π is an n-element set of sentence letters. Deduce from this that B has at most 2^{2^n} elements (cf. Exercise 19.2.1).

19.6 Connecting with the Stone Representation

The fact that every Boolean algebra is isomorphic to a field of sets is known as the *Stone representation theorem* after its discoverer, Marshall Stone. The most commonly presented proof of this consists of an embedding of the algebra into the power set of its set of *ultrafilters*.

Now, a *filter* of a Boolean algebra B is a nonempty set $F \subseteq B$ satisfying

$$x \sqcap y \in F \quad \text{iff} \quad x \in F \text{ and } y \in F$$

for all $x, y \in F$. Thus a filter *on a set* I, in the sense of Section 2.3, is the same thing as a filter of the power set algebra $\mathcal{P}(I)$. A filter is *proper* if $0 \notin F$, and is an *ultrafilter* if it is proper and has

$$x \in F \text{ or } x' \in F$$

for all $x \in B$. For example, if $a \in B$, the *principal filter*

$$\{x \in B : a \leq x\}$$

of B *generated by* a is an ultrafilter of B if and only if a is an atom of B. More generally, if a is an atom of any algebra B^+ that has B as a subalgebra, then the set

$$\{y \in B : a \leq y \text{ in } B^+\}$$

is an ultrafilter of B. This is just the restriction to B of the principal ultrafilter $\{y \in B^+ : a \leq y\}$ of B^+ generated by a. The restriction itself need not have a generator in B (see below).

Any ultrafilter F satisfies

$$x \sqcup y \in F \quad \text{iff} \quad x \in F \text{ or } y \in F,$$
$$x' \in F \quad \text{iff} \quad x \notin F,$$

so if U_x is the set of ultrafilters of B that have x as an element, then U_1 is the set of all ultrafilters of B, $U_0 = \emptyset$, and

$$U_{x \sqcap y} = U_x \cap U_y,$$
$$U_{x \sqcup y} = U_x \cup U_y,$$
$$U_{x'} = -U_x.$$

This means that the map $x \mapsto U_x$ is a homomorphism from B into $\mathcal{P}(U_1)$. To make it injective it is enough to show that

$$x \neq 0 \quad \text{implies} \quad U_x \neq \emptyset,$$

or in other words,

any nonzero element of a Boolean algebra belongs to an ultrafilter. (iii)

To *prove* this, we have to invoke Zorn's lemma to extend any proper filter to a maximal one, and then observe that a maximal proper filter is the same thing as an ultrafilter. This establishes the

Ultrafilter Theorem: *any proper filter of a Boolean algebra B can be extended to an ultrafilter of B.*

By applying this theorem to the principal filter generated by a given nonzero element, we obtain the desired result (iii) (cf. Section 2.6).

When the ultrafilter theorem is applied to power set algebras it produces the ultrafilters needed to carry out the construction of ultrapowers of superstructures as in Section 14.3. This construction then yields the *enlargement theorem*, asserting the existence, for any set $\mathbb{X}$, of an enlargement of a universe over $\mathbb{X}$. Here we can turn the tables by *assuming* the existence of enlargements and using them to give an explicit demonstration of the ultrafilter theorem lying at the heart of Stone's theory.

We saw in the last section that if B is any Boolean algebra, then in a suitable enlargement there is a *hyperfinite and atomic* Boolean algebra B^+ having B as a subalgebra. Now, if F is a proper filter of B, we can extend F to a maximal filter by showing that in B^+ there is an atom that is below all members of F, and then using this atom to define an ultrafilter of B in the way described above.

There are two main steps in this procedure:

(1) Extend F to an *internal* proper filter F^+ of B^+.

(2) Show that in B^+, every internal filter is principal.

To carry out (1), note that the enlargement $^*F$ of F will be a proper filter of $^*B$, by transfer of

$$(0 \notin F) \wedge (\forall x, y \in B)\,[x \sqcap y \in F \leftrightarrow x \in F \wedge x \in F].$$

Hence if we define $F^+ = {}^*F \cap B^+$, then F^+ will be a proper filter of B^+. Moreover, F^+ is internal, being the intersection of two internal sets. Note that $F \subseteq {}^*F \cap B \subseteq F^+$.

Result (2) holds in any hyperfinite Boolean algebra like B^+, because in a finite Boolean algebra any filter is principal: the filter $\{b_1, \ldots, b_n\}$ is generated by the element $b_1 \sqcap \cdots \sqcap b_n$. This fact transfers to show that any internal subset of a hyperfinite Boolean algebra that is a filter must have a generating element.

By (1) and (2), there is some element $b \in B^+$ that generates $F^+ = {}^*F \cap B^+$:

$$F^+ = \{y \in B^+ : b \leq y\}.$$

Now, $b \neq 0$, since F^+ is proper, and B^+ is atomic, so there is an atom a in B^+ with $a \leq b$. Then the set

$$G = \{y \in B : a \leq y \text{ in } B^+\}$$

is an ultrafilter of B, and G includes F because $F \subseteq F^+ \subseteq G$. This completes our nonstandard proof of the ultrafilter theorem.

The ultrafilter G is just the restriction to B of the principal ultrafilter

$$\{y \in B^+ : a \leq y \text{ in } B^+\}$$

of B^+. G itself may be nonprincipal in B, for instance when F is the principal filter of B generated by some $x \neq 0$ and there is no atom in B below x. Thus we may think of a nonprincipal ultrafilter as a rather complicated "ghost of a departed atom", just as a sequence of real numbers that converges to 0 is a vestige in the standard world of a missing infinitesimal.

It would be incorrect to conclude on the basis of this discussion alone that the ultrafilter theorem and the enlargement theorem are equivalent as axiomatic principles. To construct an enlargement as an ultrapower, we need to establish Łoś's theorem to verify the transfer principle, and the proof of Łoś's theorem involves some form of choice principle in handling the inductive case of the quantifiers. In fact, it has been shown that Łoś's theorem and the ultrafilter theorem together are equivalent to the axiom of choice, while the ultrafilter theorem by itself is weaker than the axiom of choice. On the other hand, there are other model-theoretic techniques for building enlargements that depend on principles no stronger than the ultrafilter theorem—but that is outside of our present scope.

19.7 Exercises on Filters and Lattices

(1) Verify that G above is an ultrafilter of B.

(2) Write out in detail the transfer argument that every internal filter of a hyperfinite Boolean algebra is principal.

(3) Work in an enlargement of a universe over a set I. Show that for any nonprincipal ultrafilter $\mathcal{F}$ on I (in the sense of Chapter 2) there is an element of $^*I$ that belongs to every member of $\mathcal{F}$.

(4) (This is really a project rather than an exercise.) A *lattice* is an algebra of the form $\langle L; \sqcap, \sqcup \rangle$ with $\sqcap$ and $\sqcup$ being binary operations on the set L that are commutative and associative and satisfy the idempotence laws
$$x \sqcap x = x = x \sqcup x$$
and the absorption laws
$$x \sqcap (x \sqcup y) = x = x \sqcup (x \sqcap y).$$
A *distributive* lattice is one satisfying
$$\begin{aligned} x \sqcap (y \sqcup z) &= (x \sqcap y) \sqcup (x \sqcap z), \\ x \sqcup (y \sqcap z) &= (x \sqcup y) \sqcap (x \sqcup z). \end{aligned}$$
These laws are satisfied by any *set lattice*, which is one of the form $\langle S, \cap, \cup \rangle$ where S is a collection of subsets of some fixed set. Any finitely generated distributed lattice is finite.

An element a of a lattice is called *join-irreducible* if
$$a = x \sqcup y \quad \text{implies} \quad a = x \text{ or } a = y$$
for all $x, y \in L$. By using join-irreducible elements in place of atoms, adapt the analysis of Boolean algebras to give proofs by hyperfinite approximation of the following facts.

(a) Every distributive lattice is isomorphic to a set lattice.

(b) Every proper filter of a distributive lattice can be extended to a proper filter G that is *prime*, meaning

$$x \sqcup y \in G \quad \text{implies} \quad x \in G \text{ or } y \in G.$$

19.8 Hyperfinite-Dimensional Vector Spaces

Our study of Boolean algebras made use of a special feature: any Boolean algebra B is *locally finite*, which means that its finitely generated subalgebras are finite. This implies that the finitely generated subalgebras are atomic, a property that can then be transferred to the approximating algebra B^+. It also leads to the conclusion that B^+ is hyperfinite.

A similar construction to this can be carried out for other kinds of algebraic structure, even if they are not locally finite. By working with the finitely generated subalgebras we can obtain an analogue of B^+ that is *hyperfinitely generated* rather than hyperfinite. This is particularly relevant to linear algebra, where the emphasis is on finite *dimensionality* rather than actual finiteness. By making an enlargement it is possible to approximate an infinite-dimensional vector space by one that is *hyperfinite-dimensional*. This provides a methodology for transferring results about finite-dimensional spaces to vector spaces in general.

We will assume familiarity with the general theory of vector spaces over fields. Recall that a *real* vector space is an Abelian group $\langle V, +, -, 0 \rangle$ with a *scalar multiplication* map $\langle \lambda, x \rangle \mapsto \lambda x$ from $\mathbb{R} \times V$ to V satisfying, for all vectors $x, y \in V$ and scalars $\lambda, \mu \in \mathbb{R}$,

$$\begin{aligned}
\lambda(x+y) &= \lambda x + \lambda y, \\
(\lambda + \mu)x &= \lambda x + \mu x, \\
\lambda(\mu x) &= (\lambda \mu)x, \\
1x &= x.
\end{aligned}$$

In a nonstandard framework for V the operations $+$ and $-$ lift to *V, and scalar multiplication becomes a map of the form *$\mathbb{R}$ × *V → *V. The vector space axioms are preserved by transfer, and so *V is a vector space over the field *$\mathbb{R}$, i.e., a *hyperreal* vector space. We can of course ignore the nonstandard scalars and restrict scalar multiplication to a map from $\mathbb{R}$ × *V to *V, thereby viewing *V as a real vector space. It is important to recognise that these two descriptions of *V, as a real space and as a hyperreal space, are descriptions of spaces with different properties. For instance, *$\mathbb{R}$ itself is a one-dimensional vector space over *$\mathbb{R}$, but is *infinite*-dimensional as a vector space over $\mathbb{R}$ (see Exercise 19.9.(2)).

A *subspace* of a real vector space V is a subset $W \subseteq V$ that is closed under vector addition and under multiplication by real scalars, and hence contains all finite *linear combinations*

$$\sum_{1}^{n} \lambda_i x_i := \lambda_1 x_1 + \cdots + \lambda_n x_n \qquad (iv)$$

for which $x_1, \ldots, x_n \in W$ and $\lambda_1, \ldots, \lambda_n \in \mathbb{R}$. W is *of dimension n*, $\dim(W) = n$, if there exists a sequence $x_1, \ldots, x_n$ of vectors in W (a *basis*) such that each member of W can be written uniquely as the linear combination $\sum_{1}^{n} \lambda_i x_i$ for some scalars λ_i.

In a nonstandard framework the vector summation operator $\sum_{1}^{n}$ can be defined for unlimited n, allowing the formation of hyperfinite sums and linear combinations in *V. To achieve this we regard the symbol "$\sum$" as denoting a function whose domain is the set $Seq(V)$ of finite sequences of elements of V, and whose range is included in V. A member of $Seq(V)$ is itself a function into V from some initial segment $\{i \in \mathbb{N} : i \leq n\}$ of $\mathbb{N}$. Thus by transfer, *$Seq(V)$ is the set of all *internal hyperfinite sequences of elements of *V*. A typical member of *$Seq(V)$ is an internal function of the form $\langle x_i : i \leq n \rangle$ defined on some initial segment $\{i \in$ *$\mathbb{N} : i \leq n\}$ of *$\mathbb{N}$, with $x_i \in$ *V and n possibly unlimited. The operator $\sum$ extends to a function from *$Seq(V)$ to *V, giving a meaning to the expression $\sum_{1}^{n} x_i$ for all internal hyperfinite sequences $\langle x_i : i \leq n \rangle$.

Let $Fin(V)$ be the set of all finite-dimensional subspaces of V. The function $\dim : Fin(V) \to \mathbb{N}$ assigns to each member of $Fin(V)$ its dimension. This extends to a function $\dim :$ *$Fin(V) \to$ *$\mathbb{N}$. By transfer, since members of $Fin(V)$ are closed under finite combinations with real scalars, a typical member of *$Fin(V)$ will be an internal subset of *V that is closed under internal hyperfinite linear combinations with hyperreal scalars. If $W \in$ *$Fin(V)$ with $\dim(W) = n$ (possibly $n \in$ *$\mathbb{N}_\infty$), then W is a hyperreal subspace of *V, and there exists an internal sequence $\langle x_i : i \leq n \rangle$ of vectors in W that forms a "basis" in the sense that each member of W is equal to $\sum_{1}^{n} \lambda_i x_i$ for a unique internal sequence $\langle \lambda_i : i \leq n \rangle$ of hyperreal numbers. Thus W is a *hyperfinite-dimensional* vector space over *$\mathbb{R}$, with hyperfinite dimension n.

Exercise 19.8.1
Write out the formal sentences whose transforms ensure that *$Seq(V)$ and *$Fin(V)$ fulfill the descriptions just given of them. (The proof of Theorem 13.17.1 may provide some guidance.) □

The approximation of a general vector space by hyperfinite-dimensional ones is given by the following result:

Theorem 19.8.2 *If V is a real vector space, then in any enlargement of a universe over V there is a hyperreal subspace V^+ of *V with*

$$V \subseteq V^+ \in {}^*Fin(V).$$

Proof. Let R be the membership relation from V to $Fin(V)$, i.e.,

$$xRW \quad \text{iff} \quad x \in W \in Fin(V).$$

Then R is concurrent, for if $x_1, \ldots, x_n$ are vectors in V, and W is the subspace they span (i.e., the set of all linear combinations $\sum_1^n \lambda_i x_i$ with real scalars), then W has dimension at most n, and so $x_i RW$ for all $1 \leq i \leq n$. This also shows that the domain of R is V.

Since $^*R \subseteq {}^*V \times {}^*Fin(V)$, it follows that in the enlargement there is some $V^+ \in {}^*Fin(V)$ with $x(^*R)V^+$ for all $x \in V$, and hence $V \subseteq V^+$. □

19.9 Exercises on (Hyper) Real Subspaces

(1) Let $\langle x_i : i \leq n \rangle$ be an internal hyperfinite sequence of elements of $^*V$. Show that there exists a $W \in {}^*Fin(V)$ such that

  (i) W is the smallest internal hyperreal subspace of $^*V$ that contains $x_1, \ldots, x_n$; and

  (ii) each member of W is equal to $\sum_1^n \lambda_i x_i$ for some internal sequence $\langle \lambda_i : i \leq n \rangle$ of hyperreal numbers.

(2) Let $r \in {}^*\mathbb{R} - \mathbb{R}$. Prove that r is not a root of any polynomial with real coefficients. (Hint: a polynomial has finitely many roots.) Deduce that the set $\{r^n : n \in \mathbb{N}\}$ of finite powers of r is linearly independent over $\mathbb{R}$, and therefore that $^*\mathbb{R}$ is infinite-dimensional as a vector space over $\mathbb{R}$.

(3) Explain why the set $\mathbb{L}$ of limited hyperreals is an infinite-dimensional *real* subspace of $^*\mathbb{R}$.

19.10 The Hahn–Banach Theorem

A function of the form $f : V \to \mathbb{R}$ is called a *functional* on a real vector space V. A *linear* functional is one that is *additive*, in the sense that

$$f(x + y) = f(x) + f(y) \qquad \text{(v)}$$

for all $x, v \in V$, and *homogeneous* in the sense that

$$f(\lambda x) = \lambda f(x) \qquad \text{(vi)}$$

for all $x \in V$ and $\lambda \in \mathbb{R}$. If f satisfies (vi) only for $\lambda \geq 0$, then it is *positively homogeneous*, and if it satisfies

$$f(x + y) \leq f(x) + f(y)$$

in place of (v), then it is *subadditive*. Finally, a function f is *dominated* by a function p if $f(x) \leq p(x)$ for all x in the domain of f.

Armed with these definitions we can state the following cornerstone result from functional analysis.

> **Hahn–Banach Theorem:** *Let p be a positively homogeneous and subadditive functional on a real vector space V, and f a linear functional on a subspace W of V with f dominated by p. Then there exists a linear functional g on V that extends f and is still dominated by p.*

They key idea for the proof is a construction that takes a vector $x \in V - W$ and extends f to the subspace generated by adjoining x to W:

Lemma 19.10.1 *Given the hypothesis of the Hahn–Banach theorem, if $x \in V - W$, then there exists a linear functional h on a subspace of V including $W \cup \{x\}$ such that h extends f and is dominated by p.*

Proof. We give only a sketch of the proof of this standard piece of linear algebra. The subspace of V generated by $W \cup \{x\}$ is the set

$$\{y + \lambda x : y \in W \text{ and } \lambda \in \mathbb{R}\}.$$

A functional h is defined on this subspace by putting $h(y+\lambda x) = f(y)+\lambda c$, where c is a real constant. Any such c will make h a linear functional extending f, but c has to be chosen suitably to ensure that h is dominated by p. It turns out that for this purpose c can any number bounded above by the infimum of the numbers $p(y+x) - f(y)$ and below by the supremum of the numbers $-p(-y-x) - f(y)$ as y ranges over W. □

The nature of this result suggests that we can prove the Hahn–Banach theorem by repeatedly applying the procedure of adjoining elements. Having extended f to an h whose domain contains x, we then choose an $x' \notin \text{dom } h$ and extend h to a linear functional h' that is dominated by p and defined at x'. We continue this until we run out of elements of V to adjoin.

Recalling the discussion in Section 2.6, we see that this process involves the axiom of choice in selecting x, x', etc. Alternatively, we could "well-order" $V - W$ into a linear list along which we iterate the construction *transfinitely* often. This was precisely how Hahn and Banach (independently) proved their theorem, and in fact, Banach expressed it very briefly. Having explained the procedure that proves Lemma 19.10.1, he completed his argument by simply stating,

> *It now suffices to well-order the set $V - W$, obtaining, by successive extensions of f, following the procedure described above, a functional g satisfying the conclusion of the theorem.*

Modern treatments of the Hahn–Banach theorem use Zorn's lemma instead to make a maximal extension of f and then show that its domain is the

19.10 The Hahn–Banach Theorem

whole of V. Here we will show that a proof can be developed in the language of enlargements. The essential idea is to iterate the construction of Lemma 19.10.1 *hyperfinitely* often, adjoining enough elements to the domain of the functional to include all of V.

By iterating the adjunctions finitely many times we can take any finite sequence $\langle x_i : i \leq n \rangle$ of elements of V and extend f to a linear functional dominated by p whose domain includes the subspace generated by the x_i's. Hence by applying transfer we can take any internal hyperfinite sequence $\langle x_i : i \leq n \rangle$ and extend f to the hyperfinite-dimensional subspace of *V that it generates. In particular, this would allow us to lift f to the space $V^+ \in {}^*Fin(V)$ including V that was obtained in Theorem 19.8.2, and then push this extension down to V itself.

In fact, an even more direct approach is to incorporate into our present situation the kind of concurrency argument used to derive V^+. Given the hypothesis of the Hahn–Banach theorem, define a binary relation R by specifying that xRh if and only if

$x \in V$ and h is a linear functional that extends f and is dominated by p and whose domain is a subspace of V that includes $W \cup \{x\}$.

If $x_1, \ldots, x_n \in V$ (where $n \in \mathbb{N}$), then applying Lemma 19.10.1 at most n times produces an h having $x_i R h$ for all $1 \leq i \leq n$. So R is current and has domain V. Working in an enlargement of a universe over V, we then obtain an f^+ such that $x({}^*R)f^+$ for all $x \in V$. By transferring properties of R we conclude that:

- f^+ is a *$\mathbb{R}$-valued function whose domain is a hyperreal subspace of *V that includes V.

- f^+ is *hyperlinear*, meaning that it is additive and is homogeneous for hyperreal scalars: $f(\lambda x) = \lambda f(x)$ whenever $\lambda \in {}^*\mathbb{R}$.

- f^+ extends f: $f^+(x) = f(x)$ for all $x \in W$.

- f^+ is dominated by the extension of p to *V: $f^+(x) \leq p(x)$ for all $x \in \text{dom } f^+$.

At this stage we cannot simply restrict f^+ to V to produce our desired functional extension of f, because f^+ may take nonstandard values on V. However, these values are always at least *limited*, so we can take their shadows. To see that this is so, note that in general,

$$-f^+(x) = f^+(-x) \leq p(-x),$$

whence

$$-p(-x) \leq f^+(x) \leq p(x)$$

for all $x \in \mathrm{dom}\, f^+$. Therefore when $x \in V$, $f^+(x)$ is sandwiched between the *real* numbers $-p(-x)$ and $p(x)$, so is limited and has a shadow. Putting

$$g(x) = \mathrm{sh}(f^+(x))$$

defines $g : V \to \mathbb{R}$ as a function that extends f. Then g is dominated by p because $g(x) \simeq f^+(x) \leq p(x)$, implying $g(x) \leq p(x)$, since both $g(x)$ and $p(x)$ are real. Finally, the shadow map $\mathrm{sh} : \mathbb{L} \to \mathbb{R}$ is a linear functional that composes with the restriction of the hyperlinear f^+ to V to make g linear.

This completes our nonstandard proof of the Hahn–Banach theorem.

19.11 Exercises on (Hyper) Linear Functionals

(1) Verify that $\mathrm{sh} : \mathbb{L} \to \mathbb{R}$ is a linear functional on the real vector space $\mathbb{L}$ of limited hyperreal numbers.

(2) Write out in full the transfer arguments showing that f^+ has the properties listed above.

(3) (Due to W. A. J. Luxemburg) Given the hypothesis of the Hahn–Banach theorem, let $\{f_i : i \in I\}$ be the set of all linear functionals that extend f, are dominated by p, and are defined on a subspace of V including W. For each $x \in V$, put $A_x = \{i \in I : x \in \mathrm{dom}\, f_i\}$, and let $r_x : I \to \mathbb{R}$ be a function satisfying $r_x(i) = f_i(x)$ for all $i \in A_x$ (for definiteness r_x can be equated to 0 outside of A_x).

  (a) Show that $\{A_x : x \in V\}$ has the finite intersection property.

  (b) Let $\mathcal{F}$ be an ultrafilter on I including $\{A_x : x \in V\}$. Taking *$\mathbb{R}$ to be the ultrapower of $\mathbb{R}$ by $\mathcal{F}$, define $f^\# : V \to {}^*\mathbb{R}$ by putting $f^\#(x) = [r_x]$. Show that $f^\#$ is hyperlinear.

  (c) Use $f^\#$ in place of f^+ to prove that there is a linear functional g on V that extends f and is dominated by p.

This gives a direct derivation of the Hahn–Banach theorem from the ultrafilter theorem.

20
Books on Nonstandard Analysis

Many monographs about nonstandard analysis and its applications have been published since the appearance in 1966 of Abraham Robinson's immortal text on the subject he founded. Most of them are listed here (Robinson's book is item 34).

There is also an extensive and growing body of articles in journals. The book by J. E. Rubio (36) has a bibliography with more than 80 pages of references to this literature.

The biography by Joseph Dauben (8) provides a comprehensive account of the life and work of a remarkable figure in the history of mathematics.

1. Sergio Albeverio, Jens Erik Fenstad, Raphael Høegh-Krohn, and Tom Lindstrøm. *Nonstandard Methods in Stochastic Analysis and Mathematical Physics.* Academic Press, 1986.

2. Sergio A. Albeverio, Wilhelm A. J. Luxemburg, and Manfred P. H. Wolff, editors. *Advances in Analysis, Probability, and Mathematical Physics: Contributions of Nonstandard Analysis.* Kluwer Academic Publishers, 1995.

3. Leif O. Arkeryd, Nigel J. Cutland, and C. Ward Henson, editors. *Nonstandard Analysis: Theory and Applications.* Kluwer Academic Publishers, 1997.

4. David Ballard. *Foundational Aspects of "Non"standard Mathematics.* American Mathematical Society, 1994.

20. Books on Nonstandard Analysis

5. Marek Capinski and Nigel J. Cutland. *Nonstandard Methods for Stochastic Fluid Mechanics.* World Scientific, 1995.

6. Nigel J. Cutland, editor. *Nonstandard Analysis and its Applications*, volume 10 of *London Mathematical Society Student Texts.* Cambridge University Press, 1998.

7. Nigel J. Cutland, Vítor Neves, Franco Oliveira, and José Sousa-Pinto, editors. *Developments in Nonstandard Mathematics*, volume 336 of *Pitman Research Notes in Mathematics Series.* Longman, 1995.

8. Joseph Dauben. *Abraham Robinson; The Creation of Nonstandard Analysis: A Personal and Mathematical Odyssey.* Princeton University Press, 1995.

9. Martin Davis. *Applied Nonstandard Analysis.* John Wiley and Sons, 1977.

10. Francine Diener and Marc Diener, editors. *Nonstandard Analysis in Practice.* Springer, 1995.

11. Francine Diener and Georges Reeb. *Analyse Non Standard.* Hermann, Paris, 1989.

12. Marc Diener and Guy Wallet, editors. *Mathématique Finitaires et Analyse Non Standard.* U. E. R. de Mathématiques de l'Université Paris VII, 1989 (in two volumes).

13. E. I. Gordon. *Nonstandard Methods in Commutative Harmonic Analysis*, volume 164 of *Translations of Mathematical Monographs.* American Mathematical Society, 1997.

14. James M. Henle and Eugene M. Kleinberg. *Infinitesimal Calculus.* MIT Press, 1979.

15. R. F. Hoskins. *Standard and Nonstandard Analysis.* Ellis Horwood, 1990.

16. Albert E. Hurd, editor. *Nonstandard Analysis–Recent Developments*, volume 983 of *Lecture Notes in Mathematics.* Springer, 1983.

17. Albert E. Hurd and Peter A. Loeb, editors. *Victoria Symposium on Nonstandard Analysis*, volume 369 of *Lecture Notes in Mathematics.* Springer, 1974.

18. Albert E. Hurd and Peter A. Loeb. *An Introduction to Nonstandard Real Analysis.* Academic Press, 1985.

20. Books on Nonstandard Analysis   281

19. D. S. Jones. *Introduction to Asymptotics: A Treatment Using Nonstandard Analysis*. World Scientific, 1997.

20. Vladimir Kanovei. *A Course on Foundations of Nonstandard Analysis*. Institute for Studies in Theoretical Physics and Mathematics, Tehran, Iran, 1994.

21. H. Jerome Keisler. *Elementary Calculus*. Prindle, Weber & Schmidt, 1976.

22. H. Jerome Keisler. *Foundations of Infinitesimal Calculus*. Prindle, Weber & Schmidt, 1976.

23. H. Jerome Keisler. *An Infinitesimal Approach to Stochastic Analysis*, volume 297 of *Memoirs of the American Mathematical Society*. American Mathematical Society, 1984.

24. Fouad Koudjeti. *Elements of External Calculus; with an application to mathematical finance*. Ph.D. thesis, Rijksuniversiteit Groningen, 1995. Published by Labyrint Publication, The Netherlands.

25. Anatoly G. Kusraev and Semen S. Kutateladze. *Nonstandard Methods of Analysis*. Kluwer Academic Publishers, 1994.

26. A. H. Lightstone and Abraham Robinson. *Nonarchimedean Fields and Asymptotic Expansions*. North-Holland, 1975.

27. Robert Lutz and Michel Goze. *Nonstandard Analysis: A Practical Guide with Applications*, volume 881 of *Lecture Notes in Mathematics*. Springer, 1981.

28. W. A. J. Luxemburg, editor. *Applications of Model Theory to Algebra, Analysis, and Probability*. Holt, Rinehart and Winston, 1969.

29. W. A. J. Luxemburg and Abraham Robinson, editors. *Contributions to Non-standard Analysis*, volume 69 of *Studies in Logic and the Foundations of Mathematics*. North-Holland, 1972.

30. Moshe Machover and Joram Hirshfeld. *Lectures on Non-standard Analysis*, volume 94 of *Lecture Notes in Mathematics*. Springer, 1969.

31. Edward Nelson. *Radically Elementary Probability Theory*. Princeton University Press, 1987.

32. Salim Rashid. *Economies With Many Agents: An Approach Using Nonstandard Analysis*. Johns Hopkins University Press, 1987.

33. Alain Robert. *Nonstandard Analysis*. Wiley, 1988.

20. Books on Nonstandard Analysis

34. Abraham Robinson. *Non-standard Analysis.* Princeton University Press, revised edition, 1996. Originally published by North-Holland 1966.

35. Abraham Robinson. *Selected Papers Volume 2: Nonstandard Analysis and Philosophy.* Edited by H. J. Keisler et al., North-Holland, 1979.

36. J. E. Rubio. *Optimisation and Nonstandard Analysis.* Dekker, 1994.

37. K. D. Stroyan and Jose Manuel Bayod. *Foundations of Infinitesimal Stochastic Analysis*, volume 119 of *Studies in Logic and the Foundations of Mathematics.* North-Holland, 1986.

38. K. D. Stroyan and W. A. J. Luxemburg. *Introduction to the Theory of Infinitesimals.* Academic Press, 1976.

39. Curtis Tuckey. *Nonstandard Methods in the Calculus of Variations*, volume 297 of *Pitman Research Notes in Mathematics Series.* Longman, 1993.

40. Imme van den Berg. *Nonstandard Asymptotic Analysis*, volume 1249 of *Lecture Notes in Mathematics.* Springer, 1987.

41. Keith R. Wicks. *Fractals and Hyperspaces*, volume 1492 of *Lecture Notes in Mathematics.* Springer, 1991.

Index

absolute convergence, 71
absolute value, 30
additive
    countably, 206
    finitely, 206
    function, 275
agreement set, 16
algebra, 204
    σ-, 204
almost all, 17, 24
almost everywhere modulo $\mathcal{F}$, 24
almost-all criterion, 26
antichain, 222
appreciable, 50
approachable point, 236
approximable, 212
Archimedes, 4
assignable quantities, 6, 11
atom, 265
atomic
    Boolean algebra, 266
    formula, 40, 167
    sentence, 41
average function value, 112
axiom of choice, 20

and Łoś's theorem, 272

Banach fixed point theorem, 84
base p expansion, 238
    hyperfinite, 255
basis, 274
Berkeley, 10
best linear approximation, 93
Bolzano–Weierstrass theorem, 66
Boolean algebra, 262
    atomic, 266
    locally finite, 273
Borel sets, 205
bound variable, 41
bounded
    quantifier, 36
    sequence, 64

Cartesian product, 158
Cauchy, 15, 75, 90
Cauchy sequence, 12, 55, 65, 233
Cauchy's convergence criterion, 65
Cauchy's principle, 85
chain (in a partial ordering), 20
chain rule, 94

closed set, 115, 116
closed term, 39, 166
closure
　point, 113
　(topological) of a set, 113
cluster point, 66
coefficient, 250
cofinite, 17
　filter, 18
colouring, 221, 260
　of a graph, 260
compact set, 116
compactness theorem, 10
comparison test, 72
complete measure, 209
complete metric space, 234
　and approachability, 237
completion
　of a metric space, 234
composition of relations, 164
comprehensive, 195
　countably, 195
　sequentially, 195
concurrent relation, 185
congruence modulo m, 237
constant polynomial, 250
constant term, 39, 166
continuous
　at a point, 75, 76, 98
　on a set, 77
　uniformly, 81
continuum hypothesis, 33
contraction mapping, 82
convergence
　absolute, 71
　in a metric space, 234
　of sequences, 61
　of series, 71
　pointwise, 86
　uniform, 86
coset, 232, 243
countable additivity, 206
countable saturation, 138, 196
countably comprehensive, 195
counting measure, 207

critical point theorem, 95
cumulative power set, 164

Dedekind completeness, 11, 12, 47, 55, 65, 130
　internal, 130
Dedekind cut, 12
defined formula, 40
degree of a coefficient, 250
density
　in a metric space, 234
　of $\mathbb{Q}$ in $\mathbb{R}$, 36
derivative, 91
　left-hand, 92
　partial, 97
　right-hand, 92
derived set, 113
differential, 93
　total, 98
dimension, 274
　hyperfinite, 274
discreteness, 37
distributive lattice, 272
diverges to infinity, 64
domain, 163
dominated, 276

edge, 260
enlargement
　as an ultrapower, 187
　of a relation, 32
　of a set, 28
　of a universe, 183
enlargement theorem, 185
Euclidean metric, 233
Eudoxus–Archimedes principle, 13, 35
Euler, 8
existential transfer, 45
extended
　tail, 61
　term, 61
extended real numbers, 206
external
　entity, 173

image, 172, 176
    set, 131
extreme value theorem, 80

field of fractions, 245
field of sets, 262
filter, 18
    of a Boolean algebra, 269
    cofinite, 18
    Fréchet, 18
    generated, 18
    maximal, 19
    prime, 273
    principal, 18, 270
    proper, 18, 269
finite
    μ-, 207
    σ-, 207
finite intersection property (fip), 19
finite Ramsey theorem, 227
finitely additive, 206
finitely generated, 268
finitely satisfiable relation, 185
fixed point, 82
fluent, 6, 91
fluxion, 6, 91
formula, 40, 167
    atomic, 40, 167
    defined, 40
    inconsistent, 264
    membership, 40, 43
    Π-, 264
    unary, 40
free variable, 41
full structure, 38
function, 160
function-value term, 166
functional, 275
    linear, 275
fundamental theorem of calculus, 112

galaxy, 52, 234
    *$\mathbb{N}$-, 56

principal, 235
generated subalgebra, 268
ghost of a departed atom, 271
graph, 260
    of a function, 32

Hahn–Banach theorem, 276
halo, 52, 234
Heine–Borel theorem, 119
homogeneous function, 275
    positively, 275
hyper-open set, 143
hyperfinite, 52
    approximant, 261
    approximation, 150
    dimension, 274
    partition, 153
    Riemann sum, 154
    set, 149, 179
    sum, 71, 152, 180
hyperfinitely generated, 273
hyperinteger, 49
hyperlinear, 277
hypernatural, 49
hyperpolynomial, 251
hyperprime, 57
hyperrational, 49
hyperreal, 3, 49
    vector space, 273
hypersequence, 31, 61

$\mathbb{I}$, 50
image
    external, 172, 176
    inverse, 164
    of a set, 172
    under a function, 164
    under a relation, 163
inconsistent formula, 264
increment, 92, 98
incremental equation, 93
    and continuity of derivative, 103–104
    for two variables, 98
    generalised, 102

individual, 158, 162
induction
    internal, 129
infinitely close, 52
    in the plane, 97
infinitely large, 3, 16
infinitesimal, 3, 15, 27, 50
    p-adic, 242, 247
    power series, 253
infix notation, 39, 167
integer
    p-adic, 239
integers modulo m, 237
interior
    of a set, 113
    point, 113
intermediate value theorem, 79
internal
    cardinality (size), 149, 180
    entity, 172
    function, 147
    function definition principle, 178
    induction, 129
    least number principle, 128
    set, 125
    set definition principle, 134, 177
    subsets of $\mathbb{R}$, 128, 135
interval topology, 121
inverse
    function theorem, 96
    image, 164
    relation, 163
    shadow, 216
isometric, 234

join, 263
join-irreducible, 272
juxtaposition of integrals, 110

κ-saturation, 198

$\mathbb{L}$, 50
Łoś's theorem, 47, 188

    and the axiom of choice, 272
Lagrange remainder, 101
language, 38
large set, 16, 24
lattice, 272
    distributive, 272
    set, 272
least number principle
    internal, 128
Lebesgue measure, 207, 210
Leibniz's rule, 5, 94
limit
    inferior, 67
    of a function, 78
    of a sequence, 61
    superior, 67
    uniqueness, 63
limit point
    of a sequence, 66
    of a set, 113
limited, 49, 235
    p-adically, 241, 247
limited distance apart, 52
linear
    combination, 274
    functional, 275
Lipschitz condition, 82
locally finite, 273
Loeb, 203
    measurable set, 210
    measure, 210
    measure space, 210
lower Riemann sum, 105

Maclaurin series, 101
maximal proper filter, 19
mean value theorem, 96
measurable set
    Lebesgue, 210
    Loeb, 210
    μ^+-, 209
measure, 206
    complete, 209
    counting, 207
    Lebesgue, 207

Loeb, 210
outer, 208
meet, 263
membership formula, 40, 43
method of exhaustion, 4
metric, 233
    Euclidean, 233
    p-adic, 239
metric space, 233
model, 10
monad, 52
monochromatic, 221, 222
monotonic, 108
μ-approximable, 212
μ-finite, 207
μ-null, 207
μ^+-measurable, 209

$\mathbb{N}$, 23
*$\mathbb{N}_\infty$, 50
near to $\mathbb{X}$, 236
nearstandard, 237
    pre-, 237
Newton, 6
Newton quotient, 92
node, 260
nondecreasing, 63, 108
nonincreasing, 63, 108
nonstandard
    entity, 170
    framework, 168
    hull, 235
    member, 29
    natural number, 29
    set, 171
norm, 233
    p-adic, 239, 240, 246
null set, 207
number
    appreciable, 50
    extended real, 206
    hyperinteger, 49
    hypernatural, 49
    hyperprime, 57
    hyperrational, 49
    hyperreal, 49
    infinitesimal, 50
    limited, 49
    p-adic, 245
    real, 11
    standard, 49
    unlimited, 27, 49

open set, 115, 116
order
    of a power series, 250
    p-adic, 238, 240, 246
order-completeness, 130
ordered pair
    definition of, 161
ordinary Riemann sum, 105
oscillation, 107
outer measure, 208
overflow, 129, 191

p-adic
    infinitesimal, 242, 247
    integer, 239
    metric, 239
    norm, 239, 240, 246
    number, 245
    order, 238, 240, 246
    unit, 240
p-adically limited, 241, 247
Paris–Harrington theorem, 229
partial derivative, 97
partial sum, 71
partition, 105
permanence, 84, 137, 191
pigeonhole principle, 152
pointwise convergence, 86
polynomial, 249
    constant, 250
    hyper, 251
positively homogeneous, 275
power series, 250
power set, 18
    algebra, 262
pre-nearstandard, 237
prime filter, 273

principal
  filter, 18, 270
  galaxy, 235
  ultrafilter, 18
proper filter, 18, 269

${}^*\mathbb{Q}^{\text{inf}}$, 232
${}^*\mathbb{Q}^{\text{inf}_p}$, 247
${}^*\mathbb{Q}^{\text{lim}}$, 232
${}^*\mathbb{Q}^{\text{lim}_p}$, 247
$\mathbb{Q}_p$, 245

$\mathbb{R}^+$, 30
$\mathbb{R}^\geq$, 233
${}^*\mathbb{R}^-_\infty$, 50
$\mathbb{R}^\mathbb{N}$, 23
${}^*\mathbb{R}^+_\infty$, 50
${}^*\mathbb{R}$, 3, 25
  uniqueness of, 33
Ramsey's theorem, 222
  finite, 227
range, 163
rank
  of a function, 187
  of an entity, 164
ratio test, 72
real comparisons, 53
real number, 11
real vector space, 273
real-open set, 120
real-radius neighbourhood, 120
reduced form, 248
relational structure, 38
relatively large set, 228
remainder, 100
  Lagrange form, 101
Riemann integrable, 106
  nonstandard criterion, 110
Riemann sum
  hyperfinite, 154
  lower, 105
  ordinary, 105
  upper, 105
ring
  σ-, 204

  of sets, 204
Robinson, 10
Robinson's
  compactness criterion, 117
  sequential lemma, 87, 193
Rolle's theorem, 96

S- as a prefix, 121
S-neighbourhood, 120
S-open set, 120
saturation, 140, 198
  κ-, 198
  countable, 138, 196
scalar multiplication, 273
sentence, 41, 167
  atomic, 41
sentence letters, 264
sequentially comprehensive, 195
series, 71
  Maclaurin, 101
  Taylor, 100
set lattice, 272
shadow, 53
  inverse, 216
  of a set, 142
shadow map, 54
σ-finite, 207
σ-algebra, 204
σ-ring, 204
sine, 30
  continuity of, 77
  derivative of, 92
smooth function, 98
standard
  entity, 170
  number, 49
  part, 54
statement value, 24
Stone representation theorem, 269
strong transitivity, 162
subadditive function, 276
subalgebra
  generated, 268
  of power set algebra, 262
subgraph, 260

subspace, 274
superstructure, 164

tautology, 264
Taylor
    formula, 101
    polynomial, 100
    series, 100
term, 39, 166
    closed, 39, 166
    undefined, 39
topology, 116
    interval, 121
    S-, 120
total differential, 98
transfer, 168
    map, 168
    principle, 11, 44, 133
transform
    $*$-transform, 35–37, 42–44, 168
transitive closure, 162
transitivity, 159
    strong, 162
triangle inequality, 233
truth values, 24
tuple, 160, 166

ultimately zero, 250
ultrafilter, 18, 269
    principal, 18
ultrafilter theorem, 270
ultrapower, 26
unary
    formula, 40
    relation (subset), 32
undefined term, 39
underflow, 136, 192
uniform convergence, 86
uniform weighting, 208
uniformly continuous, 81
unit
    of a Boolean algebra, 262
    p-adic, 240
universal transfer, 45
universe, 162

embedding, 168
over $\mathbb{X}$, 162
unlimited, 27, 49
upper Riemann sum, 105

valuation, 264
variable, 39, 166
    bound, 41
    free, 41
vector space
    hyperfinite-dimensional, 274
    hyperreal, 273
    real, 273
vertex, 260

weighting function, 208
    uniform, 208
well inside, 103, 120
well-defined, 25, 125

$\widehat{\mathbb{X}}$, 235
$^*\mathbb{X}^{\lim}$, 235
$^*\mathbb{X}^{\mathrm{ap}}$, 236

$\mathbb{Z}/m$, 237
$\mathbb{Z}^{\geq}$, 238
$^*\mathbb{Z}^{\inf_p}$, 241
$^*\mathbb{Z}^{\lim_p}$, 241
$\mathbb{Z}_p$, 239
zero of a Boolean algebra, 262
Zorn's lemma, 20

Graduate Texts in Mathematics

(continued from page ii)

62 KARGAPOLOV/MERLZJAKOV. Fundamentals of the Theory of Groups.
63 BOLLOBAS. Graph Theory.
64 EDWARDS. Fourier Series. Vol. I 2nd ed.
65 WELLS. Differential Analysis on Complex Manifolds. 2nd ed.
66 WATERHOUSE. Introduction to Affine Group Schemes.
67 SERRE. Local Fields.
68 WEIDMANN. Linear Operators in Hilbert Spaces.
69 LANG. Cyclotomic Fields II.
70 MASSEY. Singular Homology Theory.
71 FARKAS/KRA. Riemann Surfaces. 2nd ed.
72 STILLWELL. Classical Topology and Combinatorial Group Theory. 2nd ed.
73 HUNGERFORD. Algebra.
74 DAVENPORT. Multiplicative Number Theory. 2nd ed.
75 HOCHSCHILD. Basic Theory of Algebraic Groups and Lie Algebras.
76 IITAKA. Algebraic Geometry.
77 HECKE. Lectures on the Theory of Algebraic Numbers.
78 BURRIS/SANKAPPANAVAR. A Course in Universal Algebra.
79 WALTERS. An Introduction to Ergodic Theory.
80 ROBINSON. A Course in the Theory of Groups. 2nd ed.
81 FORSTER. Lectures on Riemann Surfaces.
82 BOTT/TU. Differential Forms in Algebraic Topology.
83 WASHINGTON. Introduction to Cyclotomic Fields. 2nd ed.
84 IRELAND/ROSEN. A Classical Introduction to Modern Number Theory. 2nd ed.
85 EDWARDS. Fourier Series. Vol. II. 2nd ed.
86 VAN LINT. Introduction to Coding Theory. 2nd ed.
87 BROWN. Cohomology of Groups.
88 PIERCE. Associative Algebras.
89 LANG. Introduction to Algebraic and Abelian Functions. 2nd ed.
90 BRØNDSTED. An Introduction to Convex Polytopes.
91 BEARDON. On the Geometry of Discrete Groups.
92 DIESTEL. Sequences and Series in Banach Spaces.
93 DUBROVIN/FOMENKO/NOVIKOV. Modern Geometry—Methods and Applications. Part I. 2nd ed.
94 WARNER. Foundations of Differentiable Manifolds and Lie Groups.
95 SHIRYAEV. Probability. 2nd ed.
96 CONWAY. A Course in Functional Analysis. 2nd ed.
97 KOBLITZ. Introduction to Elliptic Curves and Modular Forms. 2nd ed.
98 BRÖCKER/TOM DIECK. Representations of Compact Lie Groups.
99 GROVE/BENSON. Finite Reflection Groups. 2nd ed.
100 BERG/CHRISTENSEN/RESSEL. Harmonic Analysis on Semigroups: Theory of Positive Definite and Related Functions.
101 EDWARDS. Galois Theory.
102 VARADARAJAN. Lie Groups, Lie Algebras and Their Representations.
103 LANG. Complex Analysis. 3rd ed.
104 DUBROVIN/FOMENKO/NOVIKOV. Modern Geometry—Methods and Applications. Part II.
105 LANG. $SL_2(\mathbf{R})$.
106 SILVERMAN. The Arithmetic of Elliptic Curves.
107 OLVER. Applications of Lie Groups to Differential Equations. 2nd ed.
108 RANGE. Holomorphic Functions and Integral Representations in Several Complex Variables.
109 LEHTO. Univalent Functions and Teichmüller Spaces.
110 LANG. Algebraic Number Theory.
111 HUSEMÖLLER. Elliptic Curves.
112 LANG. Elliptic Functions.
113 KARATZAS/SHREVE. Brownian Motion and Stochastic Calculus. 2nd ed.
114 KOBLITZ. A Course in Number Theory and Cryptography. 2nd ed.
115 BERGER/GOSTIAUX. Differential Geometry: Manifolds, Curves, and Surfaces.
116 KELLEY/SRINIVASAN. Measure and Integral. Vol. I.
117 SERRE. Algebraic Groups and Class Fields.
118 PEDERSEN. Analysis Now.
119 ROTMAN. An Introduction to Algebraic Topology.

120 ZIEMER. Weakly Differentiable Functions: Sobolev Spaces and Functions of Bounded Variation.
121 LANG. Cyclotomic Fields I and II. Combined 2nd ed.
122 REMMERT. Theory of Complex Functions. *Readings in Mathematics*
123 EBBINGHAUS/HERMES et al. Numbers. *Readings in Mathematics*
124 DUBROVIN/FOMENKO/NOVIKOV. Modern Geometry—Methods and Applications. Part III.
125 BERENSTEIN/GAY. Complex Variables: An Introduction.
126 BOREL. Linear Algebraic Groups. 2nd ed.
127 MASSEY. A Basic Course in Algebraic Topology.
128 RAUCH. Partial Differential Equations.
129 FULTON/HARRIS. Representation Theory: A First Course. *Readings in Mathematics*
130 DODSON/POSTON. Tensor Geometry.
131 LAM. A First Course in Noncommutative Rings.
132 BEARDON. Iteration of Rational Functions.
133 HARRIS. Algebraic Geometry: A First Course.
134 ROMAN. Coding and Information Theory.
135 ROMAN. Advanced Linear Algebra.
136 ADKINS/WEINTRAUB. Algebra: An Approach via Module Theory.
137 AXLER/BOURDON/RAMEY. Harmonic Function Theory.
138 COHEN. A Course in Computational Algebraic Number Theory.
139 BREDON. Topology and Geometry.
140 AUBIN. Optima and Equilibria. An Introduction to Nonlinear Analysis.
141 BECKER/WEISPFENNING/KREDEL. Gröbner Bases. A Computational Approach to Commutative Algebra.
142 LANG. Real and Functional Analysis. 3rd ed.
143 DOOB. Measure Theory.
144 DENNIS/FARB. Noncommutative Algebra.
145 VICK. Homology Theory. An Introduction to Algebraic Topology. 2nd ed.
146 BRIDGES. Computability: A Mathematical Sketchbook.
147 ROSENBERG. Algebraic *K*-Theory and Its Applications.
148 ROTMAN. An Introduction to the Theory of Groups. 4th ed.
149 RATCLIFFE. Foundations of Hyperbolic Manifolds.
150 EISENBUD. Commutative Algebra with a View Toward Algebraic Geometry.
151 SILVERMAN. Advanced Topics in the Arithmetic of Elliptic Curves.
152 ZIEGLER. Lectures on Polytopes.
153 FULTON. Algebraic Topology: A First Course.
154 BROWN/PEARCY. An Introduction to Analysis.
155 KASSEL. Quantum Groups.
156 KECHRIS. Classical Descriptive Set Theory.
157 MALLIAVIN. Integration and Probability.
158 ROMAN. Field Theory.
159 CONWAY. Functions of One Complex Variable II.
160 LANG. Differential and Riemannian Manifolds.
161 BORWEIN/ERDÉLYI. Polynomials and Polynomial Inequalities.
162 ALPERIN/BELL. Groups and Representations.
163 DIXON/MORTIMER. Permutation Groups.
164 NATHANSON. Additive Number Theory: The Classical Bases.
165 NATHANSON. Additive Number Theory: Inverse Problems and the Geometry of Sumsets.
166 SHARPE. Differential Geometry: Cartan's Generalization of Klein's Erlangen Program.
167 MORANDI. Field and Galois Theory.
168 EWALD. Combinatorial Convexity and Algebraic Geometry.
169 BHATIA. Matrix Analysis.
170 BREDON. Sheaf Theory. 2nd ed.
171 PETERSEN. Riemannian Geometry.
172 REMMERT. Classical Topics in Complex Function Theory.
173 DIESTEL. Graph Theory.
174 BRIDGES. Foundations of Real and Abstract Analysis.
175 LICKORISH. An Introduction to Knot Theory.
176 LEE. Riemannian Manifolds.
177 NEWMAN. Analytic Number Theory.
178 CLARKE/LEDYAEV/STERN/WOLENSKI. Nonsmooth Analysis and Control Theory.
179 DOUGLAS. Banach Algebra Techniques in Operator Theory. 2nd ed.

180 SRIVASTAVA. A Course on Borel Sets.
181 KRESS. Numerical Analysis.
182 WALTER. Ordinary Differential Equations.
183 MEGGINSON. An Introduction to Banach Space Theory.
184 BOLLOBAS. Modern Graph Theory.
185 COX/LITTLE/O'SHEA. Using Algebraic Geometry.
186 RAMAKRISHNAN/VALENZA. Fourier Analysis on Number Fields.
187 HARRIS/MORRISON. Moduli of Curves.
188 GOLDBLATT. Lectures on the Hyperreals: An Introduction to Nonstandard Analysis.
189 LAM. Lectures on Modules and Rings.